Fruit Quality and its Biological Basis

Edited by

MICHAEL KNEE
Department of Horticulture and Crop Science
The Ohio State University
Columbus, Ohio
USA

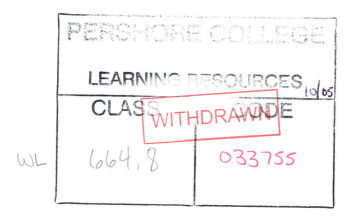

Sheffield
Academic Press

CRC
CRC Press

First published 2002
Copyright © 2002 Sheffield Academic Press

Published by
Sheffield Academic Press Ltd
Mansion House, 19 Kingfield Road
Sheffield S11 9AS, UK

ISBN 1-84127-230-2

Published in the U.S.A. and Canada (only) by
CRC Press LLC
2000 Corporate Blvd., N.W.
Boca Raton, FL 33431, U.S.A.
Orders from the U.S.A. and Canada (only) to CRC Press LLC

U.S.A. and Canada only:
ISBN 0-8493-9781-2

Printed on acid-free paper in Great Britain by
Antony Rowe Ltd, Chippenham, Wiltshire

British Library Cataloguing-in-Publication Data:
A catalogue record for this book is available from the British Library

Library of Congress Cataloging-in-Publication Data:
A catalog record for this book is available from the Library of Congress

Fruit Quality and its Biological Basis

Sheffield Biological Sciences

A series which provides an accessible source of information at research and professional level in chosen sectors of the biological sciences.

Series Editors:

Professor Jeremy A. Roberts, Plant Science Division, School of Biosciences, University of Nottingham.
Professor Peter N.R. Usherwood, Molecular Toxicology Division, School of Biosciences, University of Nottingham.

Titles in the series:

Biology of Farmed Fish
Edited by K.D. Black and A.D. Pickering.

Stress Physiology in Animals
Edited by P.H.M. Balm.

Seed Technology and its Biological Basis
Edited by M. Black and J.D. Bewley.

Leaf Development and Canopy Growth
Edited by B. Marshall and J.A. Roberts.

Environmental Impacts of Aquaculture
Edited by K.D. Black.

Herbicides and their Mechanisms of Action
Edited by A.H. Cobb and R.C. Kirkwood.

The Plant Cell Cycle and its Interfaces
Edited by D. Francis.

Meristematic Tissues in Plant Growth and Development
Edited by M.T. McManus and B.E. Veit.

Fruit Quality and its Biological Basis
Edited by M. Knee.

Preface

It is now thirty years since Hulme's monumental 'Biochemistry of Fruits and their Products' (1971, Academic Press, London) reviewed important aspects of fruit physiology and summarized the state of knowledge of major types of fruit. Since then, major changes in the conditions of trade have precipitated a restructuring of food production that continues to this day. Fruit growing has not been immune from the challenges of an increasingly globalized market. Also in those thirty years, the manipulation of fruit physiology at the gene level has moved from a dimly perceived possibility to a commercial reality. The growth in understanding of fruit development at the molecular level has contributed to the overall growth of plant science, particularly in the areas of cell wall metabolism and ethylene synthesis and action.

No subsequent book has achieved the depth and breadth of coverage of Hulme (1971). Some have dealt more generically with applied aspects of the postharvest handling of fruits and vegetables, with specific aspects of physiology or with specific types of fruit. This volume synthesizes our understanding of the biological basis for important features of fruit quality and how this can be controlled in postharvest operations. These topics are pertinent to horticulture, food science and plant science in general. Readers are likely to be actively involved in research or application of technical knowledge of fruit ripening, storage or utilization, or they may have an interest in aspects which are not restricted to fruit (such as ethylene physiology).

Throughout the book, 'fruit' is used more in its everyday sense than in its strict botanical meaning. The fruits covered are the fleshy, often sweet types that are more likely to be eaten raw for dessert than to appear cooked as an accompaniment to a main course or on a salad plate. However, it is inevitable that the tomato should figure largely in several chapters because so much has been learned about fruit physiology using the tomato as a model system. The opening chapter clarifies the botanical nature of these fruits, provides an international context and previews some of the themes of later chapters. While it is a truism that growing conditions influence the quality of plant crops, the effects are particularly evident for fruits and can often be related to mineral uptake during growth. This is the main rationale for the second chapter of the book. The next two chapters address important aspects of fruit quality as perceived by the consumer: texture and flavour. Then three chapters consider how various factors influence fruit physiology after harvest. The first two factors are the artificial manipulation of temperature and atmosphere, whereas the third is the natural regulator, ethylene, which can also be artificially controlled. Two chapters consider how fruit quality can be adversely affected by mechanical damage and pathogen attack. The book concludes with a review of the potential for genetic control of fruit physiology and development. Applications of this technology are in their infancy, but are sure to figure prominently in future books of this nature. It will be intriguing to see the extent to which they augment or provide a substitute for existing technology.

Michael Knee

Contributors

Dr Elizabeth A. Baldwin USDA Citrus and Subtropical Products Laboratory, 600 Ave S N.W., Winterhaven, FL 3388-2118, USA

Dr Raymond M. Beaudry Department of Horticulture, Michigan State University, East Lansing, MI 48824-1325, USA

Ms Linda Boyd Horticulture and Food Research Institute of New Zealand, Mt Albert Research Centre, Private Bag 92169, Auckland, New Zealand

Dr Ian B. Ferguson Horticulture and Food Research Institute of New Zealand, Mt Albert Research Centre, Private Bag 92169, Auckland, New Zealand

Dr Monica Fischer Nestlé Research Centre, Nestec Ltd, Vers-chez-les Blanc, PO Box 44, CH-1000 Lausanne 26, Switzerland

Dr Adel A. Kader Department of Pomology, University of California, One Shields Avenue, Davis, CA 95616-8683, USA

Dr Susan Lurie Department of Postharvest Science, Volcani Center, PO Box 6, Bet Dagen 50250, Israel

Dr Michael Knee Department of Horticulture and Crop Science, Ohio State University, 2001 Fyffe Court, Columbus, OH 43210-1096, USA

Mr Kenneth Manning Horticulture Research International, Wellesbourne, Warwick CV35 9EF, UK

Dr A. Raymond Miller Department of Horticulture and Crop Science, Ohio State University, Ohio Agricultural Research and Development Center, Wooster, OH 44691, USA

Dr Nazir A. Mir   Department of Plant Biology and Pathology, Rutgers, The State University of New Jersey, 59 Dudley Road, New Brunswick, NJ 08901-8520, USA

Dr Robert Redgwell   Nestlé Research Centre, Nestec Ltd, Vers-chez-les Blanc, PO Box 44, CH-1000 Lausanne 26, Switzerland

Dr Graham B. Seymour   Horticulture Research International, Welles-bourne, Warwick CV35 9EF, UK

Dr David Sugar   Southern Oregon Research and Extension Center, Oregon State University, 569 Hanley Road, Medford, OR 97502-1251, USA

Dr Chris B. Watkins   Department of Horticulture, Cornell University, Ithaca, NY 14853-5908, USA

Contents

1 Fruits in the global market

Adel A. Kader

1.1 Introduction

Fruits are not only colorful and flavorful components of the human diet, but they also serve as a source of energy, vitamins, minerals, dietary fiber and antioxidants. More than 100 fruit species are grown throughout the world. Fruits are commonly classified by growing region as follows:

1. temperate zone fruits

   - pome fruits: apple, Asian pear (nashi), European pear, quince
   - stone fruits: apricot, cherry, nectarine, peach, plum
   - small fruits and berries: grape (European and American types), strawberry, raspberry, blueberry, blackberry, cranberry

2. subtropical fruits

   - citrus fruits: grapefruit, lemon, lime, mandarin, orange, pummelo (pomelo), tangelo
   - non-citrus fruits: avocado, cherimoya, fig, kiwifruit, olive, pomegranate

3. tropical fruits

   - major tropical fruits: banana, mango, papaya, pineapple
   - minor tropical fruits: star fruit (carambola), cashew apple, durian, guava, longan, lychee, mangosteen, passion fruit, rambutan, sapota, tamarind

Botanical classification of fruits (table 1.1) shows the great diversity among fruit species, which are now widely distributed well beyond their original homes. Table 1.1 includes only the major commercially grown fruit species. There are many more fruit species that are produced on a smaller scale in terms of quantity and geographic distribution. Among the families listed, Rosaceae includes the largest number of fruit species (pome fruits, stone fruits, strawberries, brambles).

This chapter is focused on fruits in the global market, their contribution to human nutrition, composition and quality, maturity and ripening, factors influencing quality, and assurance of food safety.

Table 1.1 Botanical classification of fruit species and their original homes

Family	Genus and species	Common name(s)	Original home(s)
MONOCOTYLEDONS			
Bromeliaceae	*Ananas comosus*	Pineapple	South America
Musaceae	*Musa paradisiaca*	Banana, plantain	Southeast Asia
Palmaceae	*Cocos nucifera*	Coconut	South Pacific
	Phoenix dactylifera	Date	North Africa
DICOTYLEDONS			
Actinidiaceae	*Actinidia deliciosa*	Chinese gooseberry, kiwifruit	China
Anacardiaceae	*Anacardium occidentale*	Cashew apple, cashew nut	South America
	Mangifera indica	Mango	Southeast Asia
	Pistacia vera	Pistachio	Southwest Asia
Betulaceae	*Corylus avellana*	Filbert, hazelnut	Europe
Caricaceae	*Carica papaya*	Papaya	Central America
Ebenaceae	*Diospyros kaki*	Persimmon	Japan and China
Ericaceae	*Vaccinium* spp.	Blueberry and cranberry	USA
Fagaceae	*Castanea sativa*	European chestnut	Europe
Juglandaceae	*Carya illinoensis*	Pecan	Southeastern USA and northern Mexico
	Juglans regia	Persian (or English) walnut	Southeastern Europe, southwestern Asia, China, and Japan
Lauraceae	*Persea americana*	Avocado	Central America
Moraceae	*Ficus carica*	Fig	Southwest Asia
Oleaceae	*Olea europea*	Olive	Southwest Asia
Puniceae	*Punica granatum*	Pomegranate	Southwest Asia
Rosaceae (pome fruits)	*Cydonia oblonga*	Quince	Asia
	Malus pumila	Apple	Southeast Europe and southwest Asia
	Pyrus communis	European pear	Europe and western Asia
	Pyrus pyrifolia	Asian pear	China
Rosaceae (stone fruits)	*Prunus avium*	Sweet cherry	Europe and southwest Asia
	Prunus cerasus	Sour cherry	Europe and southwest Asia
	Prunus armeniaca	Apricot	China
	Prunus domestica	European plum, prune	Southwest Asia
	Prunus salicina	Japanese plum	China
	Prunus dulcis	Almond	Southwest Asia
	Prunus persica	Peach and nectarine	China
Rosaceae (berries)	*Fragaria x ananassa*	Strawberry	USA, Chile
	Rubus spp.	Blackberry and raspberry	USA, Europe
Rutaceae	*Citrus aurantifolia*	Lime	Southeast Asia and eastern India
	Citrus grandis	Pummelo (pomelo)	Southeast Asia and eastern India

Table 1.1 (continued)

Family	Genus and species	Common name(s)	Original home(s)
	Citrus limon	Lemon	Southeast Asia and eastern India
	Citrus paradisi	Grapefruit	Southeast Asia and eastern India
	Citrus reticulata	Mandarin	Southeast Asia and eastern India
	Citrus sinensis	Orange	Southeast Asia and eastern India
Saxifragaceae	*Ribes hirtellum*	Gooseberry	USA, Europe
	Ribes sativum	Red currant	Western Europe
Vitaceae	*Vitis labrusca*	American grape	Eastern USA
	Vitis rotundifolia	Muscadine grape	Southeastern USA
	Vitis vinifera	European grape	Southwestern Asia

Sources: Brouk, 1975; Mitra, 1997; Morton, 1987; Ryall and Pentzer, 1982.

1.2   Classification of fruits based on postharvest considerations

Fruits can be classified according to their respiration rate and pattern (table 1.2). There is a good correlation between respiration rate and degree of perishability; the higher the respiration rate the shorter the postharvest-life of fruits. While climacteric fruits can ripen on or off the tree (if picked mature), non-climacteric fruits must be picked when ripe because they do not ripen after harvest.

Table 1.2 Classification of fruits according to their respiration rate and pattern

Class	Range of respiration rates ($ml\ CO_2/kg·h$) at 5°C	Fruits Climacteric	Non-climacteric
Very low	<5	—	Date, grape, nuts, pepino, pineapple
Low	5–10	Apple, Asian pear (some cultivars), kiwifruit, papaya, pear, persimmon, plum, quince	Cactus pear, cherry, cranberry, grapefruit, jujube, lemon, lime, mandarin, olive, orange, pomegranate, pummelo (pomelo)
Moderate	11–20	Apricot, banana, blueberry, feijoa, fig, guava, mango, nectarine, peach, plantain, rambutan, sapotes	Gooseberry, lanson, longan, loquat, lychee, tamarillo, star fruit (carambola)
High	21–30	Avocado, breadfruit, passion fruit	Blackberry, boysenberry, raspberry, strawberry
Very high	>30	Cherimoya, durian, jackfruit, soursop	—

Sources: Hardenburg *et al.*, 1996; Kader, 1992; Ryall & Pentzer, 1982.

Table 1.3 Classification of fruits according to their optimal storage temperatures and potential storage-life

Potential storage-life (weeks)	Optimal storage temperatures		
	0–2°C	4–6°C	10–14°C
<2	Apricot, blackberry, fig, raspberry, strawberry	Avocado (ripe), guava, feijoa	Papaya, rambutan, sapota, soursop
2–4	Blueberry, cherry, currant, gooseberry, loquat, nectarine, peach	Cactus pear, kumquat, longan, lychee, star fruit (carambola)	Avocado, banana, breadfruit, cherimoya, jackfruit, jujube, mangosteen, passion fruit, pineapple
4–6	Cashew apple, plum, plumcot	Mandarin, pepino	Durian, mango, plantain
6–8	Coconut, grape, persimmon	Olive, orange, pomegranate, tamarillo	Grapefruit, lime, pummelo (pomelo)
>8	Apple, Asian pear, cranberry, date, kiwifruit, pear, quince, tree nuts	Apple (chilling-sensitive cultivars)	Lemon

Sources: Hardenburg *et al.*, 1986; Kader, 1992; Ryall & Pentzer, 1982.

Subtropical and tropical fruits are chilling-sensitive and their minimum safe temperature ranges from 4 to 14°C, depending on the species and cultivar. Table 1.3 includes a classification of fruits according to their optimal storage temperatures and potential storage life. This classification is useful in determining compatibility when mixing fruits during transport and storage and in developing marketing strategies for the various fruit species.

1.3   World fruit production and trade

Worldwide availability of fruits and nuts continues to increase in terms of the number of species and cultivars as well as their expanded season of availability with production in northern and southern hemisphere countries. Out of more than 400 million metric tonnes of fruits produced in 1995, world exports were about 36 million metric tonnes. As shown in table 1.4, the top nine fruits in international trade were banana, citrus, apple, grape, pear, peach and nectarine, pineapple and kiwifruit. The leading exporters were Ecuador for bananas, Spain for citrus, France for apples, Italy for grapes, Argentina for pears, Italy for peaches and nectarines, Costa Rica for pineapples, and Italy for kiwifruit (with New Zealand a close second). Based on total fruit exports, the leading five exporters were Spain, Ecuador, USA, Costa Rica and Italy (table 1.4). Ranking of Belgium and the Netherlands as number six and 11, respectively, based on

Table 1.4 World fruit exports (1000 tonnes) by product and country in 1995

Country	Bananas	Citrus fruits	Apple	Grape	Pear	Peach and nectarine	Pineapple	Kiwifruit	Other fruits	Total
1. Spain	142	2817	38	91	69	96	9	3	358	3623
2. Ecuador	3344	—	—	—	—	—	6	—	75	3425
3. USA	398	1238	635	264	146	72	16	16	222	3007
4. Costa Rica	2269	—	—	—	—	—	171	—	20	2460
5. Italy	51	212	499	471	113	382	4	192	139	2063
6. Belgium	829	260	368	46	148	3	59	80	76	1869
7. Philippines	1213	—	—	—	—	—	164	—	46	1423
8. Columbia	1360	—	—	—	—	—	—	—	3	1363
9. Chile	1	7	433	443	147	84	—	111	80	1306
10. France	105	69	768	18	80	76	44	25	111	1296
11. The Netherlands	41	360	412	64	143	14	18	19	68	1139
12. South Africa	—	449	214	90	150	3	4	—	61	971
13. Argentina	—	258	243	9	222	—	—	—	14	746
14. Guatemala	636	27	4	—	1	—	—	—	17	658
15. Honduras	522	—	—	—	—	—	44	—	9	602
16. Greece	41	307	19	108	1	70	—	19	25	590
17. Turkey	—	393	28	25	8	9	—	—	39	502
18. New Zealand	—	1	302	—	6	1	—	173	7	490
19. Morocco	—	414	—	—	5	1	—	—	18	438
20. Ivory Coast	177	—	—	—	—	—	136	—	8	321
Other countries	2664	2450	1094	264	213	39	91	14	1694	8523
World total	13,157	8869	5025	1868	1443	841	766	652	3034	35,655

Source: Food and Agriculture Organization (FAO Statistical Database at http://fao.org).

total fruit export is partially due to their role as distribution centers for other European countries (obviously the bananas and citrus exported from these two countries are not produced there).

During the past few years many countries, such as China and Turkey, have greatly expanded their fruit production areas and they are expected to become major exporters of fruits in the near future. In 1996, China's fruit production was about 60 million metric tonnes of which only about 800,000 metric tonnes were exported, but this quantity is expected to increase rapidly in the future. Some southern hemisphere countries, such as Argentina and Brazil, have expanded their fruit production and exports in recent years and this trend is expected to continue.

Continued consolidation and vertical integration among producers, shippers and marketers, plus increased use of modern communications, characterize the global marketing systems for fresh produce and related value-added products. Both retail and foodservice produce buyers are demanding more services, including the application of 'product look up' (PLU) codes and more partnerships for efficient response to consumer demands. Mass-marketers are becoming major competitors to traditional food retailers, but there is also a growing demand for superior quality produce in certain areas. Food safety has become a major concern for all those concerned with marketing fresh produce.

Consumption of fresh fruits and vegetables is increasing in most countries. People are interested in healthier diets, natural foods, organic produce, convenience and diversity of foods. The produce industry is responding to consumers' demand for convenience by increasing the quantity and range of 'ready-to-eat' and 'ready-to-cook' fresh-cut fruit and vegetable products and including them with other food products in 'home meal replacement' marketing programs.

International marketing of fruits has been facilitated by improvements in attaining and maintaining the optimum environmental conditions (temperature; relative humidity; concentrations of oxygen, carbon dioxide and ethylene) in marine transport containers and trucks. Controlled atmosphere and precision temperature management allow non-chemical insect control for markets that have quarantine restrictions against pests endemic to exporting countries and for markets that do not want produce exposed to chemical fumigants.

Many studies are under way to develop alternative methods of insect control that are effective, are not phytotoxic to the fruits, and present no hazard to the consumer (Paull and Armstrong, 1994; Sharp and Hallman, 1994). This is a high-priority research and development area because of the possible loss of methyl bromide as an option for postharvest insect control. Alternatives to methyl bromide include cold treatments, hot water or air treatments (Paull and Chen, 2000), ionizing radiation (0.15–0.30 kGy), safer and more environmentally friendly chemical fumigants, and exposure to reduced (less than 0.5%) oxygen and/or elevated (40–60%) carbon dioxide atmospheres.

1.4 Importance of fruits and nuts in human nutrition and health

Fruits and vegetables contribute about 91% of vitamin C, 48% of vitamin A, 27% of vitamin B_6, 17% of thiamine, 15% of niacin, 16% of magnesium, 19% of iron and 9% of calories in the human diet. Tree nuts are a good source of essential fatty acids, high quality proteins, fiber, vitamin E and minerals. Other important nutrients, supplied by fruits, nuts and vegetables, include folacin, riboflavin, zinc, calcium, potassium and phosphorus (USDA, 1983). An excellent source of information on food composition and nutritional value is on the Internet at http://www.nal.usda.gov/fnic/foodcomp.

Fruits, nuts, and vegetables in the daily diet have been strongly associated with reduced risk for some forms of cancer, heart disease, stroke and other chronic diseases (Prior and Cao, 2000; Produce for Better Health Foundation, 1999; Southon, 2000; Tomas-Barberan and Robins, 1997; Wargovich, 2000). Some components of fruits and vegetables are strong antioxidants and function to modify the metabolic activation and detoxification/disposition of carcinogens, or even influence processes that alter the course of the tumor cell growth (Wargovich, 2000). Although antioxidant capacity varies greatly among fruits and vegetables (Prior and Cao, 2000) it is better to consume a variety of commodities rather than limiting consumption to a few with the highest antioxidant capacity. The USDA 2000 Dietary Guidelines (USDA, 2000) encourage consumers to: 1. enjoy five a day, i.e. eat at least two servings of fruits and at least three servings of vegetables each day, 2. choose fresh, frozen, dried or canned forms of a variety of colors and kinds, and 3. choose dark-green leafy vegetables, orange fruits and vegetables, and cooked dry beans and peas often. In some countries, consumers are encouraged to eat up to ten servings of fruits and vegetables per day.

There is increasing evidence that consumption of whole foods is better than isolated food components (such as dietary supplements and nutraceuticals). For example, increased consumption of carotenoid-rich fruits and vegetables was more effective than carotenoid supplements in increasing LDL oxidation resistance, lowering DNA damage, and inducing higher repair activity in human volunteers who participated in a study conducted in France, Italy, the Netherlands and Spain (Southon, 2000). Similar comparative studies are needed on other constituents of fruits and vegetables and on the bioavailability of nutrients taken as dietary supplements or as foods that contain these nutrients.

Constituents of fruits and nuts that have a positive impact on human health due to their antioxidant activity include the following (along with important sources among fruits and nuts):

1. Vitamin C (citrus fruits, guava, kiwifruit, pineapple, strawberry)
2. Vitamin E (almonds, cashew nuts, filberts, macadamias, pecans, pistachios, walnuts)
3. Carotenoids including provitamin A (apricot, mango, nectarine, orange, papaya, peach, persimmon, pineapple)
4. Flavonoids (apple, blackberry, blueberry, cranberry, grape, nectarine, peach, plum and prune, pomegranate, raspberry, strawberry)

1.5   Fruit composition and quality

Carbohydrates are the most abundant and widely distributed food component derived from plants. Fresh fruits vary greatly in their carbohydrate content, with a general range being between 10 and 25%. The structural framework, texture, taste and food value of a fresh fruit are related to its carbohydrate content (Kader and Barrett, 1996).

Sucrose, glucose and fructose are the primary sugars found in fruits and their relative importance varies among commodities. Sugars are found primarily in the vacuole and range from about 0.9% in limes to 16% in fresh figs. Sucrose content ranges from a trace in cherries, grapes and pomegranates to more than 8% in ripe bananas and pineapples. Such variation influences taste since fructose is sweeter than sucrose and sucrose is sweeter than glucose.

Starch occurs as small granules within the cells of immature fruits. Starch is converted to sugar as the fruits mature and ripen. Other polysaccharides present in fruits include cellulose, hemicellulose and pectin, which are found mainly in cell walls and vary greatly among commodities. These large molecules are broken down into simpler and more soluble compounds resulting in fruit softening. The transformation of insoluble pectins into soluble pectins is controlled, for the most part, by the enzymes pectinesterase and polygalacturonase.

Fruits contain less than 1% protein (as opposed to 9–20% protein in nuts such as almond, pistachio and walnut). Changes in the level and activity of proteins resulting from permeability changes in cell membranes may be involved in chilling injury. Enzymes, which catalyze metabolic processes in fruits, are proteins that are important in the reactions involved in fruit ripening and senescence (Seymour et al., 1993).

Lipids constitute only 0.1–0.2% of most fresh fruits, except for avocados, olives and nuts; however, lipids are very important because they make up the surface wax, which contributes to fruit appearance, and cuticle, which protects the fruit against water loss and pathogens. Lipids are also important constituents of cell membranes.

Most fresh fruits are acidic. Some fruits, such as lemons and limes, contain as much as 2–3% of their total fresh weight as acid. Total titratable acidity, specific organic acids present and their relative quantities, and other factors influence the buffering system and affect pH. Acid content usually decreases during ripening due to the utilization of organic acids during respiration or their conversion to sugars. Malic and citric acids are the most abundant in fruits except grapes (tartaric acid is the most important in most cultivars) and kiwifruits (quinic acid is the most abundant).

Pigments, which are the chemicals responsible for skin and flesh colors, undergo many changes during the maturation and ripening of fruits; these include: 1. Loss of chlorophyll (green color), which is influenced by pH changes, oxidative conditions and chlorophyllase action; 2. synthesis and/or revelation of carotenoids (yellow and orange colors), and 3. development of anthocyanins (red, blue and purple colors), which are fruit-specific.

Total phenolic content is higher in immature fruits than in mature fruits and typically ranges between 0.1 and 2 g per 100 g fresh weight (Macheix *et al.*, 1990). Fruit phenolics include chlorogenic acid, catechin, epicatechin, leucoanthocyanidins, flavonols, cinnamic acid derivatives and simple phenols. Chlorogenic acid (ester of caffeic acid) occurs widely in fruits and is the main substrate involved in enzymatic browning of cut, or otherwise damaged, fruit tissues when exposed to air.

Astringency is directly related to phenolic content and it usually decreases with fruit ripening because of conversion of astringent phenolic compounds from the soluble to the insoluble non-astringent form. Loss of astringency occurs via: 1. binding or polymerization of phenolics; 2. change in molecular size of phenolics; and/or 3. change in hydroxylation pattern of phenolic compounds.

Volatiles are responsible for the characteristic aroma of fruits. They are present in extremely small quantities ($<100\,\mu g/g$ fresh weight). The total amount of carbon involved in the synthesis of volatiles is $<1\%$ of that expelled as CO_2. The major volatile formed in climacteric fruits is ethylene (50–70% of the total carbon content of all volatiles). Ethylene does not have a strong aroma and does not contribute to typical fruit aromas.

Volatile compounds are largely esters, alcohols, acids, aldehydes, and ketones (low-molecular weight compounds). Very large numbers of volatile compounds have been identified in fruits, and more are identified as advances in separation and detection techniques and gas chromatographic methods are made; however, only a few key volatiles are important for the particular aroma of a given fruit (Eskin, 1991). Their relative importance depends upon threshold concentration (which can be as low as 1 ppb), potency, and interaction with other compounds.

1.6 Quality attributes of fruits

Appearance quality factors include size, shape, color, gloss and freedom from defects and decay. Defects can originate before harvest as a result of damage by insects, diseases, birds and/or hail; chemical injuries; and various blemishes (such as scars, scabs, russeting, rind staining). Postharvest defects may be physical, physiological or pathological (Snowdon, 1990). Textural quality factors include firmness, crispness, juiciness and mealiness. Flavor or eating quality depends upon sweetness (types and concentrations of sugars), sourness or acidity (types and concentrations of acids, buffering capacity), astringency (phenolic compounds) and aroma (concentrations of odor-active volatile compounds). Off-flavors may result from accumulation of fermentative metabolites (acetaldehyde, ethanol, ethyl acetate). Nutritional quality is related to contents of vitamins, minerals, dietary fiber and phytochemicals (Eskin, 1991; Kader, 1992; Seymour *et al.*, 1993).

Consumers consider good quality fruits to be those that look good, are firm and offer good flavor and nutritive value. Although consumers buy on the basis of appearance and textural (based on feel) quality, their satisfaction and repeat purchases are dependent upon good eating (flavor) quality. In contrast, producers and handlers are concerned first with appearance and textural quality along with long postharvest life (storage-life plus shelf-life). The challenge is to encourage the producers and handlers to pay more attention to flavor and nutritional quality and to encourage buyers and consumers to be willing to pay a higher price for fruit varieties and picking maturities that are superior in flavor because they often have lower yield and require more careful handling.

It is important to determine postharvest life and 'best if used by date' on the basis of flavor and nutritional quality rather than appearance and textural quality of intact and fresh-cut fruits. Several attempts have been made to develop non-destructive methods for measuring flavor quality and other quality attributes (Abbott, 1999). Examples include: use of near infrared spectroscopy for estimation of sugar content in fruits; removal of a very small amount of fruit tissue and measurement of soluble solids content; and use of electronic noses to determine concentration of odor-active volatiles. Until such methods become widely available, we will continue to depend on destructive techniques, such as soluble solids determination by refractometer and titratable acidity measurements by titration, to evaluate flavor quality of fruits.

1.7   Fruit maturity, ripening and quality relationships

Maturity at harvest is the most important factor that determines storage-life and final fruit quality. Immature fruits are more subject to shriveling and mechanical damage and are of inferior quality when ripe. Overripe fruits are likely to become soft and mealy with insipid flavor soon after harvest. Any fruit picked either too early or too late in its season is more susceptible to physiological disorders and has a shorter storage-life than fruit picked at the proper maturity (Kader, 1999).

All fruits, with a few exceptions (such as European pears, avocados and bananas), reach their best eating quality when allowed to ripen on the tree or plant. However, some fruits are usually picked mature but unripe so that they can withstand the postharvest handling system when shipped long-distance. Most currently used maturity indices are based on a compromise between those indices that would ensure the best eating quality to the consumer and those that provide the needed flexibility in marketing. Examples of these indices are total solids concentration (avocado), soluble solids concentration (cherry, grape, grapefruit, kiwifruit, mandarin, orange, and pear), titratable acidity concentration (pomegranate), ratio of soluble solids to titratable acidity (citrus fruits, grape) and flesh firmness (apple, pear). Color is used as the index of maturity for apricot, nectarine, peach, persimmon, plum, raspberry and strawberry. It would be better to add a minimum soluble solids concentration, a maximum acid concentration, and/or a minimum soluble solids/acid ratio as maturity indices for these fruits to assure better flavor quality to the consumers.

Ripening is the composite of the processes that occur from the latter stages of growth and development through the early stages of senescence and that results in the characteristic aesthetic and/or food quality, as evidenced by changes in composition, color, texture or other sensory attributes.

Fruits can be divided into two groups:

1. non-climacteric fruits that are not capable of continuing their ripening process once removed from the plant. Examples include berries (such as blackberry, raspberry, strawberry), cherry, citrus (grapefruit, lemon, lime, orange, mandarin and tangerine), grape, lychee, pineapple, pomegranate, tamarillo.
2. climacteric fruits that can be harvested mature and ripened off the plant. Examples include apple, pear, quince, persimmon, apricot, nectarine, peach, plum, kiwifruit, avocado, banana, mango, papaya, cherimoya, sapodilla, sapote, guava, passion fruit.

Fruits of the first group produce very small quantities of ethylene and do not respond to ethylene treatment except in terms of de-greening (removal of chlorophyll); these should be picked when fully ripe to ensure good flavor quality. Fruits in group 2 produce much larger quantities of ethylene in association with their ripening, and exposure to ethylene treatment will result in faster and more uniform ripening. Bananas must be treated with 10–100 ppm ethylene to initiate their ripening during transport or at destination handling facilities. Ripening of avocado, kiwifruit, mango and pear fruits before marketing is increasingly being used to provide consumers with the choice of purchasing ready-to-eat, ripe fruits or mature fruits that can be ripened at home. This practice has, in many cases, resulted in increased sales and profits (Kader, 1999).

Once fruits are ripened they require quick marketing and careful handling to minimize bruising. If delays cannot be avoided, the ripe fruits should be cooled to their minimum safe temperature and kept at that temperature until ready for retail display. Ripe, chilling-sensitive fruits tolerate lower temperatures than unripe fruits. Removal of ethylene and decreasing oxygen concentration to the 3–5% range can be useful supplements to maintaining the optimal temperature and relative humidity for delaying further ripening and deterioration of partially ripe fruits.

1.8   Factors influencing quality

Many preharvest and postharvest factors influence the composition and quality of fruits and nuts. These include: genetic factors (selection of cultivars and rootstocks); preharvest climatic conditions and cultural practices; maturity at harvest and harvesting method; postharvest handling procedures; and processing methods (Goldman *et al.*, 1999; Lee and Kader, 2000).

Within each fruit species there is a range of genotypic variation in composition, quality and postharvest-life potential. There are many opportunities using plant breeding and biotechnology methods to develop new cultivars that have good flavor and nutritional quality and to introduce resistance to physiological disorders and/or decay-causing pathogens to reduce use of fungicides. A cost–benefit analysis (including consumer acceptance issues) should be used to determine priorities for genetic improvement programs. In some cases, increasing the consumption of certain cultivars that are high in nutritive value and flavor may be more effective and less expensive than breeding for higher contents of phytonutrients.

The effects of preharvest climatic conditions and cultural practices on postharvest quality of fruits have been reviewed by several authors (Arpaia, 1994; Crisosto *et al.*, 1997; Ferguson *et al.*, 1999; Goldman *et al.*, 1999; Lee and Kader, 2000; Mattheis and Fellman, 1999; Prange and DeEll, 1997). In general, lower light intensity during growth leads to lower contents of ascorbic acid and sugars in fruits. Temperature influences the uptake and metabolism of mineral nutrients by plants. Rainfall affects the water supply to the plant and may cause fruit cracking. Soil type, rootstock, mulching, irrigation and fertilization influence the water and nutrient supply to the plant. High calcium content in fruits has been related to longer postharvest-life (Ferguson *et al.*, 1999). High nitrogen content is often associated with shorter postharvest-life of fruits (Arpaia, 1994; Crisosto *et al.*, 1997; Prange and DeEll, 1997).

There are numerous physiological disorders associated with mineral deficiencies. For example, bitter pit of apples, cork spot in apples and pears, and red blotch of lemons are associated with calcium deficiency in these fruits. Boron deficiency results in corking of apples, apricots and pears, lumpy rind of citrus fruits and cracking of apricots. Poor color of stone fruits may be related to iron and/or zinc deficiencies. Excess sodium and/or chloride (due to salinity) results in reduced fruit size and higher soluble solids content.

Severe water stress results in increased sunburn of fruits, irregular ripening of pears, tough and leathery texture in peaches and incomplete kernel development in nuts. Moderate water stress reduces fruit size and increases contents of soluble solids, acidity and ascorbic acid. On the other hand, excess water supply to the plants results in cracking of fruits (such as cherries and prunes), excessive turgidity leading to increased susceptibility to physical damage, reduced firmness, delayed maturity and reduced soluble solids content.

Cultural practices such as pruning and thinning determine the crop load and fruit size, which can influence nutritional composition of fruit. The use of pesticides and growth regulators does not directly influence fruit composition but may indirectly affect it due to delayed or accelerated fruit maturity.

Harvesting method can determine the extent of variability in maturity and physical injuries and consequently influence composition and quality of fruits. Mechanical injuries (bruising, surface abrasions, cuts, etc.) can accelerate loss of water and vitamin C and increase susceptibility to decay-causing pathogens. The incidence and severity of such injuries are influenced by the method of harvest (hand versus mechanical) and management of the harvesting and handling operations.

Keeping fruits within their optimal ranges of temperature and relative humidity is the most important factor in maintaining their quality and minimizing postharvest losses. Above the freezing point (for non-chilling-sensitive commodities) and the minimum safe temperature (for chilling-sensitive commodities), every $10°C$ increase in temperature accelerates deterioration and the rate of loss in nutritional quality by two- to threefold (Kader, 1992). Delays between harvesting and cooling or processing can result in direct losses (due to water loss and decay) and indirect losses (losses in flavor and nutritional quality). The extent of these losses depends upon the fruit's condition at harvest and its temperature, which can be several degrees higher than ambient temperatures, especially when exposed to direct sunlight. The distribution chain rarely has the facilities to store each commodity under ideal conditions and requires handlers to make compromises as to the choice of temperature and relative humidity. These choices can lead to physiological stress and loss of shelf-life and quality (Paull, 1999).

Responses to atmospheric modification vary greatly among plant species, organ type and developmental stage, and duration and temperature of exposure (Beaudry, 1999; Kader *et al.*, 1989). Maintaining the optimal ranges of oxygen, carbon dioxide and ethylene concentrations around the commodity extends its postharvest-life by about 50 to 100% relative to air control. Exposure to ethylene induces faster and more uniform ripening of climacteric fruits (Saltveit, 1999).

Postharvest opportunities for enhancing the quantity and quality of essential nutrients present in fruits and vegetables include: 1. increasing overall consumption of fruits and vegetables; 2. improving bioavailability of nutrients; 3. increasing levels of essential nutrients through fortification methods; and 4. reducing nutrient losses (Buescher *et al.*, 1999). All these strategies require effective interdisciplinary cooperation in research, establishing policy and educating consumers and all those involved in produce production and marketing.

1.9   Food safety assurance

During the past few years, food safety became and continues to be the number one concern of the fresh produce industry. The US Food and Drug Administration published in October 1998 a 'Guide to Minimize Microbial Food Safety Hazards for Fresh Fruits and Vegetables', which is based on the following principles: '1. Prevention of microbial contamination of fresh produce is favored over reliance on corrective actions once contamination has occurred; 2. To minimize microbial food safety hazards in fresh produce, growers, packers, or shippers should use good agricultural and management practices in those areas over which they have control; 3. Fresh produce can become microbiologically contaminated at any point along the farm-to-table food chain. The major source of microbial contamination with fresh produce is associated with human or animal feces; 4. Whenever water comes in contact with produce, its quality dictates the potential for contamination. Minimize the potential of microbial contamination from water used with fresh fruits and vegetables; 5. Practices using animal manure or municipal biosolid wastes should be managed closely to minimize the potential for microbial contamination of fresh produce; and 6. Worker hygiene and sanitation practices during production, harvesting, sorting, packing, and transport play a critical role in minimizing the potential for microbial contamination of fresh produce.'

Chlorine dioxide and ozone-based systems for water sanitization are increasingly being used instead of chlorine-based systems. Researchers continue to evaluate the efficacy of various ultraviolet radiation treatments in reducing microbial contamination of fresh produce. Other active research areas include identification of relevant indices of microbial quality of intact and fresh-cut fruits and investigation of how various postharvest handling treatments and conditions influence survival of human pathogens on produce.

References

Abbott, J.A. (1999) Quality measurement of fruits and vegetables. *Postharvest Biology and Technology*, **15**, 207-225.

Arpaia, M.L. (1994) Preharvest factors influencing postharvest quality of tropical and subtropical fruit. *HortScience*, **29**, 982-985.

Beaudry, R.M. (1999) Effect of O_2 and CO_2 partial pressure on selected phenomena affecting fruit and vegetable quality. *Postharvest Biology and Technology*, **15**, 293-303.

Brouk, B. (1975) *Plants Consumed by Man*, Academic Press, New York.

Buescher, R., Howard, L. and Dexter, P. (1999) Postharvest enhancement of fruits and vegetables for improved human health. *HortScience*, **34**, 1167-1170.

Crisosto, C.H., Johnson, R.S. and DeJong, T. (1997) Orchard factors affecting postharvest stone fruit quality. *HortScience*, **32**, 820-823.

Eskin, N.A.M. (ed.) (1991) *Quality and Preservation of Fruits*, CRC Press, Boca Raton, FL.

Ferguson, I., Volz, R. and Woolf, A. (1999) Preharvest factors affecting physiological disorders of fruit. *Postharvest Biology and Technology*, **15**, 255-262.

Goldman, I.L., Kader, A.A. and Heintz, C. (1999) Influence of production, handling, and storage on phytonutrient contents of foods. *Nutrition Reviews*, **57(9)**, S46-S52.

Hardenburg, R.E., Watada, A.E. and Wang, C.Y. (1986) *The Commercial Storage of Fruits, Vegetables, and Florist and Nursery Stocks*, US Department of Agriculture, Agriculture Handbook 66.

Kader, A.A. (ed.) (1992) *Postharvest Technology of Horticultural Crops*, 2nd edn. University of California Division of Agriculture and Natural Resources, Publication 3311.

Kader, A.A. (1999) Fruit maturity, ripening, and quality relationships. *Acta Horticulurae*, **485**, 203-208.

Kader, A.A. and Barrett, D.M. (1996) Classification, composition of fruits, and postharvest maintenance of quality, in *Processing Fruits: Science and Technology, Vol. 1. Biology, Principals, and Applications* (eds L.P. Somogyi *et al.*), Technomic Publishing Co., Lancaster, PA, pp. 1-24.

Kader, A.A., Zagory, D. and Kerbel, E.L. (1989) Modified atmosphere packaging of fruits and vegetables. *CRC Critical Reviews in Food Science and Nutrition*, **28**, 1-30.

Lee, S.K. and Kader, A.A. (2000) Preharvest and postharvest factors influencing vitamin C content of horticultural crops. *Postharvest Biology and Technology*, **20**, 207-220.

Macheix, J., Fleuriet, A. and Billot, J. (1990) *Fruit Phenolics*. CRC Press, Boca Raton, FL.

Mattheis, J.P. and Fellman, J.K. (1999) Preharvest factors influencing flavor of fresh fruits and vegetables. *Postharvest Biology and Technology*, **15**, 227-232.

Mitra, S. (ed.) (1997) *Postharvest Physiology and Storage of Tropical and Subtropical Fruits*, CAB International, Wallingford.

Morton, J.F. (1987) *Fruits of Warm Climates*, Creative Resource Systems, Winterville, NC.

Paull, R.E. (1999) Effect of temperature and relative humidity on fresh commodity quality. *Postharvest Biology and Technology*, **15**, 263-277.

Paull, R.E. and Armstrong, J.W. (eds) (1994) *Insect Pests and Fresh Horticultural Products: Treatments and Responses*, CAB International, Wallingford.

Paull, R.E. and Chen, N.J. (2000) Heat treatment and fruit ripening. *Postharvest Biology and Technology*, **21**, 21-37.

Prange, R. and DeEll, J.R. (1997) Preharvest factors affecting quality of berry crops. *HortScience*, **32**, 824-830.

Prior, R.L. and Cao, G. (2000) Antioxidant phytochemicals in fruits and vegetables: diet and health implications. *HortScience*, **35**, 588-592.

Produce for Better Health Foundation (1999) *Dietary Guidelines: The Case for Fruits and Vegetables First*, Produce for Better Health Foundation, Wilmington, DE (http://www.5aday.com).

Ryall, A.L. and Pentzer, W.T. (1982) *Handling, Transportation, and Storage of Fruits and Vegetables, Vol. 2. Fruits and Tree Nuts*, AVI Publishing Co., Westport, CT.

Saltveit, M.E. (1999) Effect of ethylene on fresh fruits and vegetables. *Postharvest Biology and Technology*, **15**, 279-292.

Seymour, G.B., Taylor, J.E. and Tucker, G.A. (eds) (1993) *Biochemistry of Fruit Ripening*, Chapman & Hall, London.

Sharp, J.L. and Hallman, G.J. (eds) (1994) *Quarantine Treatments for Pests of Plant Foods*, Westview Press, Boulder, CO.

Snowdon, A.L. (1990) *A Color Atlas of Postharvest Diseases and Disorders of Fruits and Vegetables, Vol. 1. General Introduction and Fruits*, CRC Press, Boca Raton, FL.

Southon, S. (2000) Increased fruit and vegetable consumption within the EU: potential health benefits. *Food Research International*, **33**, 211-217.

Tomas-Barberan, F.A. and Robins, R.J. (eds) (1997) *Phytochemistry of Fruits and Vegetables.* Oxford Science, Oxford.

USDA (1983) *Composition of Fruits and Fruit Juices, Raw, Processed, Prepared.* US Department of Agriculture, Agricultural Handbook 8-9 (http://www.nal.usda.gov/fnic/foodcomp).

USDA (2000) *Nutrition and Your Health: Dietary Guidelines for Americans*, Home and Garden Bulletin 232, US Department of Agriculture, Washington, DC (www.usda.gov/cnpp).

Wargovich, M.J. (2000) Anticancer properties of fruits and vegetables. *HortScience*, **35**, 573-575.

2 Inorganic nutrients and fruit quality

Ian B. Ferguson and Linda M. Boyd

2.1 Introduction

Inorganic nutrients influence fruit quality in many ways, the most substantial being direct effects of specific minerals on fruit disorders. Other effects are more subtle and often harder to quantify. Nevertheless, the concentration and distribution of inorganic nutrients in fruit is part of the physiological framework that determines quality.

In relating inorganic nutrients to fruit quality, we are focusing on consumer acceptance of fruit. This includes direct consumer-based properties such as texture, flavour and colour, and the absence of storage disorders and rots. In addition, inorganic nutrients influence fruit maturation and ripening. These processes underlie consumer acceptance, and are important equally to the grower and marketer.

Inorganic nutrients include the major cations potassium, calcium and magnesium, trace elements, and inorganic forms of phosphorus and nitrogen. These affect the nutrition of the developing fruit, postharvest quality, consumer preference and human nutrition. The requirements for fruit to develop to maturity and ripeness may be in conflict with our demands to maintain quality over an extended postharvest life and to provide a source of human nutrition. We will discuss these issues, together with some comments on the genetic basis of inorganic nutrition of fruit.

Of the mineral nutrients, calcium has the biggest impact on fruit quality. It is difficult not to write a chapter devoted to this mineral. In order to cover a wider range of nutrient effects, we have restricted our discussion of calcium; the cited literature will allow a greater exploration of this nutrient.

2.2 Inorganic contents of fruit and how they are achieved

If specific mineral concentrations in fruit are required for optimal postharvest quality, we need to understand how these are achieved and how we can influence them. This requires knowledge of typical levels of fruit minerals, the pattern of mineral flow into developing fruit and the factors that influence this.

2.2.1   Concentrations in mature fruit

2.2.1.1   Whole fruit

Concentrations of inorganic nutrients in mature fruit vary widely with species and fruit type (table 2.1). However, there are some general features that apply to most fruits. Nutrients that are less mobile in the plant (e.g. calcium) tend to have lower concentrations in the fruit than in leaves. Input of a relatively immobile nutrient depends on direct water flow rather than recycling from leaves via the phloem (see section 2.2.2). Concentrations of relatively immobile cations such as calcium and magnesium typically have fruit to leaf ratios of less than 0.2, whereas more mobile nutrients such as potassium may reach close to 1.0 (Ferguson, 1980). Nitrogen fruit to leaf ratios are also usually close to 1.0 (e.g. in tomatoes, Anderson *et al.*, 1999). Consequently leaf concentrations of nutrients often cannot be used to predict fruit concentrations.

2.2.1.2   Within fruit variation

Most nutrients are unevenly distributed within fruit; general patterns are shown in the examples in table 2.2. Typically, higher concentrations are found in the skin and seeds, and the lowest in the flesh. These patterns will reflect both the distribution of vascular tissue and sink characteristics. For example, sides of avocado fruit exposed to direct sunlight, where there is potentially greater water loss, tend to have higher concentrations of nutrients such as calcium that are dependent on water flow (Witney *et al.*, 1990a,b; Woolf *et al.*, 1999).

Nutrient concentrations within a fruit may be related to cellular function. Tissues which have high metabolic rates, such as the core, and cells under the skin are likely to have higher requirements of nitrogen and phosphorus (Hulme *et al.*, 1968). Rapidly expanding fruit parts, such as cortical flesh of apples and melons, are unlikely to have high calcium concentrations: their cell walls are elastic and less rigid than in non-expanding tissues. Cell density is also a factor.

Table 2.1 Mineral composition of a range of fruits. Results are in mg/100 g fresh weight of edible portion (US Department of Agriculture, 1992)

Nutrient	Apple	Apricot	Avocado	Banana	Orange	Pineapple	Strawberry
% Water	83.9	86.4	74.3	74.3	86.8	86.5	91.6
Ca	7	14	11	6	40	7	14
Mg	5	8	39	29	10	14	10
K	115	296	599	396	181	113	166
P	7	19	41	20	14	7	19
Fe	0.18	0.54	1.02	0.31	0.10	0.37	0.38
Na	0	1	10	1	0	1	1
Zn	0.04	0.26	0.42	0.16	0.07	0.08	0.13
Cu	0.04	0.09	0.26	0.10	0.05	0.11	0.05
Mn	0.05	0.08	0.23	0.15	0.035	1.65	0.29

Table 2.2 Distribution of some major nutrients in different tissues and positions in fruit (note that the mango data are on a fresh weight basis)

Tissue	Kiwifruit[a]		Mango[b]			Strawberry[c]		
	K	Ca	K	P	Ca	K	P	Ca
	mg/g DW		mg/100 g FW			mg/g DW		
Basal	49.74	1.70	149.3	8.89	11.56	23.43	2.17	1.77
Mid	47.12	1.25	164.6	10.08	8.27	—	—	—
Distal	45.08	1.06	165	11.19	7.25	20.73	2.11	1.29
Seeds	37.61	1.69	—	—	—	1.09	3.74	5.09
Skin	63.69	2.02	—	—	—	—	—	—
Inner flesh	28.74	1.08	202.3	13.75	4.61	9.76	0.91	0.77
Outer flesh	57.59	0.64	236.8	15	8.10	22.08	2.14	1.53

[a] Data recalculated from Ferguson (1980); inner flesh is core, seeds includes surrounding tissue.
[b] Data from Burdon *et al.* (1991).
[c] Data averaged for 2 cvs recalculated from Makus and Morris (1998).
FW = fresh weight; DW = dry weight.

The large cells in mid-cortical tissues in fruits such as in apples or melons are likely to have lower concentrations of the less mobile cations than inner cortical or core tissue. These poorly mobile nutrients are often located in cell walls and the extracellular fluid, which occupy a smaller volume of the mid-cortex than elsewhere. For instance, the area of sections of apple tissue occupied by air spaces doubled while the area occupied by cell walls decreased by half during growth to maturity (Harker and Ferguson, 1988a). Fruit such as avocados, which have smaller cells that continue to divide over the whole development period, may have higher mineral concentrations (table 2.1). Intracellular exclusion of cations such as calcium is another reason for lower concentrations in large-celled tissues.

Differences in distribution have postharvest implications. For example, even though total calcium levels in a fruit may be within a normal range, concentrations in the cortical flesh may be critical with respect to storage disorders such as bitter pit in apples (Ferguson and Watkins, 1989), or vulnerability to low temperature (e.g. avocados, Bower and Cutting, 1988). Commercial use of particular tissues or components, such as juice, highlights the importance of localized composition (e.g. high nitrate in pineapple juice is undesirable in canning). In these cases, the nutrient contents of the particular tissue are more important than those of the fruit as a whole.

2.2.1.3 Mineral ratios

Mineral concentrations are not always as important as the balance between them. This is recognized practically where ratios among calcium, magnesium and potassium are used as part of an index to predict quality of particular lines of apple fruit in storage (Waller, 1980). Calcium/nitrogen ratios have also been related to storage quality of apple fruit. High nitrogen soil treatments can result in higher bitter pit incidence in apple fruit (Faust and Shear, 1968; Bramlage *et al.*, 1980), and calcium and nitrogen have been related in diagnostic schemes (Fallahi and Righetti, 1984). However, calcium/nitrogen ratios are difficult to understand from a physiological or functional viewpoint, since effects of nitrogen on fruit can be indirect, for instance an increase in shoot/fruit ratios.

There may not be a direct physiological link between components of a ratio. Kiwifruit that develop a pitting disorder consistently have a low calcium to phosphorus ratio (Ferguson *et al.*, 2001). A high phosphorus level results from a predominance of phloem transport into the fruit when xylem transport (carrying calcium) is low. However, components of a ratio may be more directly linked. In apples, magnesium and potassium are often high when calcium is low because these cations compete in root uptake, xylem transport and at cell wall binding sites in the fruit.

2.2.2 Patterns of mineral input

Mineral nutrient flow into developing fruit over time is usually not linear. There is a rapid uptake phase associated with the early period of cell division. This has been observed in a range of crops (e.g. kiwifruit, figure 2.1). The pattern of uptake depends on the nature of the mineral and the transport pathways. A relatively immobile cation such as calcium may cease moving into fruit at the later stages of development (figure 2.1), although this pattern is not found in every season. Calcium contents of apple fruit are sometimes observed to increase continually over the whole season (Haynes and Goh, 1980). A more mobile cation such as potassium will continue to flow into the fruit over the season, often accumulating at a greater rate than fresh weight gain. These uptake patterns usually result in decreasing concentrations of minerals, such as calcium, and stable or increasing concentrations of nutrients, such as potassium and phosphorus, as fruit develop (e.g. see curves for mango fruit, Raymond *et al.*, 1998). High calcium/magnesium and calcium/potassium ratios can be important for postharvest quality and the drop in calcium concentration during the later stages of development increases the risk of calcium-related storage disorders.

A change in mineral ratios can be seen if fruit of different sizes but of the same maturity are analysed. In apple fruit, calcium concentrations usually decline with increasing fruit size, whereas potassium and magnesium concentrations may be constant, or increase (Ferguson and Watkins, 1992).

It is important with any fruit crop to understand the developmental changes in mineral accumulation, particularly those that might occur at the end of the growing season, if we are to harvest fruit with optimal concentrations for postharvest quality. An example is avocado, where fruit of greater maturity

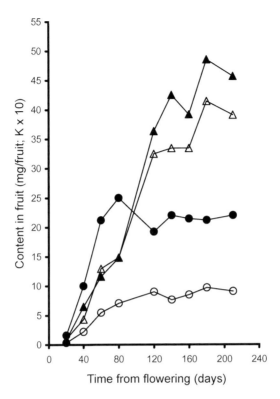

Figure 2.1 Accumulation of inorganic nutrients in developing kiwifruit. Data are recalculated from Ferguson (1980). —●— Ca; —○— Mg; —△— K; —▲— P.

generally have lower calcium and magnesium concentrations and are more susceptible to low temperature injury than less mature fruit (Cutting *et al.*, 1992).

There are two main phases of fruit mineral accumulation, which are, respectively, xylem-based and phloem-based. Initially a young fruit with active stomata and relatively large surface area to volume ratio will accumulate inorganic nutrients mostly through xylem transport. In this phase phloem supply to the fruit may be relatively low. In many fruit crops, leaf development predominates at this time, and leaves are the main sinks for phloem supply. Later, with rapid dry matter accumulation in fruit and a declining surface area to volume ratio, phloem tends to provide the predominant flow into the fruit. This is a long-standing hypothesis to explain patterns of nutrient accumulation by fruit but there is now physiological evidence for it from apple (Lang, 1990). Under conditions of high evaporative demand xylem allows flow out of the fruit and the presence of spur leaves promotes outflow during the day (Lang and Volz, 1998).

These patterns suggest that inorganic nutrient accumulation in fruit relies heavily on water flow from soil drawn by transpirational pull at the fruit surface.

Consequently there is interest in relating water economy of the plant with fruit development. For instance, grape berry transpiration has been strongly correlated with accumulation of nitrogen, phosphorus, potassium and calcium by the fruit; there was little or no correlation between leaf transpiration and these nutrients (Boselli et al., 1998).

A direct route from soil to fruit is supported by data showing increases in fruit mineral contents when soil treatments are applied. [15]N partitioning shows that nitrogen recently applied to the soil is directly allocated to fruit such as blackberry (Malik et al., 1991). Although there has been substantial research on nitrogen cycling in fruit trees, there is little information on the transport of re-mobilized versus soil-based nitrogen into developing fruit (Millard, 1995). Soil applications of potassium can result in higher fruit contents (e.g. in apple, Neilsen et al., 1998), consistent with the idea that potassium is a relatively mobile element in both phloem and xylem. Even a cation such as zinc, which is not recognized as particularly mobile, can move directly from soil to fruit, as shown in control of zinc deficiency in mango trees (Bahadur et al., 1998); foliar sprays were also effective in increasing fruit pulp zinc levels. A direct effect of soil calcium treatments resulting in increased calcium levels in fruit has rarely been observed (Ferguson and Watkins, 1989). However, Raese (1998) showed some increase in fruit calcium (and phosphorus) from soil fertilization of apples and pears. The paucity of direct effects from soil calcium has led to development of other techniques, such as spraying, to increase fruit calcium levels.

It is important to understand the relationship between mineral accumulation and fruit growth. Inorganic nutrients are expressed in the literature as concentrations on the basis of fresh weight, dry weight or total fruit content. All three provide useful information but they are not interchangeable. For instance the more mobile minerals will keep up with dry matter increase whilst the less mobile fall in concentration as dry matter accumulates. Rapid fruit expansion is usually associated with increased water in-flow, which may lower concentrations on a fresh-weight basis but not on a dry-weight basis. There is a need for more research relating mineral nutrient input into fruit to growth curves. Such curves are sigmoidal (e.g. mango, Saini et al., 1971), or double sigmoidal (e.g. cherries, Tukey, 1964). The period or periods of rapid growth may be modified by environmental conditions, particularly temperature and rainfall. The ability of inorganic nutrients to keep up with fruit growth may determine the concentrations of important nutrients such as calcium at harvest and consequent postharvest responses.

2.2.3   Factors influencing mineral input

2.2.3.1   Water relations

Most of the factors influencing nutrient accumulation are associated with the transport systems feeding the fruit. Any alteration of water economy, such as through modification of the relative humidity (RH) environment of the fruit, will affect mineral input. For instance, bagging fruit increases RH and seems to result in lower calcium concentrations in apples (Witney *et al.*, 1991) and mangoes (Hofman *et al.*, 1997). Incidence of blossom-end rot in tomatoes is often greater with higher humidity conditions in glasshouses; treatments that increase the transpirational flow reduce disorder incidence (e.g. Ho *et al.*, 1993; Paiva *et al.*, 1998).

2.2.3.2 Fruit position and pollination

Nutrient contents differ according to position of the fruit on the tree or vine. These differences are often caused by the effects of leaves on associated fruit development. In apple trees, fruit from upper parts of the tree tend to have lower calcium contents than fruit from lower down. Similar patterns can be found along a branch, where unloading seems to decrease with distance from the trunk (Schumacher *et al.*, 1980; Ferguson and Watkins, 1992).

Apple fruit with low seed numbers (poor pollination) tend to have lower calcium contents and higher bitter pit risk (Bramlage *et al.*, 1990; Brookfield *et al.*, 1996; Volz *et al.*, 1996b). The relationship between calcium contents of fruit and seed development has been linked with auxin supply by the seed and subsequent transport (Bangerth, 1976), but no adequate physiological explanation has yet been given.

2.2.3.3 Crop load

Light cropping apple trees have fruit with lower calcium and higher potassium concentrations than fruit from relatively heavy cropping trees (Ferguson and Watkins, 1992). These cropping differences may partly reflect a predominance of different fruiting wood types. Volz *et al.* (1994) found that fruit on terminal positions tend to have higher calcium and magnesium concentrations. The effect of leaves on the fruiting wood can be seen where removal of primary and bourse shoot leaves early in the season reduced calcium and magnesium contents of apple fruit, but this had no effect on potassium levels (Volz *et al.*, 1996a). These data suggest that leaves associated with fruiting wood attract mineral nutrients into the fruit vascular supply.

2.2.4 Horticultural practices influencing fruit mineral contents

2.2.4.1 Preharvest practices

Tree or vine management can be used to optimize mineral contents of fruit. For instance, alternate-cluster thinning can result in increased calcium contents of apple fruit, although within-cluster thinning has no such affect (Volz and Ferguson, 1999). Greater tree vigour in crops such as avocados has been associated with lower calcium and magnesium levels in fruit (Witney *et al.*, 1990a,b). Unfortunately, methods designed to increase fruit size usually have negative effects on critical nutrients such as calcium, and growers need to compromise on such practices.

Modification of soil nutrition is the usual method for altering nutrient contents of plant parts. Direct effects of soil applications on fruit contents are not consistent, as discussed above. Effects of changes in soil pH and available nutrients have been observed in fruit, but usually only after more than one season. In more intensive crops, such as glasshouse tomatoes, the growing medium can markedly influence fruit nutrient contents. For example, tomato fruit had higher calcium and lower magnesium, sodium and potassium concentrations when grown on organic (compost/soil mix) versus hydroponic substrates (Premuzic et al., 1998).

Where soil treatments have failed to increase fruit concentrations of nutrients, such as calcium, preharvest sprays have been developed and are widely used (Ferguson and Watkins, 1989). A calcium spray programme may greatly increase the fruit concentration (e.g. 13% in pears, Gerasopoulis and Richardson, 1997; >30% in apples, Turner et al., 1977; >100% in kiwifruit, Gerasopoulis et al., 1996). Uptake depends on coverage of the fruit with the spray and characteristics of the fruit skin. In a comparison of foliar sprays and soil applications, Makus and Morris (1998) found increases in fruit concentrations from sprays for one strawberry cultivar but not another. Herbaceous plants may respond more readily to soil and foliar calcium than woody fruit trees.

Calcium uptake through the fruit skin is dependent on natural openings such as stomata, unsuberized lenticels and cracks (Harker and Ferguson, 1988b). Thus, to increase the efficiency of uptake, movement through a small pore needs to be improved. Recent advances in the use of calcium sprays have centred on the development of compounds which minimize skin damage. The traditional chemicals, calcium chloride ($CaCl_2$) and calcium nitrate ($Ca(NO_3)_2$) can burn leaves and increase russet. Proprietary calcium chemicals tend to be buffered (at a higher pH) to reduce this damage and, in some cases, include chelated calcium to assist uptake.

2.2.4.2 Postharvest practices

Failure to grow fruit consistently with the desired mineral concentrations has led to the widespread industry practice of postharvest treatments. These have focused on increasing calcium concentrations in pome fruit, although they have also been used with other crops such as mangoes (Gunjate et al., 1977) and kiwifruit (Hopkirk et al., 1990). The simplest method is to dip or drench fruit and substantial benefits have been shown. A postharvest dip in 3% $CaCl_2$ significantly reduced the rate of softening of 'Hayward' kiwifruit (Hopkirk et al., 1990); data summarized by Ferguson and Watkins (1989), from about 50 apple calcium dipping experiments carried out in six countries, showed an average reduction in bitter pit of more than 50% with calcium dips. The effectiveness of such dips or drenches can be increased with additives such as lecithin

(Sharples *et al.*, 1979) or xanthan gum (Johnson, 1979), although results are not always consistent. As with sprays, various commercial calcium formulations are used in addition to $CaCl_2$, and the need to reduce potential calcium damage has driven this development as much as the effort to increase calcium uptake. Vacuum infiltration and pressure treatments with calcium salts have also been used. Vacuum infiltration can reduce bitter pit more than dipping alone (Scott and Wills, 1979) and pressure treatments can increase calcium concentrations in apples manyfold (Conway and Sams, 1987; Saftner *et al.*, 1999). Ultrastructure of fruit cuticles will determine the success of calcium uptake during postharvest infiltration; surface cracks in apple cuticles can increase and deepen during low temperature storage and infiltration can be greater if applied after storage (Roy *et al.*, 1999).

Postharvest calcium treatments have effects beyond the potential alleviation of storage disorders. Respiration, ethylene production and internal O_2 levels can be reduced, and internal CO_2 levels increased, with postharvest calcium dips or pressure treatments (Conway and Sams, 1987; Saftner *et al.*, 1999). Flavour volatiles are also reduced but recover with time in storage. These results suggest that calcium infiltration has a direct effect on gas exchange, perhaps by filling airspaces with solution. Part of the response may be caused by the calcium salt or its ionic properties, since water alone had a smaller effect.

The development of postharvest treatments is not necessarily the best way of solving storage or postharvest quality problems. Such treatments are expensive, increase risk of fruit damage through extra handling, and will not remove all risk in some fruit. They also encourage growers to pay less attention to on-tree quality, since a postharvest treatment is seen as a panacea. Growers should be producing fruit with a low risk of disorders, and properties that result in optimal postharvest quality, and which do not require special postharvest treatment. This requires more attention to fruit development and how various desired properties are determined.

2.3 Inorganic nutrients and fruit quality at harvest and during postharvest storage

Inorganic nutrients can influence most quality attributes of fruit during storage, retail handling and shelf life. The influence can be direct, such as when calcium in cell walls affects the rate of softening, or indirect, such as when nitrogen increases tree vigour resulting in less exposure of fruit to light, and reduced skin colour.

2.3.1 Fruit colour

Fruit colour changes can involve combinations of chlorophyll breakdown and the synthesis and degradation of carotenoids and phenolic pigments such as anthocyanins (Lancaster *et al.*, 1997). For example ripening in 'Tommy Atkins' mangoes is accompanied by chlorophyll and anthocyanin breakdown and increased carotenoid levels (Medlicott *et al.*, 1986). Red colour development in 'Fuji' apples has been associated with chlorophyll decline unmasking anthocyanins (Marsh *et al.*, 1996) and in tomatoes to increased synthesis of lycopene (Hobson and Grierson, 1993).

Colour changes can be affected by nitrogen and potassium nutrition. Nitrogen has been directly associated with maintaining green colour in fruit such as lemons (Koo *et al.*, 1974), stone fruit (Crisosto *et al.*, 1997) and pears (Raese, 1977). Skin chlorophyll concentration in 'Fuji' apples has been positively associated with fruit nitrogen concentrations (Marsh *et al.*, 1996). The possibility that the effect of nitrogen on fruit colour is an indirect result of an effect on tree vigour was mentioned above. There is no clear physiological or biochemical explanation for a more direct effect.

Manganese application has been positively associated with green background colour of 'Jonagold' apples (Deckers *et al.*, 1997). Red colour development in apples can be enhanced by potassium fertilizers, although not consistently (Neilsen *et al.*, 1998; Raese, 1998). In tomato fruit, there may be a direct effect of potassium on carotenoid synthesis, where potassium deficiency is associated with higher levels of β-carotene and a decrease in lycopene (Trudel and Ozbun, 1971). Increased nitrogen, phosphorus and potassium fertilization lowered the anthocyanin content of cranberries (Francis and Atwood, 1965).

Fruit flesh characteristics are important for processed products as well as whole fresh fruit. Increased nitrogen fertilization of peaches can result in improved 'a', 'b' and chroma values for fruit purée (more intense yellow colour) and a higher sensory rating (Olienyk *et al.*, 1997).

2.3.2 Flavour

Fruit flavour is a combination of taste (such as sweet or sour) and aroma. Total soluble solids, titratable acidity and aroma volatile composition are all associated with flavour and are commonly measured as part of fruit quality assessment.

Potassium has been associated with increased acidity in many fruits including citrus (e.g. Koo *et al.*, 1974), pears (Johnson *et al.*, 1998), some cultivars of apple (Marcelle, 1995) and blueberries (Ballinger and Kushman, 1969). While increased rates of potassium fertilization can result in increased titratable acidity in berryfruit, excessive potassium can also result in higher pH in grape tissue (Prange and DeEll, 1997). It is likely that there are direct effects of potassium on organic acid levels and on H^+ exchange, which may have opposite effects on the eventual quality of the fruit.

Soluble solids levels in fruit may respond to soil or foliar treatments of nutrients such as potassium and nitrogen, although not consistently. High nitrogen status has been associated with lowered soluble solids levels in pears (Raese, 1977) and apples (Dris *et al.*, 1999). Soluble solids increased in tomatoes with increasing fertilizer nitrogen levels (Barringer *et al.*, 1999), but did not respond to nitrogen fertilizers in studies on strawberries and blackberries (Alleyne and Clark, 1997; Miner *et al.*, 1997). Soil application of zinc sulphate increased total soluble solids in mango (Bahadur *et al.*, 1998). As with colour, it is difficult to determine whether these responses are caused by a direct physiological effect in the fruit or are indirect through increased health or growth of the tree. Nitrogen application usually enhances plant vigour and fruit yield and lower soluble solids concentrations with such treatments may be caused by dilution or shading effects.

Nitrogen fertilization has been associated positively with volatile production in apples (Somogyi *et al.*, 1964) and negatively with flavour ratings in 'd'Anjou' pears (Raese, 1977). Amino acid precursors are required in the synthesis of some aroma volatiles (Hansen and Poll, 1993), yet Fellman *et al.* (2000) found no marked effect of nitrogen application on the availability of these precursors and little or no effect of nitrogen on aroma production in apples.

2.3.3 Flesh firmness

Flesh firmness is a prime indicator of fruit quality, being a ripening index in fruit such as apples, kiwifruit, stone fruit, mangoes and avocados. Much postharvest research effort is dedicated to ways of maintaining fruit firmness during postharvest storage and shelf-life. Greater flesh firmness (slower rate of softening), in fruits such as apples, pears and kiwifruit, has been associated with higher flesh calcium concentrations (Poovaiah *et al.*, 1988; Ferguson and Watkins, 1989; Hopkirk *et al.*, 1990; Gerasopoulis *et al.*, 1996; Gerasopoulos and Richardson, 1997). Berryfruit also seem responsive to postharvest calcium dips, which help maintain firmness in strawberries (Garcia *et al.*, 1996) and blueberries (Hanson *et al.*, 1993). Pressure infiltration of $CaCl_2$ into apples can result in firmer fruit after storage, and this has been related to lower levels of soluble polyuronides (Sams and Conway, 1984).

This retention of firmness is probably caused by binding of calcium to pectic polymers in cell walls (reviewed in Ferguson, 1984; Poovaiah *et al.*, 1988; Fallahi *et al.*, 1997). Indirect evidence for this comes from tensile strength measurements. Tissue with relatively high calcium concentrations shows greater tensile strength, a measure of cell–cell cohesion (Poovaiah *et al.*, 1988; Stow, 1989). Calcium binding probably reduces the susceptibility of the pectic substrate to enzymic attack. The extent of a direct effect of calcium on activity of pectin-degrading enzymes such as polygalacturonase (PG) and pectinmethylesterase (PME), however, is still not certain. Both PME and exo-PG activity is enhanced with increasing calcium levels in tomato (Wills and Rigney, 1979) and peach (Pressey and Avants, 1973) tissues. In the same work with peach, endo-PG activity was reduced by calcium. Buescher and Hobson (1982), have also shown in tomato that EDTA treatment (presumably removing some fraction of bound calcium) reduced endogenous resistance to PG. Since then, there has been little clarification of the role of calcium in cell wall breakdown and reduction of softening. The published data show that effects of calcium on firmness are not often substantial and are usually observed in fruit at late stages of storage when flesh softening is advanced.

Both high and low levels of boron have been linked to fruit firmness. The relationship may be largely indirect and related to the effects of boron on fruit calcium levels (Faust and Shear, 1968; Shear, 1975). Results are not always consistent and may depend on cultivar. For example, Wojcik and Cieslinski (2000) reported that boron sprayed onto apple trees after flowering increased fruit calcium concentrations in the two cultivars studied, yet fruit firmness after storage increased in one cultivar and decreased in the second. Excess boron has been associated with premature softening in kiwifruit during storage. This has been linked to lower concentrations of calcium in fruit where boron status was high (Smith and Clark, 1989).

The other major nutrient associated with fruit firmness is nitrogen. Generally, high nitrogen contents of fruit have been negatively associated with firmness, as clearly shown for berry fruits such as strawberries, kiwifruit and cranberries (Prange and DeEll, 1997). However, there has been little or no explanation for such effects. Since firmness is related to cell turgor and cell wall characteristics, perhaps the major effect of nitrogen is on rate of fruit growth, with consequent effects on cell properties.

2.3.4   Ripening

Fruit ripening characteristics include changes to flesh firmness, carbohydrate metabolism, respiration rate and ethylene production. The effect of a mineral nutrient on ripening suggests an effect on a specific ripening characteristic. For example calcium sprays can increase pear flesh calcium concentrations by about 13%, increase chilling requirements (from 55 to 70 days; Gerasopoulis and Richardson, 1997), and reduce rates of softening (Gerasopoulis and Richardson, 1999). Similar results have been shown in kiwifruit (Gerasopoulis et al., 1996), avocados (Eaks, 1985) and mangoes where postharvest calcium dips increased shelf-life and delayed colour change, loss of firmness and weight loss (Mootoo, 1991).

These effects of calcium, found in a range of fruit, are not clearly explained but are likely to reside in the interaction of calcium with cell wall polymers and enzymes, and with membrane function. There is little indication of a direct effect of calcium on specific enzymes. Lipoxygenase, 1-aminocyclopropane-1-carboxylic acid (ACC) content and ethylene production can all be lower in apple fruit disks high in calcium (Marcelle, 1991). It is likely that effects of calcium on enzymes and metabolism are indirect, associated with calcium maintaining membrane function under conditions where permeability changes are naturally occurring. Calcium also has a role in signal transduction and we might speculate that reduced levels of extracellular calcium in fruit flesh ultimately lead to a breakdown in signalling processes, which may result in premature senescence (Ferguson and Watkins, 1989).

Positive effects of calcium on membrane function can often be observed by measuring relatively crude responses such as electrolyte leakage. Leakage was decreased in muskmelon mesocarp tissue by dipping fruit in calcium compounds (Lester and Grusak, 1999). A number of physiological changes associated with the plasma membrane, such as loss of phospholipid and protein contents, and H^+-ATPase activity, have also been reduced with calcium treatments of melon disks (Lester, 1996). This role of calcium in maintaining membrane function could be a major anti-senescence effect applicable to fruits and other plant organs. The lowering of respiration rates with increased calcium (e.g. in avocado fruit, Eaks, 1985) is likely to be linked to maintenance of membrane function. Similarly the association of calcium with a reduction in browning disorder symptoms, such as flesh and vascular browning in avocado fruit, is likely to be associated with maintaining membrane function and cellular compartmentation (Thorp et al., 1997).

Phosphorus may have a direct effect on respiration and metabolism through involvement in energy charge and ATP metabolism, in addition to its role in membrane lipid chemistry. Low-phosphorus cucumber fruit mesocarp had lower concentrations of phospholipids and respiration rates than tissues with relatively high phosphorus levels (Knowles et al., 2001). Low respiration rates in apples with high phosphorus levels were observed by Letham (1969) but not by Davenport and Peryea (1989). The potential importance of phosphorus in fruit lipid composition may be seen in the increase of total phospholipid in apple fruit at low postharvest temperatures (Lurie et al., 1987)

Uneven ripening is a problem in many fruits such as avocados and tomatoes, causing quality problems both with fresh fruit and processing. In tomatoes, this is manifest as yellow and white coloration of the flesh and can be reduced by increasing potassium supply to the plant (Hartz et al., 1999). Fruit with such symptoms may have lower potassium contents (Picha, 1987). Potassium seems to have a strong effect on tomato quality; apart from an effect on lycopene synthesis it is not clear how this occurs (Trudel and Ozbun, 1971).

2.3.5 Rots

Fruit nutrient contents, particularly calcium levels, can affect the disposition of fruit to fungal rots. Increased calcium levels in apple fruit, from either preharvest sprays or postharvest dip or infiltration treatments, can reduce storage rot incidence (Sams et al., 1993; Fallahi et al., 1997; Conway et al., 1999). However, the levels necessary may be in the range where calcium damage occurs. Calcium applications to strawberry plants can reduce postharvest rot incidence, although not always with noticeable increases in fruit calcium contents (Makus and Morris, 1989). Dipping strawberry fruit in $CaCl_2$ can increase fruit calcium and prolong postharvest life through effects on firmness, soluble solids and decay (Garcia *et al.*, 1996). In most of these studies, the association between maintenance of firmness and reduced incidence of rots suggests that calcium affects both processes through an inhibition of cell wall breakdown (Sams and Conway, 1984; Sams *et al.*, 1993). However, rots in stone fruit were not affected by either pre- or postharvest calcium treatments (Crisosto *et al.*, 1997). Stone fruit also provide evidence of indirect nutritional effects; high leaf nitrogen has been associated with greater rot susceptibility in nectarines, partly because of differences in fruit cuticle thickness (Crisosto *et al.*, 1997).

2.3.6 Specific nutrient-related disorders and postharvest responses

Low temperature breakdown, scald, bitter pit, cork spot and water core in apples, and jelly seed and soft nose in mangoes, are all examples of physiological disorders. These differ from damage to fruit caused by mechanical or pathological means. An association between physiological disorders and mineral nutrients has been developed over many years. In most cases disorders are associated with deficiency, often in some specific region rather than the whole fruit. Some storage responses, for example to low temperature, can be modified by varying levels of nutrients such as calcium.

Pitting is a major storage symptom manifest as the breakdown and browning of groups of cells either in the flesh, in the cells underlying the epidermis, or in cells associated with lenticels. The prime example, bitter pit in apples, usually occurs during low temperature storage, although occasionally it is found on the tree (Ferguson and Watkins, 1989). The discrete nature of pitting suggests that there may be groups of tissue or skin cells that have particularly low levels of nutrients such as calcium or boron. There are two types of evidence supporting the involvement of these elements:

- the relationship between low tissue concentrations of the nutrients and disorder incidence,
- the efficacy of artificial means of increasing the particular nutrient levels, with associated reduction in the disorder incidence.

For bitter pit and lenticel blotch in apples, the calcium/disorder relationship helps prediction based on tissue analysis (section 2.3.7) and decisions to apply preharvest calcium sprays or postharvest dips (e.g. Ferguson and Watkins, 1989; Watkins *et al.*, 1989).

Pitting in pears and apples has also been related to low fruit boron levels and can be can be treated by boron application (Faust and Shear, 1968; Raese, 1989).

Boron-related pitting, often known as 'cork spot', has different symptoms from calcium-related bitter pit. These include a drier appearance of the tissue, often described as 'corkiness', and sometimes cell division has been associated with the symptom (Faust and Shear, 1968). Boron application can sometimes help reduce the incidence of disorders which are traditionally associated with low levels of calcium, such as bitter pit and internal breakdown in apples (Wojcik and Cieslinski, 2000) and black end in 'Bartlett' pears (Raese, 1994). This highlights a relationship between calcium and boron which is often discussed but rarely clarified. Both nutrients are associated with binding to cell wall polymers.

In addition to pitting disorders, there are relationships between fruit inorganic nutrient contents and a wider breakdown or softening of fruit flesh. A major calcium-related disorder, which reflects uneven calcium distribution in the fruit, is blossom-end rot in tomatoes. This is flesh breakdown at the distal end of the fruit and has been strongly related to low calcium concentrations in the susceptible flesh. The causes for these low levels and the uneven distribution are a composite of high light and temperature stimulating fruit growth and poor calcium and water supply because of inadequate xylem development and competition from leaves and shoot growth (Ho *et al.*, 1993, 1999). The soft nose disorder in mango is also associated with breakdown of the flesh at the distal end of the fruit. Although the disorder is not exclusively associated with calcium content of the fruit, the distal region of the fruit can have the lowest calcium concentrations (Burdon *et al.*, 1991). Early work by Young and Miner (1961) suggested that the tendency to the disorder was exacerbated by high nitrogen levels in the tree. Another mango disorder, resulting in spongy tissue in the flesh, has been related to low calcium and high potassium contents, although this is not always observed (Gautam and Lizada, 1984), and Raymond *et al.* (1998) failed to find significant relationships between mineral contents and internal breakdown in mango fruit. Analysis of disordered tissue is not always reliable in giving an indication of a mineral relationship since nutrients can move into such sites after the cell breakdown occurs. The most useful relationships are those of mineral levels in tissue measured before the disorder occurs, and subsequent disorder incidence. Understandably, this is often difficult to quantify.

While some physiological disorders, such as soft nose in mango and blossom-end rot in tomato can be observed at harvest or upon immediate ripening, many are associated with long-term low temperature storage. Tropical and subtropical fruit are particularly susceptible to chilling injury, whose symptoms often include cell collapse, water loss and flesh browning. Chilling injury tolerance in muskmelons is decreased with low calcium nutrition and can be improved with boron application (Combrink *et al.*, 1995). Low levels of calcium have been associated with chilling injury susceptibility in avocado fruit and vacuum infiltration of calcium can reduce chilling injury symptoms, but with reduced external fruit quality (Chaplin and Scott, 1980; Eaks, 1985).

In pome fruit, internal breakdown symptoms are associated with, and may be a response to, low temperature, or a reflection of tissue senescence. Low temperature and senescent breakdown in apples have been related to low calcium concentrations in the flesh and to low phosphorus levels (Johnson and Yogaratnam, 1978; Yogaratnam and Sharples, 1982), although there is no obvious explanation for an effect of phosphorus.

The low levels of a nutrient such as calcium may not necessarily directly cause the symptoms of tissue breakdown. It is more likely that at certain stages of ripening or postharvest duration, cells collapse or become dysfunctional where calcium is at a lower level. In other words, there is some ripening or senescence trigger to which predisposed tissue responds differently. A low level of calcium, for example, may make the cell plasma membranes more susceptible to water loss or osmotic changes. The positive effects of calcium in reducing chilling injury may also depend on protection against loss of selective membrane permeability.

Watercore is a common disorder in many apple cultivars (although it is regarded as a natural and sometimes desirable feature in cultivars such as 'Fuji'). It is the result of accumulation of soluble sugars such as sorbitol in the free space of the fruit tissue (Marlow and Loescher, 1984). Both high levels of nitrogen and low levels of calcium in the fruit have been associated with the disorder (Bowen and Watkins, 1997; Fallahi *et al.*, 1997), although there is no evidence for a direct effect of either nutrient on the physiology of watercore development. Excess boron application has also been associated with watercore and internal breakdown in apples (Bramlage and Thompson, 1962; Martin *et al.*, 1976).

Fruit cracking is a further disorder that is commercially important in crops such as cherries and pome fruit. It has been associated with deficiency of both boron and calcium; incidence can be reduced by spray treatments with these nutrients (Opara *et al.*, 1997). The deficiencies are probably associated with weakening of skin structure, leading to breakdown under conditions of high turgor pressure in the flesh tissue.

2.3.7 *Prediction of disorders*

Prediction schemes are based on measurements that can be made preharvest or at harvest and will give an indication of how well a line of fruit will store. One way of increasing the predictive accuracy and confidence is to incorporate more than one factor into the prediction scheme. The calcium/bitter pit relationship has been expanded to include fruit size and other nutrients such as magnesium and potassium (Bramlage *et al.*, 1985; Autio *et al.*, 1986; Ferguson and Watkins, 1989). Up to 12 variables have been included in predictive models for bitter pit and low temperature breakdown in apple fruit, even though fruit calcium and potassium were the strongest predictors (Johnson and Ridout, 1998). Leaf analysis has not proved successful in predicting postharvest fruit quality or, in most cases, fruit mineral concentrations (e.g. Fallahi *et al.*, 1985; Ferguson and Watkins, 1989; Fallahi and Simons, 1996). Where regression analysis has shown positive correlations between leaf and fruit properties, investigations have rarely encompassed many orchards, districts and seasons.

Environmental influences on disorders or quality properties need to be factored into predictions; Johnson and Ridout (1998) found that their predictive models accounted for 67% of the variance in bitter pit incidence in fruit stored in air, but only 39% for fruit in controlled atmospheres. This suggests that other factors associated with long-term storage played a larger role in disorder incidence under controlled atmospheric conditions.

Successful prediction depends on a number of factors:

- the fruit property (or properties) needs to be accurately measured
- there needs to be a strong and significant relationship between the fruit property and the targeted quality characteristics
- the relationship must hold over different orchards and seasons

If the relationship is based on fruit mineral concentrations, then sampling and analytical techniques become important; variability needs to be minimized. For example, Ferguson and Triggs (1990) showed for bitter pit prediction that fruit to fruit variability in calcium concentrations within a tree was greater than that between trees or within fruit. In order to represent an orchard block and to minimize variation it was more important to sample from many trees than to take many fruit from one tree. Calcium/bitter pit relationships can differ also depending on the type of fruit sample; they are usually stronger for bulk rather than individual fruit analyses, for example (Perring, 1968; Perring and Sharples, 1975).

2.4 Inorganic nutrients in relation to human nutrition and the consumer

There is increasing appreciation that quality means more than just taste, texture and appearance. Nutritional properties of fruit (e.g. vitamins, minerals) and perceived health benefits (e.g. antioxidants) are becoming factors in consumer preference.

2.4.1 Health and fruit consumption

Epidemiological data show that a diet rich in fruit and vegetables may provide protection against chronic diseases prevalent in western society, particularly some forms of cancer (Steinmetz and Potter, 1996) and, to a lesser extent, cardiovascular disease and stroke (Joshipura *et al.*, 1999; Van't Veer *et al.*, 2000). The protective effect of these foods is not fully understood. Supplementation with specific food components rather than whole fruit and vegetables does not confer the same protective effects (Greenberg *et al.*, 1994) and may even be deleterious (Omenn *et al.*, 1996).

Dietary guidelines in many countries encourage the consumption of at least two servings of fruit and three of vegetables a day (WHO, 1990). Nutrition surveys reveal wide variation in daily fruit and vegetable consumption. In New Zealand, for example, 34% of adult males and 56% of adult females consume at least two daily servings of fruit whilst 34% of males and 19% of females consume less than one serving a day (Russell *et al.*, 1999). In the USA around 17 to 20% of adults meet target consumption of five or more servings of fruit and vegetables a day (Thompson *et al.*, 1999; Li *et al.*, 2000).

While health is a motivating factor for purchasing fruit, there are many reasons why a high proportion of people do not meet recommended daily intakes. These include socioeconomic and social factors. In a society where time available for food preparation and consumption is decreasing, consumers have reported that fruits have less reliable or predictable eating quality than manufactured snacks (Jack *et al.*, 1997). Other things which may dissuade people from buying and eating fruit include poor appearance, taste, texture and unpredictable ripening.

Fruit quality means different things to different people: for the grower, achieving high yield and large fruit size; for the transporter, long storage potential and continuity of supply; for the consumer, nutritional value and eating quality. These requirements are sometimes in conflict. Nitrogen fertilization for example may increase fruit yield and size. However high nitrogen status can result in higher fruit losses in storage (Tagliavini *et al.*, 1995) and lower vitamin C, acidity and soluble sugar concentrations (Nagy, 1980; Marcelle, 1995). These are all negative factors from the consumer point of view. High fruit potassium and low fruit calcium are often associated with high fruit acidity and sugars (Baldry *et al.*, 1982; Marcelle, 1995) and potassium fertilization can increase vitamin C levels (Nagy, 1980), yet high calcium and low potassium fruit tend to have better storage potential.

Consumer acceptability tends to be based on appearance and sensory properties associated with texture and flavour rather than on laboratory measurements such as titratable acidity, dry matter and firmness. For example, consumers may have a preference for a sweet, hard apple or a juicy, acidic apple (Daillant-Spinnler *et al.*, 1996). Some preharvest nutritional conditions can result in altered consumer preferences. With tomatoes, increased salinity can result in increased dry matter, and high concentrations of sugars, acids, carotene and vitamin C (Adams, 1991; Petersen *et al.*, 1998); the increased sweetness can be detected by sensory analysis. However, there is not always such a correlation between chemical analysis of components and consumer preference. For example, Johnson *et al.* (1998) found that there was no sensory discrimination (flavour, sweetness, firmness and juiciness) between fruit with enhanced potassium levels and control fruit, even though the high potassium fruit had higher acidity, lower sugar levels and were firmer.

Consumer acceptability is also being measured as part of the quality attributes of fruit treated to enhance postharvest life. For example, apples infiltrated with 2% $CaCl_2$ to enhance storage were perceived as crisper, sweeter and overall more acceptable than untreated fruit (Klein *et al.*, 1998). Postharvest $CaCl_2$ dips reduced strawberry decay during shelf-life without affecting the sensory quality of the fruits (Garcia *et al.*, 1996).

2.4.2   *Fruit as a direct source of minerals*

Mayer (1997) has studied food composition tables over the past 50 years and reports a trend towards lower mineral and dry matter content in fruits. The most significant reductions were in the levels of magnesium, iron, copper and potassium. Possible explanations for this trend include changes in sampling and analytical methodologies through to agricultural and breeding practices.

There is a drive to increase nutrient levels in fruit and vegetables, both to counter nutrient deficiencies in developing countries and as consumers voice a preference for high nutrient produce. Buescher *et al.* (1999) list the following postharvest strategies (some of which are being applied to grains and cereal products) for improving the micronutrient intake from fruits and vegetables: 1. increasing consumption of fruits and vegetables; 2. improving bioavailability of nutrients; 3. increasing levels of essential nutrients through fortification methods; and 4. improving nutrient retention.

From a nutrition standpoint, fruit are not considered to be a primary source of minerals (Fairweather-Tait and Hurrell, 1996). However, fruit can be a good source of minerals, in terms of the US FDA (1993) definition that a serving of the fruit would provide 10–20% of the daily value of that nutrient (Miller-Ihli, 1996). According to this definition, avocado, strawberry, banana and melons are good sources of potassium and avocado, coconut, banana and pineapple are good sources of manganese.

2.5   Genetic approaches

Modifying fruit concentrations of inorganic nutrients to optimize postharvest quality incurs considerable cost. This is embodied in preharvest fruit sprays, postharvest treatments, such as dips and drenches, and soil and foliar fertilizer applications. When market losses are added, costs substantially multiply. Another approach is to select or breed cultivars that have mineral fruit concentrations in the required ranges. Although little has been done in this regard, there have been attempts to improve calcium contents of fruit and vegetables genetically.

It was first necessary to determine whether calcium content of organs such as fruit is a heritable trait. Experience with different fruit cultivars suggests that there is a genetic component to some calcium-related disorders. Various comparative studies of apple cultivars consistently show a range of susceptibility to bitter pit (Ferguson and Watkins, 1989). While this is suggestive, more direct evidence comes from analysis of specific breeding lines or populations. In apple, analysis of 30 half-sib families from open-pollinated seed showed that bitter pit was a moderately heritable trait (Volz *et al.*, 2000) and calcium content of the fruit strongly heritable (Volz *et al.*, 2001). Segregation analysis in apple seedlings has also led to suggestions of 'bitter pit genes' which control calcium accumulation and distribution in the fruit (Korban and Swiader, 1984).

The concept of genes controlling calcium flow into developing fruit does not necessarily demand a direct calcium function. For example, genes controlling vascular development, leaf and fruit balance, skin permeance, cell division and expansion, may all influence calcium accumulation. Once heritability is established, gene mapping can be used to develop marker-assisted selection. The marker genes in such a progamme need not be directly calcium-associated.

Blossom-end rot in tomatoes has a genetic component (Greenleaf and Adams, 1969). Efficiency of calcium utilization varies among different genetic lines, both in terms of root uptake and internal distribution (Giordano *et al.*, 1982; Li and Gabelman, 1990). There appears to be sufficient strength in this genetic control to allow selection of calcium efficient lines that would be less susceptible to blossom-end rot.

A further genetic study showed significant differences in calcium concentrations in snap bean pods (Quintana *et al.*, 1996, 1999a) and these have been associated with differences in xylem flow rates (Quintana *et al.*, 1999b). Rates appeared more important than calcium concentration in the sap. This highlights the potential impact of environment and plant processes (in this case transpiration and water availability) on eventual calcium contents of fruits and fruit-like organs. In the above genetic studies on tomatoes and apples, strong environmental influences on calcium and magnesium concentrations were also observed. While such environmental or site interactions can complicate genetic improvement of fruit mineral contents, they do not necessarily preclude it.

Selection of improved fruit cultivars and breeding of new ones has been driven by a mixture of yield, disease resistance and consumer requirements. There has been little attention paid to mineral-based attributes such as storage disorders. However, this strategy must be an increasingly attractive option, particularly with the availability of genomic DNA sequences of fruit crop species that will make available potential candidate genes for a gene mapping and marker-assisted selection approach to cultivar development.

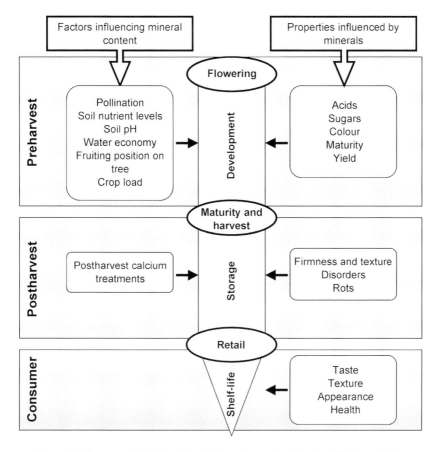

Figure 2.2 Summary of the interactions between inorganic nutrients and fruit production.

2.6 Conclusions

Despite the enormous amount of research carried out on the relationships between inorganic nutrient composition of fruit and postharvest quality, the impact of modifying mineral nutrients on quality in most cases is small, compared with the impact of ethylene, modified O_2 and CO_2 concentrations, low temperature and fruit maturity. The exception is with specific disorders such as bitter pit, where calcium is ameliorative. The value in understanding the relationships between quality and nutrients is that fruit nutrition must be a recognized input in the overall management of fruit crop production (figure 2.2). Further, this management must be directed to postharvest and consumer-based fruit quality rather than just to production. Nutritional composition should be the framework for developing other quality characters of fruit, such as texture, flavour and appearance. As the summary chart in figure 2.2 shows, this framework has to be developed initially in the orchard and then maintained and improved upon in the later stages of the fruit production pipeline. Postharvest quality is determined on the tree and only modified by postharvest practice.

References

Adams, P. (1991) Effects of increasing the salinity of the nutrient solution with major nutrients or sodium chloride on the yield, quality and composition of tomatoes grown in rockwool. *Journal of Horticultural Science*, **66**, 201-207.

Alleyne, V. and Clark, J.R. (1997) Fruit composition of 'Arapaho' blackberry following nitrogen fertilization. *HortScience*, **32**, 282-283.

Anderson, P.C., Rhoads, F.M., Olson, S.M. and Hill, K.D. (1999) Carbon and nitrogen budgets in Spring and Fall tomato crops. *HortScience*, **34**, 648-652.

Autio, W.R., Bramlage, W.J. and Weis, S.A. (1986) Predicting poststorage disorders of 'Cox's Orange Pippin' and 'Bramley's Seedling' apples by regression equations. *Journal of the American Society for Horticultural Science*, **111**, 738-742.

Bahadur, L., Malhi, C.S. and Singh, Z. (1998) Effect of foliar and soil applications of zinc sulphate on zinc uptake, tree size, yield, and fruit quality of mango. *Journal of Plant Nutrition*, **21**, 589-600.

Baldry, J., Dougan, J. and Howard, G.E. (1982) The effect of nutritional treatments on postharvest quality and flavour of Valencia oranges. *Journal of Horticultural Science*, **57**, 239-242.

Ballinger, W.E. and Kushman, L.J. (1969) Relationship of nutrition and fruit quality of Wolcott blueberries grown in sand culture. *Journal of Horticultural Science*, **94**, 329-335.

Bangerth, F. (1976) A role for auxin and auxin transport inhibitors on the Ca content of artificially induced parthenocarpic fruits. *Physiologia Plantarum*, **37**, 191-194.

Barringer, S.A., Bennett, M.A. and Bash, W.D. (1999) Effect of fruit maturity and nitrogen fertilizer levels on tomato peeling efficiency. *Journal of Vegetable Crop Production*, **5**, 3-11.

Boselli, M., Di Vaio, C. and Pica, B. (1998) Effect of soil moisture and transpiration on mineral content in leaves and berries of Cabernet Sauvignon grapevine. *Journal of Plant Nutrition*, **21**, 1163-1178.

Bowen, J.H. and Watkins, C.B. (1997) Fruit maturity, carbohydrate and mineral content relationships with watercore in 'Fuji' apple. *Postharvest Biology and Technology*, **11**, 31-38.

Bower, J.P. and Cutting, J.G.M. (1988) Avocado fruit development and ripening physiology. *Horticultural Reviews*, **10**, 229-271.

Bramlage, W.J. and Thompson, A.H. (1962) The effects of early-season sprays of boron on fruit set, colour, finish and storage life of apples. *Proceedings of the American Society for Horticultural Science*, **80**, 64-72.

Bramlage, W.J., Drake, M. and Lord, W.J. (1980) The influence of mineral nutrition on the quality and storage performance of pome fruits grown in North America, in *Mineral nutrition of fruit trees* (eds D. Atkinson, J.E. Jackson, R.O. Sharples and W.M. Waller). Butterworths, London, pp. 29-39.

Bramlage, W.J., Weis, S.A. and Drake, M. (1985) Predicting poststorage disorders of 'McIntosh' apples from preharvest mineral analyses. *Journal of the American Society for Horticultural Science*, **110**, 493-498.

Bramlage, W.J., Weis, S.A. and Greene, D.W. (1990) Observations on the relationships among seed number, fruit calcium, and senescent breakdown in apples. *HortScience*, **25**, 351-353.

Brookfield, P.L., Ferguson, I.B., Watkins, C.B. and Bowen, J.H. (1996) Seed number and calcium concentrations of 'Braeburn' apple fruit. *Journal of Horticultural Science*, **71**, 265-271.

Buescher, R.W. and Hobson, G.E. (1982) Role of calcium and chelating agents in regulating the degradation of tomato fruit tissue by polygalacturonase. *Journal of Food Biochemistry*, **6**, 147-160

Buescher, R., Howard, L. and Dexter, P. (1999) Postharvest enhancement of fruits and vegetables for improved human health. *HortScience*, **34**, 1167-1170.

Burdon, J.N., Moore, K.G. and Wainwright, H. (1991) Mineral distribution in mango fruit susceptible to the physiological disorder soft-nose. *Scientia Horticulturae*, **48**, 329-336.

Chaplin, G.R. and Scott, K.J. (1980) Association of calcium in chilling injury susceptibility of stored avocados. *HortScience*, **15**, 514-515.

Combrink, N.J.J., Jacobs, G. and Maree, P.C.J. (1995) The effect of calcium and boron on the quality of muskmelons. *Journal of the Southern African Society for Horticultural Sciences*, **5**, 33-38.

Conway, W.S. and Sams, C.E. (1987) The effects of postharvest infiltration of calcium, magnesium or strontium on decay, firmness, respiration and ethylene production in apples. *Journal of the American Society for Horticultural Science*, **112**, 300-302.

Conway, W.S., Janisiewicz, W.J., Klein, J.D. and Sams, C.E. (1999) Strategy for combining heat treatment, calcium infiltration, and biocontrol to reduce postharvest decay of 'Gala' apples. *HortScience*, **34**, 700-704.

Crisosto, C.H., Johnson, R.S., DeJong, T. and Day, K.R. (1997) Orchard factors affecting postharvest stone fruit quality. *HortScience*, **32**, 820-823.

Cutting, J.G.M., Wolstenholme, B.N. and Hardy, J. (1992) Increasing relative maturity alters the base mineral composition and phenolic concentration of avocado fruit. *Journal of Horticultural Science*, **67**, 761-768.

Daillant-Spinnler, B., MacFie, H.J.H., Beyts, P.K. and Hedderley, D. (1996) Relationships between sensory properties and major preference directions of 12 varieties of apples from the Southern Hemisphere. *Food Quality and Preference*, **7**, 113-126.

Davenport, J.R. and Peryea, F.J. (1989) Whole fruit mineral element content and respiration rates of harvested 'Delicious' apples. *Journal of Plant Nutrition*, **12**, 701-713.

Deckers, T., Daemen, E., Lemmens, K., Missotten, C. and Val, J. (1997) Influence of foliar applications on Mn during summer on the fruit quality of Jonagold. *Acta Horticulture*, **448**, 467-473.

Dris, R., Niskanen, R. and Fallahi, E. (1999) Relationship between leaf and fruit minerals and fruit quality attributes of apples grown under northern conditions. *Journal of Plant Nutrition*, **22**, 1839-1851.

Eaks, I.L. (1985) Effect of calcium on ripening, respiratory rate, ethylene production, and quality of avocado fruit. *Journal of the American Society for Horticultural Science*, **110**, 145-148.

Fairweather-Tait, S. and Hurrell, R.F. (1996) Bioavailability of minerals and trace elements. *Nutrition Research Reviews*, **9**, 295-324.

Fallahi, E. and Righetti, T.L. (1984) Use of Diagnosis and Recommendation Integrated System (DRIS) in apple. *HortScience*, **19**, 54.

Fallahi, E. and Simons B.R. (1996) Interrelations among leaf and fruit minerals nutrients and fruit quality in 'Delicious' apples. *Journal of Tree Fruit Production*, **1**, 15-25.

Fallahi, E., Righetti, T.L. and Richardson, D.G. (1985) Predictions of quality by preharvest fruit and leaf mineral analyses in 'Starkspur Golden Delicious' apple. *Journal of the American Society for Horticultural Science*, **110**, 71-74.

Fallahi, E., Conway, W.S., Hickey, K.D. and Sams, C.E. (1997) The role of calcium and nitrogen in postharvest quality and disease resistance of apples. *HortScience*, **32**, 831-835.

Faust, M. and Shear, C.B. (1968) Corking disorders of apples: a physiological and biochemical review. *Botanical Reviews*, **34**, 441-469.

Fellman, J.K., Miller, T.W., Mattinson, D.S. and Mattheis, J.P. (2000) Factors that influence biosynthesis of volatile flavour compounds in apple fruit. *HortScience*, **35**, 1026-1033.

Ferguson, I.B. (1980) Movement of mineral nutrients into the developing fruit of the kiwifruit (*Actinidia chinensis* Planch.) *New Zealand Journal of Agricultural Research*, **23**, 349-353.

Ferguson, I.B. (1984) Calcium in plant senescence and fruit ripening. *Plant, Cell and Environment*, **7**, 477-489.

Ferguson, I.B. and Triggs, C.M. (1990) Sampling factors affecting the use of mineral analysis of apple fruit in the prediction of bitter pit. *New Zealand Journal of Crop and Horticultural Science*, **18**, 147-152.

Ferguson, I.B. and Watkins, C.B. (1989) Bitter pit in apple fruit. *Horticultural Reviews*, **11**, 289-355.

Ferguson, I.B. and Watkins, C.B. (1992) Crop load affects mineral concentrations and incidence of bitter pit in 'Cox's Orange Pippin' apple fruit. *Journal of the American Society for Horticultural Science*, **117**, 373-376.

Ferguson, I.B., Thorp, G., Barnett, A. and Boyd, L. (2001) Physiological pitting in Hayward kiwifruit. *New Zealand Kiwifruit Journal*, **Jan/Feb**, 6-8.

Francis, F.J. and Atwood, W.M. (1965) The effect of fertilizer treatments on the pigment of cranberries. *Proceedings of the American Society of Horticultural Science*, **77**, 351-357.

Garcia, J.M., Herrara, S. and Morilla, A. (1996) Effects of postharvest dips in calcium chloride on strawberry. *Journal of Agricultural and Food Chemistry*, **44**, 30-33.

Gautam, D.M. and Lizada, M.C.C. (1984) Internal breakdown in 'Carabao' mango subjected to modified atmospheres IV. Ca and K levels in the spongy tissue. *Postharvest Research Notes*, **1**, 90-94.

Gerasopoulis, D. and Richardson, D.G. (1997) Fruit maturity and calcium affect chilling requirements and ripening of 'd'Anjou' pears. *HortScience*, **32**, 911-913.

Gerasopoulis, D. and Richardson, D.G. (1999) Storage temperature and fruit calcium alter the sequence of ripening events of 'd'Anjou' pears. *HortScience*, **34**, 316-318.

Gerasopoulis, D., Chouliaras, V. and Lionakis, S. (1996) Effects of preharvest calcium chloride sprays on maturity and storability of Hayward kiwifruit. *Postharvest Biology and Technology*, **7**, 65-72.

Giordano L. de B., Gabelman, W.H. and Gerloff, G.C. (1982) Inheritance of differences in calcium utilization by tomatoes under low-calcium stress. *Journal of the American Society for Horticultural Science*, **107**, 664-669.

Greenberg, E.R., Baron, J.A., Tosteson, T.D. *et al.* (1994) A clinical trial of antioxidant vitamins to prevent colorectal adenoma. *New England Journal of Medicine*, **331**, 141-147.

Greenleaf, W.H. and Adams, F. (1969) Genetic control of blossom-end rot disease in tomatoes through calcium metabolism. *Journal of the American Society for Horticultural Science*, **94**, 248-250.

Gunjate, R.T., Tare, S.J., Rangwala, A.D. and Limaye, V.P. (1977) Effect of pre-harvest and post-harvest calcium treatments on calcium content and occurrence of spongy tissue in Alphonso mango fruits. *Indian Journal of Horticulture*, **37**, 140-144.

Hansen, K. and Poll, L. (1993) Conversion of L-isoleucine into 2-methylbut-2-enyl esters in apples. *Lebensmittel-wissenschaft und-Technologie*, **26**, 178-180.

Hanson, E.J., Beggs, J.L. and Beaudry, R.M. (1993) Applying calcium chloride postharvest to improve highbush blueberry firmness. *HortScience*, **28**, 1033-1034.

Harker, F.R. and Ferguson, I.B. (1988a) Calcium ion transport across discs of the cortical flesh of apple fruit in relation to fruit development. *Physiologia Plantarum*, **74**, 695-700.

Harker, F.R. and Ferguson, I.B. (1988b) Transport of calcium across cuticles isolated from apple fruit. *Scientia Horticulturae*, **36**, 205-217.

Hartz, T.K., Miyao, G., Mullen, R.J., Cahn, M.D., Valencia, J. and Brittain, K.L. (1999) Potassium requirements for maximum yield and fruit quality of processing tomato. *Journal of the American Society for Horticultural Science*, **124**, 199-204.

Haynes, R.J. and Goh, K.M. (1980) Distribution and budget of nutrients in a commercial apple orchard. *Plant and Soil*, **56**, 445-457.

Ho, L.C., Belda, R., Brown, M., Andrews, J. and Adams, P. (1993) Uptake and transport of calcium and the possible causes of blossom-end rot in tomato. *Journal of Experimental Botany*, **44**, 509-518.

Ho, L.C., Hand, D.J. and Fussell, M. (1999) Improvement of tomato fruit quality by calcium nutrition. *Acta Horticulturae*, **481**, 463-469.

Hobson, G. and Grierson, D. (1993) Tomato, in *Biochemistry of fruit ripening*. (eds Seymour, G.B., Taylor, J.E. and Tucker, G.A.), Chapman & Hall, London, pp. 403-442.

Hofman, P.J., Smith, L.G., Joyce, D.C., Johnson, G.I. and Meiburg, G.F. (1997) Bagging of mango (*Mangifera indicia* cv. 'Keitt') fruit influences fruit quality and mineral composition. *Postharvest Biology and Technology*, **12**, 83-91.

Hopkirk, G., Harker, F.R. and Harman, J.E. (1990) Calcium and firmness of kiwifruit. *New Zealand Journal of Crop and Horticultural Science*, **18**, 215-219.

Hulme, A.C., Rhodes, M.J.C., Galliard, T. and Wooltorton, L.S.C. (1968) Metabolic changes in excised fruit tissue. IV. Changes occurring in discs of apple peel during the development of the respiration climacteric. *Plant Physiology*, **43**, 1154-1161.

Jack, F.R., O'Neill, J., Piacentini, M.G. and Schroder, M.J.A. (1997) Perception of fruit as a snack: a comparison with manufactured snack foods. *Food Quality and Preference*, **8**, 175-182.

Johnson, D.S. (1979) New techniques in the post-harvest treatment of apple fruits with calcium salts. *Communications in Soil Science and Plant Analysis*, **10**, 373-382.

Johnson, D.S. and Ridout, M.S. (1998) Prediction of storage quality of 'Cox's Orange Pippin' apples from nutritional and meteorological data using multiple regression models selected by cross validation. *Journal of Horticultural Science and Biotechnology*, **73**, 622-630.

Johnson, D.S. and Yogaratnam, N. (1978) The effects of phosphorus sprays on the mineral composition and storage quality of 'Cox's Orange Pippin' apples. *Journal of Horticultural Science*, **53**, 171-176.

Johnson, D.S., Samuelson, T.J., Pearson, K. and Taylor, J. (1998) Effect of soil applications of potassium sulphate on the mineral composition and eating quality of stored 'Conference' and 'Doyenne du Comice' pears. *Journal of Horticultural Science and Biotechnology*, **73**, 151-157.

Joshipura, K.J., Ascherio, A., Manson, J.E., Stampfer, M.J., Rimm, E.B., Speizer, F.E., Hennekens, C.H., Spiegelman, D. and Willett, W.C. (1999) Fruit and vegetable intake in relation to risk of ischemic stroke. *Journal of the American Medical Association*, **282**, 1233-1239.

Klein, J.D., Abbott, J.A., Basker, D., Conway, W.S., Fallik, E. and Lurie, S. (1998) Sensory evaluation of heated and calcium-treated fruits. *Acta Horticulturae*, **464**, 467-471.

Knowles, L., Trimble, M.T. and Knowles, N.R. (2001) Phosphorus status affects postharvest respiration, membrane permeability and lipid chemistry of European seedless cucumber fruit (*Cucumis sativus* L.). *Postharvest Biology and Technology*, **21**, 179-188.

Koo, R.J.C., Young, T.W., Reese, R.L. and Kesterson, J.W. (1974) Effect of nitrogen, potassium and irrigation on yield and quality of lemon. *Journal of the American Society for Horticultural Science*, **99**, 289-291.

Korban, S.S. and Swiader, J.M. (1984) Genetic and nutritional status in bitter pit-resistant and -susceptible apple seedlings. *Journal of the American Society for Horticultural Science*, **109**, 428-432.

Lancaster, J.E., Lister, C.E., Reay, P.F. and Triggs, C.M. (1997) Influence of pigment composition on skin colour in a wide range of fruit and vegetables. *Journal of the American Society for Horticultural Science*, **122**, 594-598.

Lang, A. (1990) Xylem, phloem and transpiration flows in developing apple fruits. *Journal of Experimental Botany*, **41**, 645-651.

Lang, A. and Volz, R.K. (1998) Spur leaves increase calcium in young apples by promoting xylem inflow and outflow. *Journal of the American Society for Horticultural Science*, **123**, 956-960.

Lester, G. (1996) Calcium alters senescence rate of postharvest muskmelon fruit disks. *Postharvest Biology and Technology*, **7**, 91-96.

Lester, G.E. and Grusak, M.A. (1999) Postharvest application of calcium and magnesium to honeydew and netted muskmelons: effects on tissue ion concentrations, quality and senescence. *Journal of the American Society for Horticultural Science*, **124**, 545-552.

Letham, D.S. (1969) Influence of fertilizer treatment on apple fruit composition and physiology. II. Influence on respiration rate and contents of nitrogen, phosphorus and titratable acidity. *Australian Journal of Agricultural Research*, **20**, 1073-1085.

Li, Y-M. and Gabelman, W.H. (1990) Inheritance of calcium use efficiency in tomatoes grown under low-calcium stress. *Journal of the American Society for Horticultural Science*, **115**, 835-838.

Li, R., Serdula, M., Bland, S., Mokdad, A., Bowman, B. and Nelson, D. (2000) Trends in fruit and vegetable consumption among adults in 16 US states: Behavioral Risk Factor Surveillance System, 1990-1996. *American Journal of Public Health*, **90**, 777-781.

Lurie, S., Sonego, L. and Ben-Arie, R. (1987) Permeability, microviscosity and chemical changes in the plasma membrane during storage of apple fruit. *Scientia Horticulturae*, **32**, 73-83.

Makus, D.J. and Morris, J.R (1989) Influence of soil and foliar applied calcium on strawberry nutrients and post-harvest quality. *Acta Horticulturae*, **265**, 443-446.

Makus, D.J. and Morris, J.R. (1998) Preharvest calcium applications have little effect on mineral distribution in ripe strawberry fruit. *HortScience*, **33**, 64-66.

Malik, H., Archibold, D.D. and McKown, C.T. (1991) Nitrogen partitioning by 'Chester Thornless' blackberry in pot culture. *HortScience*, **26**, 1492-1494.

Marcelle, R.D. (1991) Relationships between mineral content, lipoxygenase activity, levels of 1-aminocyclopropane-1-carboxylic acid and ethylene emission in apple fruit flesh disks (cv. Jonagold) during storage. *Postharvest Biology and Technology*, **1**, 101-109.

Marcelle, R.D. (1995) Mineral nutrition and fruit quality. *Acta Horticulturae*, **383**, 219-226.

Marlow, G.C. and Loescher, W.H. (1984) Watercore. *Horticultural Reviews*, **6**, 189-251.

Marsh, K.B., Volz, R.K., Cashmore, W. and Reay, P. (1996) Fruit colour, leaf nitrogen level, and tree vigour in 'Fuji' apples. *New Zealand Journal of Crop and Horticultural Science*, **24**, 393-399.

Martin, D., Lewis, T.L., Cerny, J. and Ratkowsky, D.A. (1976) The effect of tree sprays of calcium, boron, zinc and naphthaleneacetic acid, alone and in all combinations on the incidence of storage disorders in Merton apples. *Australian Journal of Agricultural Research*, **27**, 391-398.

Mayer, A.M. (1997) Historical changes in the mineral content of fruits and vegetables, *British Food Journal*, **99**, 207-21.

Medlicott, A.P., Bhogol, M. and Reynolds, S.B. (1986) Changes in peel pigmentation during ripening of mango fruit (*Mangifera indica* var. Tommy Atkin). *Annals of Applied Biology*, **109**, 651-656.

Millard, P. (1995) Internal cycling of nitrogen in trees. *Acta Horticulturae*, **383**, 3-14.

Miller-Ihli, N.J. (1996) Atomic absorption and atomic emission spectrometry for the determination of the trace element content of selected fruits consumed in the United States. *Journal of Food Composition and Analysis*, **9**, 301-311.

Miner, G.S., Poling, E.B., Carroll, D.E., Nelson, L.A. and Campbell, C.R. (1997) Influence of fall nitrogen and spring nitrogen–potassium applications on yield and fruit quality of 'Chandler' strawberry *Journal of the American Society for Horticultural Science*, **122**, 290-295.

Mootoo, A. (1991) Effect of post-harvest calcium chloride dips on ripening changes in 'Julie' mangoes. *Tropical Science*, **31**, 243-248.

Nagy, S. (1980) Vitamin C contents of citrus fruit and their products; a review. *Journal of Agricultural and Food Chemistry*, **28**, 8-18.

Neilsen, G., Parchomuk, P., Meheriuk, M. and Neilsen, D. (1998) Development and correction of K-deficiency in drip-irrigated apple. *HortScience*, **33**, 258-261.

Olienyk, P., Gonzalez, A.R., Mauromoustakos, A., Patterson, W.K., Rom, C.R. and Clark, J. (1997) Nitrogen fertilization affects quality of peach puree. *HortScience*, **32**, 284-287.

Omenn, G.S., Goodman, G.E., Thornquist, M.D., Balmes, J., Cullen, M.R., Glass, A., Keogh, J.P., Meyskens, F.L., Valanis, B., Williams, J.H., Barnhart, S. and Hammar, S. (1996) Effects of a combination of beta carotene and vitamin A on lung cancer and cardiovascular disease. *New England Journal of Medicine*, **334**, 1150-1155.

Opara, L.U., Studman, C.J. and Banks, N.H. (1997) Fruit skin splitting and cracking. *Horticultural Reviews*, **19**, 217-262.

Paiva, E.A.S., Martinez, H.E.P., Casali, V.W.D. and Padilha, L. (1998) Occurrence of blossom-end rot in tomato as a function of calcium dose in the nutrient solution and air relative humidity. *Journal of Plant Nutrition*, **21**, 2663-2670.

Perring, M.A. (1968) Mineral composition of apples. VII. The relationship between fruit composition and some storage disorders. *Journal of the Science of Food and Agriculture*, **19**, 186-192.

Perring, M.A. and Sharples, R.O. (1975) The mineral composition of apples. Composition in relation to disorders for fruit imported from the Southern Hemisphere. *Journal of the Science of Food and Agriculture*, **26**, 681-689.

Petersen, K.K., Willumsen, J. and Kaack, K. (1998) Composition and taste of tomatoes as affected by increased salinity and different salinity sources. *Journal of Horticultural Science & Biotechnology*, **73**, 205-215.

Picha, D.H. (1987) Physiological factors associated with yellow shoulder expression in tomato fruit. *Journal of the American Society for Horticultural Science*, **112**, 798-801.

Poovaiah, B.W., Glenn, G.M. and Reddy, A.S.N. (1988) Calcium and fruit softening: physiology and biochemistry. *Horticultural Reviews*, **10**, 107-152.

Prange, R.K. and DeEll, J.R. (1997) Preharvest factors affecting postharvest quality of berry crops. *HortScience*, **32**, 824-830.

Premuzic, Z., Bargiela, M., Garcia, A., Rendina, A. and Iorio, A. (1998) Calcium, iron, potassium, phosphorus, and vitamin C content of organic and hydroponic tomatoes. *HortScience*, **33**, 255-257.

Pressey, R. and Avants, J.K. (1973) Separation and characterization of endopolygalacturonase and exopolygalacturonase from peaches. *Plant Physiology*, **52**, 1349-1351.

Quintana, J.M., Harrison, H.C., Nienhuis, J., Palta, J.P. and Grusak, M.A. (1996) Variation in calcium concentration among sixty S$_1$ families and four cultivars of snap beans (*Phaseolus vulgaris* L.). *Journal of the American Society for Horticultural Science*, **121**, 789-793.

Quintana, J.M., Harrison, H.C., Nienhuis, J., Palta, J.P., Kmiecik, K. and Miglioranza, E. (1999a) Comparison of pod calcium concentration between two snap bean populations. *Journal of the American Society for Horticultural Science*, **124**, 273-276.

Quintana, J.M., Harrison, H.C., Nienhuis, J., Palta, J.P., Kmiecik, K. and Miglioranza, E. (1999b) Xylem flow rate differences are associated with genetic variation in snap bean pod calcium concentration. *Journal of the American Society for Horticultural Science*, **124**, 488-491.

Raese, J.T. (1977) Response of young 'd'Anjou' pear trees to triazine and triazole herbicides and nitrogen. *Journal of the American Society for Horticultural Science*, **102**, 215-218.

Raese, J.T. (1989) Physiological disorders and maladies of pear fruit. *Horticultural Reviews*, **11**, 357-411.

Raese, J.T. (1994) Effect of calcium sprays on control of black end, fruit quality, yield and mineral composition of 'Bartlett' pears. *Acta Horticulturae*, **367**, 314-322.

Raese, J.T. (1998) Response of apple and pear trees to nitrogen, phosphorus, and potassium fertilizers. *Journal of Plant Nutrition*, **21**, 2671-2696.

Raymond, L., Schaffer, B., Brecht, J.K. and Hanlon, E.A. (1998) Internal breakdown, mineral element concentration, and weight of mango fruit. *Journal of Plant Nutrition*, **21**, 871-889.

Roy, S., Conway, W.S., Watada, A.E., Sams, C.E., Erbe, E.F. and Wergin, W.P. (1999) Changes in the ultrastructure of the epicuticular wax and postharvest calcium uptake in apples. *HortScience*, **34**, 121-124.

Russell, D.G. *et al.* (1999) *NZ Food: NZ People. Key results of the 1997 National Nutrition Survey*, Ministry of Health, Wellington, New Zealand.

Saftner, R.A., Conway, W.S. and Sams, C.E. (1999) Postharvest calcium infiltration alone and combined with surface coating treatments influence volatile levels, respiration, ethylene production, and internal atmospheres of 'Golden Delicious' apples. *Journal of the American Society for Horticultural Science*, **124**, 553-558.

Saini, S.S., Singh, R.N. and Paliwal, G.S. (1971) Growth and development of mango (*Mangifera indica* L.) fruit. I. Morphology and cell division. *Indian Journal of Horticulture*, **28**, 247-256.

Sams, C.E. and Conway, W.S. (1984) Effect of calcium infiltration on ethylene production, respiration rate, soluble polyuronide content, and quality of 'Golden Delicious' apple fruit. *Journal of the American Society for Horticultural Science*, **109**, 53-57.

Sams, C.E., Conway, W.S., Abbott, J.A., Lewis, R.J. and Ben-Shalom, N. (1993) Firmness and decay of apples following postharvest pressure infiltration of calcium and heat treatment. *Journal of the American Society for Horticultural Science*, **118**, 623-627.

Schumacher, R., Fanhauser, F. and Stadler, W. (1980) Influence of shoot growth, average fruit weight and daminozide on bitter pit, in *Mineral Nutrition of Fruit Trees* (eds D. Atkinson, J.E. Jackson, R.O. Sharples and W.M. Waller), Butterworths, London, pp. 83-92.

Scott, K.J. and Wills, R.B.H. (1979) Effects of vacuum and pressure infiltration of calcium chloride and storage temperature on the incidence of bitter pit and low temperature breakdown of apples. *Australian Journal of Agricultural Research*, **30**, 917-926.

Sharples, R.O., Reid, M.S. and Turner, N.A. (1979) The effects of postharvest mineral element and lecithin treatments in the storage disorders of apples. *Journal of Horticultural Science*, **54**, 299-304.

Shear, C.B. (1975) Calcium-related disorders in fruits and vegetables. *HortScience*, **10**, 361-365.

Smith, G.S. and Clark, C.G. (1989) Effect of excess boron on yield and postharvest storage of kiwifruit. *Scientia Horticulturae*, **38**, 105-115.

Somogyi, L.P., Childers, N.F. and Chang, S.S. (1964) Volatile constituents of apple fruits as influenced by fertiliser treatments. *Proceedings of the American Society of Horticultural Science*, **84**, 51-58.

Steinmetz K.A. and Potter J.D. (1996) Vegetables, fruit, and cancer prevention; a review. *Journal of the American Dietetic Association*, **96**, 1027-1039.

Stow, J. (1989) The involvement of calcium ions in maintenance of apple fruit tissue structure. *Journal of Experimental Botany*, **40**, 1053-1057.

Tagliavini, M., Toselli, M., Marangoni, B., Stampi, G. and Pelliconi, F. (1995) Nutritional status of kiwifruit affects yield and fruit storage. *Acta Horticulturae*, **383**, 227-237.

Thompson, B., Demark-Wahnefried, W., Taylor, G., McClelland, J.W., Stables, G., Havas, S., Feng, Z., Topor, M., Heimendinger, J., Reynolds, K.D. and Cohen, N. (1999) Baseline fruit and vegetable intake among adults in seven 5 a day study centers located in diverse geographic areas. *Journal of the American Dietary Association*, **99**, 1241-1248.

Thorp, T.G., Hutching, D., Lowe, T. and Marsh, K.B. (1997) Survey of fruit mineral concentrations and postharvest quality of New Zealand-grown 'Hass' avocado (*Persea americana* Mill.). *New Zealand Journal of Crop and Horticultural Science*, **25**, 251-260.

Trudel, M.J. and Ozbun, J.L. (1971) Influence of potassium on carotenoid content of tomato fruit. *Journal of the American Society for Horticultural Science*, **96**, 763-765.

Tukey, L.D. (1964) A linear electronic device for continuous measurement and recording of fruit enlargement and contraction. *Proceedings of the American Society for Horticultural Science*, **84**, 653-660.

Turner, N.A., Ferguson, I.B. and Sharples, R.O. (1977) Sampling and analysis for determining the relationship of calcium concentration to bitter pit in apple fruit. *New Zealand Journal of Agricultural Research*, **20**, 525-532.

US Department of Agriculture (1992) *Composition of Food, Fruits and Fruit Juices: Raw, Processed, Prepared*. Agriculture Handbook No. 8–9 United States Department of Agriculture, August 1992.

US FDA (1993) Section 101.54 Nutrient content claims for 'good source', 'high' and 'more'. *Code of Federal Regulations*, title 21, 84-85.

Van't Veer, P., Jansen, M.C., Klerk, M. and Kok, F.J. (2000) Fruits and vegetables in the prevention of cancer and cardiovascular disease. *Public Health and Nutrition*, **3**, 103-107.

Volz, R.K. and Ferguson, I.B. (1999) Flower thinning method affects mineral composition of 'Braeburn' and 'Fiesta' apple fruit. *Journal of Horticultural Science and Biotechnology*, **74**, 452-457.

Volz, R.K., Ferguson, I.B., Hewett, E.H. and Woolley, D.J. (1994) Wood age and leaf area influence fruit size and mineral composition of apple fruit. *Journal of Horticultural Science*, **69**, 385-395.

Volz, R.K., Tustin, D.S. and Ferguson, I.B. (1996a) Mineral accumulation in apple fruit as affected by spur leaves. *Scientia Horticulturae*, **65**, 151-161.

Volz, R.K., Tustin, D.S. and Ferguson, I.B. (1996b) Pollination effects on fruit mineral composition, seeds and cropping characteristics of 'Braeburn' apple trees. *Scientia Horticulturae*, **66**, 169-180.

Volz, R.K., Alspach, P.A., White, A.G. and Ferguson, I.B. (2000) Genetic variability in apple fruit storage disorders. *Acta Horticulturae*, **553**, 241-244.

Volz, R.K., Alspach, P.A., White, A.G. and Ferguson, I.B. (2001) Genetic variability in mineral accumulation in apple fruit, in *Plant Nutrition – Food Security and Sustainability of Agro-ecosystems* (eds Horst, W.J. *et al.*), Kluwer Academic Publishers, The Netherlands, pp. 92-93.

Waller, W.M. (1980) Use of apple analysis, in *Mineral Nutrition of Fruit Trees* (eds D. Atkinson, J.E. Jackson, R.O. Sharples and W.M. Waller). Butterworths, London, pp. 383-394.

Watkins, C.B., Hewett, E.W., Bateup, C., Gunson, A. and Triggs, C.M. (1989) Relationships between maturity, starch pattern index and storage disorders in 'Cox's Orange Pippin' apples as influenced by preharvest calcium or ethephon sprays. *New Zealand Journal of Crop and Horticultural Science*, **17**, 283-292.

WHO (1990) *Diet, Nutrition and the Prevention of Chronic Diseases*, Technical Report Series 797, Geneva, World Health Organisation.

Wills, R.B.H. and Rigney, C.J. (1979) Effect of calcium on activity of mitochondria and pectic enzymes isolated from tomato fruits. *Journal of Food Biochemistry*, **3**, 103-110.

Witney, G.W., Hofman, P.J. and Wolstenholme, B.N. (1990a) Effect of cultivar, tree vigour and fruit position on calcium accumulation in avocado fruits. *Scientia Horticulturae*, **44**, 269-278.

Witney, G.W., Hofman, P.J. and Wolstenholme, B.N. (1990b) Mineral distribution in avocado trees with reference to calcium cycling and fruit quality. *Scientia Horticulturae*, **44**, 279-291.

Witney, G.W., Kushad, M.M. and Barden, J.A. (1991) Induction of bitter pit in apple. *Scientia Horticulturae*, **47**, 173-176.

Wojcik, P.P. and Cieslinski, G. (2000) Effect of boron fertilization on yield and fruit quality of 'Elstar' and 'Sampion' apple cultivars. *Acta Horticulturae*, **512**, 189-190.

Woolf, A.B., Ferguson, I.B., Requejo-Tapia, L.C., Boyd, L., Laing, W.A. and White, A. (1999) Impact of sun exposure on harvest quality of 'Hass' avocado fruit, in *Proceedings of the IV World Avocado Congress, Uruapan, Mexico. Revista Chaingo Serie Horticultura 5*, pp. 352-358.

Yogaratnam, N. and Sharples, R.O. (1982) Supplementing the nutrition of Bramley's Seedling apple with phosphorus sprays. II. Effects on fruit composition and storage quality. *Journal of Horticultural Science*, **57**, 53-59.

Young, T.W. and Miner, J.T. (1961) Relationship of nitrogen and calcium to 'soft-nose' disorder in mango fruits. *Journal of the American Society for Horticultural Science*, **78**, 201-208.

3 Fruit texture, cell wall metabolism and consumer perceptions

Robert J. Redgwell and Monica Fischer

And so from hour to hour we ripe and ripe
and then from hour to hour we rot and rot
and thereby hangs a tale

W. Shakespeare

3.1 Introduction

During the 1980s, a wealth of data accumulated on the biochemistry of fruit softening, based on chemical, enzymatic and histochemical approaches. These attempted to make correlations between some textural change and a specific alteration to cell wall chemistry, or the activity of a cell-wall associated enzyme. Many of these correlations were compelling, but none was conclusive. In tomato, endo-polygalacturonase (endo-PG, EC 3.2.1.15) seemed to be responsible for softening. Endo-PG activity increased dramatically during softening (Brady *et al.*, 1983), there was a concomitant increase in soluble polyuronide (Huber, 1983) and a loss of middle lamella integrity (Crookes and Grierson, 1983). However the expression of the endo-PG gene in a non-ripening tomato mutant did not increase the rate of softening (Giovannoni *et al.*, 1989).

The mechanisms of fruit softening have interest beyond the practical requirements of fruit marketing. Understanding how an integrated, compact and largely insoluble structure is transformed into a dispersed, swollen and partially solubilized matrix (John and Dey, 1986) will contribute to our basic understanding of the architecture of the cell wall.

In the 12 years since the work of Giovannoni *et al.* (1989) molecular techniques and advances in biochemical methods for wall analysis have lead to steady progress in elucidating the mechanisms of fruit softening. During this period there have been reviews of fruit texture (Jackman and Stanley, 1995; Harker *et al.*, 1997c), softening (MacRae and Redgwell, 1992; Wakabayashi, 2000) and cell wall enzymes (Fischer and Bennett, 1991; Seymour and Gross, 1996). Reviews on cell wall architecture are also relevant to our understanding of changes in developing fruit (Carpita and Gibeaut, 1993; Rose and Bennett,

1999; Cosgrove, 2001). This review will focus on research on fruit softening and plant cell walls in the last ten years.

3.2   Texture and the cell wall

Food texture is defined as: 'All the rheological and structural (geometrical and surface) attributes of a food product perceptible by means of mechanical, tactile and where appropriate, visual and auditory receptors' (ISO, 1981; Jowitt, 1974). A lexicon of sensory texture attributes can be found in the reviews of Harker *et al.* (1997c) and Lapsley (1989). Terms such as crispness, juiciness, hardness and mealiness are a few of a broad spectrum of attributes that define the feel of fruit in the mouth. They are experienced during mastication, which causes the breakdown of the tissue. In many fruits this tissue is made up of parenchyma cells. Although these cells contain a variety of inclusions such as starch, it is the presence and structural integrity of the cell wall that plays a major role in the perception of texture.

The integrity of the fruit cell can be ascribed to wall-to-wall adhesion between cells and the strength of the primary wall. Wall-to-wall adhesion has been described as the most critical factor influencing the perception of fruit texture (Diehl and Hamann, 1980; Pitt and Chen, 1983) and is believed to be dictated by the strength of the middle lamella, the area of cell-to-cell contact and the extent of plasmodesmatal connections (Harker and Hallett, 1994). It has also been related to the different textures of temperate fruit (Harker *et al.*, 1997b). Bourne (1979) separated temperate fruit into two groups: those that softened considerably to a melting texture (e.g. plum, tomato and kiwifruit) and those that softened only moderately and retained a crisp fracturable texture (e.g. apple, nashi pear). In the first group cell-to-cell adhesion is poor, and the tissue can separate with minimal cell rupture. In the second group cell-to-cell adhesion is strong, so that the cell walls rupture giving the feeling of crispness. This does not explain all textural differences between fruits. Batisse *et al.* (1996) showed that the middle lamella in cherry cultivars which had a soft texture at maturity was better preserved than in those cultivars that remained crisp. These results suggest that differences in the texture of the two groups of fruit also require changes to the internal wall structure.

Redgwell *et al.* (1997b) examined the swelling of cell walls in intact fruit, by microscopy and *in vitro*, using isolated cell wall material (CWM). In fruit which ripened to a soft melting texture (group 1), wall swelling was pronounced, particularly *in vitro*. Fruit which ripened to a crisp fracturable texture (group 2) did not show either *in vivo* or *in vitro* wall swelling. Kiwifruit was unusual in that it demonstrated both *in vivo* and *in vitro* swelling to a marked degree (figure 3.1). Tomato, strawberry, plum and persimmon, all group 1 fruit, showed only swelling of the CWM *in vitro*.

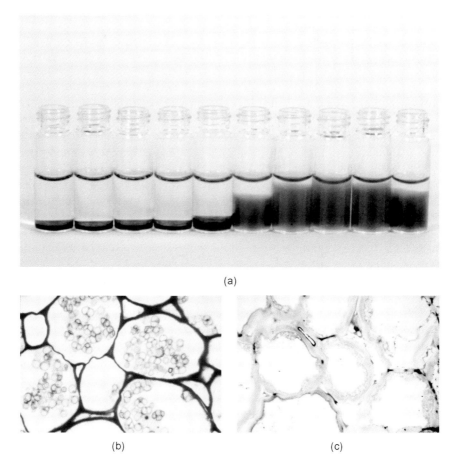

Figure 3.1 *In vitro* and *in vivo* cell wall swelling in ripening kiwifruit (Redgwell *et al.*, 1997b). (a) Isolated cell wall material (10 mg) allowed to swell in water. From left to right, kiwifruit 3, 2 and 1 week before harvest, at harvest and 1, 2, 4, 6, 8 and 10 days after ethylene treatment. Bottom pictures: Light microscopy of tissue showing cell walls at maturity (b) and after fruit softening (c), 6 days after ethylene treatment. Walls stained with toluidine blue.

3.3 Chemistry of the fruit cell wall

Fruit flesh is predominantly composed of thin walled parenchyma cells, which for the most part are unlignified. The cell wall is a dynamic structure, changing its composition and physicochemical properties continuously in response to its development and environment. Primary walls comprise cellulose microfibrils suspended in a matrix of pectic and hemicellulose polysaccharides, glycoproteins and water. The polysaccharides in fruit walls are hydrated, hydrophilic molecules. Many are water-soluble molecules, *yet* form a coherent water-insoluble structure within the wall. Why this should be and how the molecules interact to accommodate all the changes necessary for growth, development and senescence of the fruit remain to be explained.

Research has concentrated on defining the primary structure of individual polysaccharides, which account for 90% of the cell wall. Major obstacles have been the complexity of the polysaccharides and the difficulty in isolating and purifying individual polymers. This has also led to a confusing nomenclature, particularly with regard to the pectic polysaccharides. It is common to refer to polysaccharide fragments (rhamnogalacturonans, RG I, RG II) isolated by enzymatic and chemical degradation as separate classes of polysaccharides, comparable to the xyloglucans and cellulose. However, whereas the latter exist as such *in vivo*, RG II, RG I, arabinans and galactans have not been found as separate molecular entities in fruit cell walls, unless the cell wall has been subjected to a prior enzymatic treatment (e.g. RG II in wine, Pellerin *et al.*, 1996). On the basis of their isolation as individual molecular forms there are five polysaccharides which have been unequivocally identified in fruit cell walls: rhamnogalacturonans, xyloglucans, galactoglucomannans, glucuronoarabinoxylans and cellulose. These seem to be ubiquitous and although a general formula for each can be given, there is heterogeneity within each category with regard to molecular weight, primary structure, linkage type, degree of branching, pattern of distribution of side-chains and substituents (acetylation, esterification).

3.3.1   Rhamnogalacturonans

The rhamnogalacturonans (RGs) or pectic polysaccharides account for 40–60% of the fruit cell wall and are the most complex group. Their molecular weight can exceed 2000 kDa (Chapman *et al.*, 1987). They consist of a linear backbone of 4-linked α-D-galacturonosyl residues, interspersed with 2-linked α-L-rhamnosyl residues at infrequent and irregular intervals. Side-chains of sugars, mostly galactose and arabinose, are attached to the *O*-4 position of some of the rhamnosyl residues and it is the degree of substitution of the RG backbone that provides the greatest structural diversity among fruit cell wall RGs (Redgwell *et al.*, 1988; Seymour *et al.*, 1990; Renard *et al.*, 1993; Fischer *et al.*, 1994; Schols *et al.*, 1995b).

RGs are solubilized from CWM by extraction in solutions of chelators, chaotropic agents or alkali. Even after such treatment the residue of the CWM contains amounts of RG. There is evidence in kiwifruit that the more substituted with galactose–arabinose the RG backbone is, the more difficult it is to extract from CWM (table 3.1). A similar pattern has been found in other fruit (Seymour *et al.*, 1990; Redgwell *et al.*, 1997a). The spectrum of chemical reagents required to solubilize RGs is an indication that their location within the wall, or mode

Table 3.1 Ease of extraction of rhamnogalacturonans from kiwifruit cell walls is related to their degree of branching (Redgwell, 1991)

Extraction medium	% of total rhamnogalacturonan	Polyuronide/ neutral sugar ratio
CDTA	10	97:3
Na$_2$CO$_3$	46	90:10
6 M GTC[a]	10	58:42
4 M KOH	16	42:58
Residue	18	36:64

[a] Guanidinium thiocyanate.

of attachment, varies considerably. How this variation relates to the degree of substitution is not known, but physical entanglement of side-chains with cellulose microfibrils may be a factor.

Within the RG molecule discrete regions exist that have distinctive structural forms. They have been isolated as separate polysaccharides only following treatment of CWM or a solubilized pectic polysaccharide with pectin-degrading enzymes. They include RG I, RG II, xylogalacturonans and the arabinogalactans.

3.3.1.1 Rhamnogalacturonan II

RG II is released as a low molecular weight (5–10 kDa) polysaccharide by treating plant polysaccharides with endo-PG (O'Neill *et al.*, 1990). It accounts for ca. 2% of the primary wall and contains 12 different monosaccharides and 20 different linkage types. RG II has been identified in the pectin of several fruits (Pellerin *et al.*, 1996; Vidal *et al.*, 2000) and appears to be ubiquitous in plant cell walls. In kiwifruit (Redgwell *et al.*, 1988), it was present in the CDTA (*trans*-1,2-diaminocyclohexane-*N*,*N*,*N'*,*N'*-tetraacetic acid) and Na$_2$CO$_3$-soluble fractions as well as the pectic fraction that remained associated with the CWM residue.

3.3.1.2 Rhamnogalacturonan I

RG I was first identified as a significant component (14%) of sycamore primary cell walls following endo-PG treatment and is characterized by a backbone with a regular alternating sequence of galacturonic acid and rhamnose in a ratio of 1:1 (O'Neill *et al.*, 1990). It has a molecular weight of approximately 200 kDa. The backbone is substituted with a series of side-chains made up of predominantly galactosyl and arabinosyl residues, although xylosyl, 4-*O*-methylglucuronosyl and glucuronosyl residues are also present (An *et al.*, 1994). RG I is used by some as a term for highly branched RGs other than RG II. RG I as originally defined in sycamore cells has not been identified in fruit cell walls. RG I must be present at lower levels in fruit than in sycamore cell walls. Pectin can account for 60% by weight of the fruit cell wall and the galacturonic acid/rhamnose ratio is ca. 20:1. This is too high for RG I to account for 14% of the cell wall.

3.3.1.3   Xylogalacturonans

A xylogalacturonan (20–30 kDa) was isolated from apple cell walls following treatment of the hairy region polysaccharides with rhamnogalacturonan hydrolase (Schols *et al.*, 1995a). Single xylosyl residues were attached by a β-1,3 linkage to *O*-3 of a galacturonosyl residue. The degree of substitution of the galacturonosyl residues with xylose was as high as 0.7, suggesting that many contiguous galacturonosyl residues were xylosylated. In contrast, in watermelon the xylose residues were far enough apart in some of the xylogalacturonan regions that endo-PG digested them to produce small fragments (Yu and Mort, 1996). How extensively xylogalacturonans are distributed among different fruit has yet to be determined.

3.3.1.4   Arabinans, galactans and arabinogalactans

It is unlikely that arabinans, galactans or type I arabinogalactans exist free in fruit cell walls. If endogenous enzyme activity is prevented during the preparation of CWM and the extraction of the pectic fraction is done in such a way that β-elimination is minimized, these polymeric sequences are only found covalently linked to the RG backbone. Nevertheless, all pectic polysaccharide fractions isolated from fruit cell walls contain some galactose and arabinose in their side-chains. The galactans occur for the most part as linear chains of β-1,4-linked residues. The arabinans exist as α-1,5-linked arabinosyl residues. In some cases the arabinosyl residues are substituted at the *O*-2 and *O*-3 position giving highly branched structures (Aspinall and Fanous, 1984).

The cell walls of eight fruit species were found to contain arabinogalactans with a very high molecular mass (Redgwell *et al.*, 1997a). The polymers resisted extraction from the CWM in 4 M KOH and were partially solubilized following endo-PG treatment of the KOH-insoluble residue, indicating that they were covalently linked to an RG. The fragments which possessed a molecular weight of approximately 400 kDa contained between 75–90% of their structure as arabinosyl and galactosyl residues.

3.3.2   Xyloglucans

Xyloglucans (XGs) in dicotyledons account for 20–25% of the primary cell wall (Bacic *et al.*, 1988). They possess a β-1,4-glucan backbone with side-chains such as α-D-Xyl*p*-(1→6), β-D-Gal*p*-(1→2)-α-D-Xyl*p*-(1→6)- or α-L-Fuc*p*-(1→2)-β-D-Gal*p*-(1→2)-α-D-Xyl*p*-(1→6) attached to up to 75% of the β-(1→4)-D-Glc*p* residues. The degree of fucosylation varies among fruit. The fucose contents of xyloglucan from tomato (Tong and Gross, 1988; Seymour *et al.*, 1990), kiwifruit (Fischer *et al.*, 1996), and persimmon (Cutillas *et al.*, 1998b) were reported as 0, 1.8 and 6 mol % respectively. Xyloglucan is hydrogen bonded to the surface of the cellulose fibrils and has been postulated to function as intermolecular tethers between adjacent microfibrils, thus contributing to the overall integrity of the structure (Hayashi, 1989). This has led to the concept of a biphasic cell wall consisting of a cellulose–xyloglucan skeleton immersed in a matrix of pectic polysaccharides and other hemicelluloses. However, immuno-logical labelling has demonstrated that the xyloglucans are not restricted to a close association with the cellulose but are spread throughout the wall, including the middle lamella (Moore *et al.*, 1986; Sutherland *et al.*, 1999). However, very little xyloglucan (or any hemicellulose) is co-extracted with the pectic polysac-charides of the cell wall by solutions of CDTA or Na_2CO_3 which can solubilize most of the cell wall pectin. This means that all the cell wall hemicellulose is either hydrogen-bonded (or otherwise attached) to the cellulose fibrils, or is retained in the wall by cross-links to pectic polymers that are insoluble in CDTA or Na_2CO_3.

3.3.3 Galactoglucomannans

In contrast to XG, the galactoglucomannans (GGMs) have been less well char-acterized in fruit, and no clear role for them has been established within the overall architecture of the wall. GGM has been reported in suspensions of cultured cells of *Rubus* (Cartier *et al.*, 1988) and has been characterized from kiwifruit (Schroeder *et al.*, 2001). The ratio of galactose (Gal), glucose (Glc) and mannose (Man) varies with plant source and developmental stage of the tissue (Sims *et al.*, 1997). In kiwifruit, the GGM backbone is predominantly alternating β-D-Glc$p$-(1→4) and β-D-Man$p$(1→4) with the latter branched at O-6 with either single α-D-Gal$p$-(1→ residues or the disaccharide β-D-Gal$p$-(1→2)-α-D-Gal$p$-(1→. The molecular weight range is 10–40 kDa. GGM in kiwifruit is more strongly hydrogen bonded to the cellulose fibrils than XG. Following extraction in 4 M KOH, which solubilized all the XGs and some GGMs, a second extraction with 8 M KOH solubilized additional GGMs (Redgwell, 1991).

It is likely that GGMs will also contribute to wall architecture by their interaction with the cellulose fibrils. A study with bacterial cellulose has demon-strated that related polymers (galactomannans) form cross-links of varying lengths between cellulose fibrils and that networks with distinct architectural and molecular features can be formed by varying the Gal:Man ratio of the galactomannans (Whitney *et al.*, 1998).

3.3.4 Glucuronoarabinoxylans

Glucuronoarabinoxylans are the least researched of the three hemicellulose groups and have been reported in olive and kiwifruit (Gil-Serrano *et al.*, 1986; Redgwell *et al.*, 1991a). They are characterized by a backbone of β-1,4-linked xylosyl residues with substitutions on the O-2 and O-3 position of the xylan backbone with single residues of arabinose and glucuronosyl and/or

4-O-methylglucuronosyl residues. In kiwifruit nearly 90% of the glucuronic acid was 4-O-methylated (Redgwell et al., 1988).

3.3.5 Cellulose

Cellulose is a linear polymer of β-(1,4)-D-glucan that self-associates to form crystalline bundles called microfibrils. Although this is believed to occur by intermolecular hydrogen bonding, computations have indicated that van der Waals forces provide the dominant cohesive forces in a cellulose crystallite (Cousins and Brown, 1995). The arrangement confers great tensile strength which allows the wall to withstand tensions up to 5000 bar (Carpita and Gibeaut, 1993). Mechanical properties of the primary wall may also be affected by the degree of molecular ordering of cellulose (crystalline versus amorphous cellulose). Solid state ^{13}C nuclear magnetic resonance (NMR) has shown cellulose in apple cell walls to have a high degree of ordering, with no amorphous cellulose (Newman et al., 1994). The ordered state, together with the narrow cross-sections of the crystallites in the apple, would translate into properties of hardness and weakness in individual crystallites and may contribute to the brittle, crisp texture of apple. There is too little data from other fruit to know whether there is a correlation between fruit texture and the pattern of molecular ordering of cellulose.

3.3.6 Proteins

In addition to numerous enzymes, structural glycoproteins are located in the cell wall (Cassab, 1998). The best known is extensin in which hydroxyproline accounts for 40% of the amino acids and is glycosylated with tri- and tetra-arabinosyl residues. It is difficult to assign a given protein to the wall in vivo, when the potential for absorption of intracellular proteins to the wall during tissue homogenization is so high. The level of covalently bound cell wall protein can be very low in fruit. CWM prepared from kiwifruit contained only 1.4% protein of which extensin was not a major component (Redgwell et al., 1988). Nevertheless, 'Cox' apples from an orchard which produced consistently firmer fruit had higher levels of hydroxyproline than found for apples from a second orchard (Huxman et al., 1999). Cell wall proteins are potential cross-linking agents and most of the cell wall extensin appears to be covalently liked to RG I in cotton suspension cultures (Qi et al., 1995).

3.4 Changes to polysaccharides during fruit softening

There is no obvious correlation between the primary structure of cell wall polysaccharides in different fruit and their texture at maturity. Chemical, microscopic, enzymatic and molecular studies indicate that alterations to the physicochemistry of the primary wall and the middle lamella are major factors in textural changes. But how cell wall metabolism leads to softening is still not clear. Methodological differences are often an obstacle. Methods used to isolate and purify the CWM and component polysaccharides vary. The composition of isolated polymers may not reflect the situation *in vivo* unless steps are taken to minimize residual endogenous enzyme activity and chemically induced depolymerization of the pectic polysaccharides during extraction. Methods used to overcome one form of inadvertent degradation may promote another. For example, use of hot alcohol to inactivate endogenous enzyme activity may induce a chemical depolymerization of methyl esterified pectic polysaccharides by β-elimination.

Experiments have attempted to correlate softening with the depolymerization of a polysaccharide by a cell-wall associated hydrolase. Almost every fruit has obliged and a reduction in the molecular weight of a pectic polysaccharide fraction, accompanied by the loss of galactosyl or arabinosyl residues, has been documented countless times as accompanying fruit softening. By the end of the 1980s it had become a belief bordering on dogma, that softening resulted from enzyme-mediated hydrolysis of cell wall polymers. It was only with the advent of molecular techniques that enabled wall hydrolases to be selectively down-regulated that this began to be questioned, promoting fresh avenues of investigation.

3.4.1 Pectic polysaccharides

In vitro analysis of cell wall pectic polysaccharides has demonstrated that they undergo solubilization and depolymerization during fruit ripening. Increased solubilization is shown by the enhanced capacity of some pectic polysaccharides to be extracted from CWMs by aqueous solvents. The degree of *in vitro* solubilization may not be strictly analogous to the *in vivo* situation. Nevertheless, it gives a measure of a phase change in the physicochemical properties of the pectic polysaccharides which is indicative of wall dissolution. Depolymerization is a lowering of the molecular mass by cleavage of either the rhamnogalacturonan backbone and/or the neutral galactose–arabinose side-chains. It may also represent a de-aggregation of pectic polysaccharide complexes held together by non-covalent bonds, in which case hydrolases may not be involved. Pectin solubilization and depolymerization of cell wall polymers during ripening have been documented in many fruit. It is reasonable to assume that the two processes are interdependent, but a consideration of the results of several studies suggest that this is not necessarily so.

3.4.1.1 Pectin solubilization
A study of *in vivo* and *in vitro* swelling of the cell walls of eight fruit types showed a relationship between swelling and the degree of pectin solubilization

that occurred during softening (Redgwell *et al.*, 1997b). In fruit which softened to a melting texture (tomato, kiwifruit, plum, avocado, blackberry, persimmon, strawberry), cell wall swelling was pronounced and was accompanied by moderate to heavy pectin solubilization. Fruit, which stayed crisp during ripening (apple, nashi pear, watermelon), demonstrated no cell wall swelling and negligible amounts of pectin solubilization. With kiwifruit ripened on the vine (Redgwell and Percy, 1992), or after a postharvest ethylene treatment (Redgwell *et al.*, 1990), softening, cell wall swelling and pectin solubilization were in synchrony.

The mechanism(s) for pectin solubilization remain unclear and they may not be the same in all fruits. While enzyme-induced depolymerization may be a general mechanism, the extent of this process varies widely. In plum and blackberry, the molecular weight of the solubilized pectin in ripe fruit was similar to that in unripe fruit (Redgwell *et al.*, 1997a). In contrast, avocado showed a dramatic reduction in the molecular weight of solubilized pectin (Huber and O'Donoghue, 1993). Tomato and kiwifruit were between the two extremes.

Molecular weight changes in polysaccharide wall fractions during ripening, as indicated by gel-permeation chromatography, do not provide unequivocal proof of *in vivo* depolymerization by hydrolases. A CDTA or Na_2CO_3-soluble fraction extracted late in ripening, when wall dissolution is advanced, may contain different polymers to those extracted by the same reagent early in ripening. The extreme polydispersity of the pectic polysaccharides could mean that different sized polymers are extracted by the same reagent at different times in ripening. Likewise, lack of change in molecular weight of a CDTA-soluble fraction during ripening does not necessarily mean that *in vivo* depolymerization of some pectin polymers has not occurred. Ripening studies with kiwifruit and tomato have shown evidence of a reallocation of fractions during sequential extraction, resulting in clear molecular weight shifts (Redgwell *et al.*, 1992; Brummell and Labavitch, 1997).

3.4.1.2   *Galactose loss*

One mechanism for depolymerization of pectic polysaccharides is the removal of galactose and/or arabinose from the side-chains, a process that seems to be even more widespread than pectin solubilization. In a survey of 17 different fruits (Gross and Sams, 1984), 15 showed a net loss of non-cellulosic neutral sugars during ripening. The sugars need not be lost as a consequence of enzymatic cleavage. They could also be lost from the cell wall along with solubilized pectin, leading to a similar pattern of loss for arabinose and rhamnose residues (Redgwell *et al.*, 1992; Carrington and Pressey, 1996). Enzymatic hydrolysis of galactose is better documented than loss of arabinose from pectic side-chains.

While galactose loss from cell wall polymers often accompanies fruit ripening it may not be involved in pectin solubilization or textural changes. Galactose was lost equally from fruit that softened to a melting texture (tomato) and from those which maintained a crisp hard texture at maturity (apple) (Redgwell *et al.*, 1997a). Plum, which showed moderate wall swelling, pectin solubilization and softened considerably, showed no net loss of galactose. In kiwifruit allowed to soften on the vine, 80% of the galactose was lost from the cell wall prior to the onset of significant softening or pectin solubilization (Redgwell and Percy, 1992). Retardation of galactose loss from the cell walls of kiwifruit discs had no effect on either the rate of softening or pectin solubilization (Redgwell and Harker, 1995). Pectin solubilization and galactose loss can occur on different polysaccharides. In eight different fruit most of the cell wall galactose was not associated with the bulk of the polyuronide that could be solubilized with CDTA and 0.05 M Na_2CO_3 solutions, but with pectic polymers that remained attached to the CWM residue, even after extraction with 4 M KOH (Redgwell *et al.*, 1997a).

Despite these data, it seems likely that neutral sugar loss from the pectic polysaccharides does affect the physicochemical properties of the fruit wall. The treatment of a chelator-soluble pectin from avocado with β-galactosidase apparently affected molecular aggregation so that solubility increased and apparent molecular weight decreased more than predicted from the galactose loss (de Veau *et al.*, 1993).

3.4.1.3 Cross-links

Some form of covalent cross-link, between either adjacent pectin molecules or pectin and other polysaccharides, could be broken during ripening, leading to a destabilization of the matrix. The nature of these linkages is unknown. Because 0.05 M Na_2CO_3 has been used routinely to solubilize a pectic fraction from fruit cell walls (Jarvis *et al.*, 1981), alkali-labile ester bonds were thought to be responsible for retaining the fraction in the wall (Fry, 1986) but this has not been confirmed. The elucidation of the mode of attachment of the Na_2CO_3-soluble fraction would be a key contribution to our understanding of cell wall architecture and its disassembly during fruit softening.

RG II exists in the primary wall as a dimer covalently linked by a 1:2 borate diol ester (Matoh *et al.*, 1993; O'Neill *et al.*, 1996). The RG II-borate complex is believed to result in the formation of a covalently cross-linked pectic matrix which may regulate the pore size and physical stability of the primary cell wall. Whether the borate-linked RG II dimer is broken during fruit ripening, and hence is a factor in pectin solubilization, is not known. Boron sprays, used after bloom in apple, increased fruit firmness and decreased the incidence of bitter bit (Wojcik *et al.*, 1999).

In theory, pectic polysaccharides can cross-link with adjacent polymers by forming an ester bond between the carboxyl group of one pectic chain and the hydroxyl group of an adjacent chain. No such cross-link has yet been established, but Brown and Fry (1993) identified novel *O*-galacturonosyl esters in the pectic polysaccharides of suspension cultured cells and suggested that they are linked via O-D-galacturonosyl bonds to relatively hydrophobic constituents of the cell wall. Kim and Carpita (1992) found a lack of stoichiometry between methyl groups and esterified uronic acid which pointed to the presence of non-methyl esterified pectic polysaccharides.

Studies with Chinese water chestnut have implicated diferulic acid cell wall cross-links (Waldron *et al.*, 1997; Waldron, 1998) in the thermal stability of cell adhesion and the maintenance of a crisp texture during cooking. Low levels of ferulic acid and its dimers have been detected in apple (Waldron *et al.*, 1997) and carrot (Parr *et al.*, 1997) and an unidentified phenolic fraction has been reported in the cell walls of kiwifruit (Redgwell *et al.*, 1988). While phenolics may be cross-linking agents in the cell walls of some plants, their extent and importance as agents for wall stability in fruit have yet to be established.

3.4.2   *Xyloglucans*

A decrease in molecular weight of xyloglucan (XG) during fruit softening has been reported for tomato (Huber, 1983), strawberry (Huber, 1984; Nogata *et al.*, 1996), kiwifruit (Redgwell *et al.*, 1991a), pepper (Gross *et al.*, 1986), melon (Rose *et al.*, 1998), avocado (Sakurai and Nevins, 1997), peaches (Hegde and Maness, 1998) and persimmon (Cutillas *et al.*, 1994). The molecular weight of tomato XG has been reported to remain unchanged (Tong and Gross, 1988; Seymour *et al.*, 1990), to decrease slightly (Sakurai and Nevins, 1993) or to decrease greatly (Huber, 1983; Maclachlan and Brady, 1994) during ripening.

Although XG is depolymerized during ripening, the side-chains are not as susceptible to degradation as they are in the pectic polysaccharides. In tomato (Maclachlan and Brady, 1994) and kiwifruit (Redgwell, 1991), the amount and composition of XG did not change during ripening despite a decrease in molecular weight. This is consistent with fission of the β-1,4-glucan backbone by cellulases, xyloglucanases or xyloglucan endotransglycosylases (XETs), all of which are present in fruit and act only on the XG backbone. There was also no detectable solubilization of XG during kiwifruit ripening, even when the wall was in a state of almost complete dissolution (Redgwell, 1991). Partial depolymerization would not completely disrupt hydrogen bonding of XG to the cellulose fibrils but hydrolysis of XG could have an effect on the integrity of the wall if it occurred in regions spanning adjacent microfibrils. XG depolymerization occurs early in softening and could have a role in cell wall loosening (Rose *et al.*, 1998; Wakabayashi, 2000).

In vitro hydrolysis of xyloglucan produces oligosaccharides with biological activity (Creelmann and Mullet, 1997). The oligosaccharides are also found in the medium of suspension-cultured cells (Baydoun and Fry, 1989) and are present *in vivo* in ripe kiwifruit (Schroeder and Redgwell, unpublished data). They could be involved in cell signalling or as substrates for the depolymerization of xyloglucan by XET-mediated endo-transglycosylation.

3.4.3 Galactoglucomannans and glucuronoarabinoxylans

There is no change in either composition or molecular size of the galactoglu-comannan (GGM) and glucuronoarabinoxylan in kiwifruit during ripening (Redgwell *et al.*, 1991a). Earlier work with tomato had reported a dramatic increase in the relative amount of a low molecular weight 8 M KOH-soluble hemicellulose fraction in the late stages of ripening (Tong and Gross, 1988). A similar phenomenon was found in the 4 M KOH-soluble fraction of kiwifruit cell walls (Redgwell *et al.*, 1991b). Recent evidence suggests that this fraction is a GGM. The increase in tomato was attributed to *de novo* synthesis, but in kiwifruit to an increase in solubility in 4 M KOH as the fruit ripens.

3.4.4 Cellulose

Ultrastructural studies have reported the apparent dissolution during ripening of the microfibrillar network in pear, apple (Ben Arie *et al.*, 1979) and avocado (Platt-Aloia *et al.*, 1980) but there is no evidence that this involves changes in the cellulose fibrils themselves. Immunolocalization studies in kiwifruit demonstrated that cellulose remained intact (Sutherland *et al.*, 1999) and cellulose levels were unaffected by fruit ripening (Fischer and Bennett, 1991). Limited hydrolysis would not necessarily cause loss of cellulose, but could affect the physical form and appearance of fibrils (e.g. swelling). Expansins, small proteins which have the ability to weaken the structure of pure cellulose paper, have been implicated as agents for loosening the xyloglucan–cellulose network (McQueen-Mason *et al.*, 1992).

Cross-polarization NMR combined with magic angle spinning was used to monitor changes to the relatively rigid polymers including crystalline cellulose during ripening of kiwifruit (Newmann and Redgwell, 2001). There was no evidence for changes in the nature of the cellulose crystallites or the polysaccharides adhering to crystallite surfaces, even in CWM from fruit in which cell wall dissolution was extreme.

3.5 Cell-wall associated enzymes: role in texture change

The last 10 years have seen a concerted effort to understand the biochemical and molecular mechanisms governing the action of cell-wall associated enzymes during fruit ripening. Tomato, which is suited to a molecular approach, has been the focus of several laboratories for the study of endo-PG and the level of understanding of its mode of action has outstripped other enzymes. At the same time, the discovery that it may not be the prime determinant of pectin solubilization in tomato has fuelled the study of other enzymes as candidates for promoting wall dissolution.

3.5.1 Polygalacturonase in tomato

Endo-polygalacturonase (endo-PG, EC 3.2.1.15) catalyses the hydrolysis of the linear α-1,4-D-galacturonan backbone of pectic polysaccharides. The enzyme requires several contiguous galacturonosyl residues in the free acid form to initiate and maintain its action (Daas et al., 2000). Pectic polysaccharides that contain methyl esterified or acetylated galacturonosyl residues, have a high rhamnose content in the backbone, or are heavily substituted with neutral sugar side-chains, are poor substrates for endo-PG.

Endo-PG activity increases dramatically during tomato ripening owing to de novo synthesis of the enzyme, the level of mRNA increasing 2000-fold (DellaPenna et al., 1986). There are at least two isoforms, PG1 and PG2 in ripe tomato. PG1 accumulates early in ripening but PG2 activity can be detected soon afterwards and becomes the dominant isoform in ripe fruit (Tucker et al., 1980). It consists of two subforms (PG2A and PG2B) but these contain the same polypeptide chains differing only in degree of glycosylation (Ali and Brady, 1982; DellaPenna et al., 1987). PG1 is composed of two peptide chains of ca. 43 and 38 kDa. The 43 kDa peptide appears to be the same as that in PG2 and presumably represents the catalytic subunit. The 38 kDa peptide has been termed the β-subunit (Zheng et al., 1992). It can be extracted from the cell walls of both green and ripe tomato fruit but in itself possesses no enzyme activity. PG2 can be converted to PG1 in vitro by incubation with an extract from green tissue (containing the β-subunit) (Tucker et al., 1981), but there is a suggestion that PG1 is an artefact of extraction and does not exist in vivo (Moore and Bennett, 1994).

Endo-PG activity in ripe tomato arises from a single gene encoding a fruit specific endo-PG protein (Bird et al., 1998) and therefore the down-regulation of endo-PG affects all isoforms that arise from a post-translational modification of a common polypeptide (Tucker and Zhang, 1996). At the molecular level, the most significant findings in relation to endo-PG activity and softening were recently summarized:

> 'The suppression of PG gene expression in wild type tomato and the ectopic expression of PG in the ripening-impaired pleiotropic mutant ripening inhibitor (rin) showed that PG mediated pectin depolymerization was not necessary for normal ripening and softening... Collectively the results obtained with transgenic tomatoes having altered PG levels are consistent with the hypothesis that PG mediated pectin disassembly does not contribute to early fruit softening but contributes significantly to tissue deterioration in the late stages of fruit ripening' (Hadfield and Bennett, 1998).

Since tissue deterioration in the later stages of tomato ripening is 'normal', there is a contradiction in these statements. Nevertheless, they encapsulate the most salient conclusions of the transgenic tomato work, and pose the question— what is the role of endo-PG in tomato cell wall pectin metabolism during ripening? Analysis of transgenic tomato plants that constitutively express an antisense PG2 transgene has shown that the residual 1% endo-PG enzyme activity extracted from ripe fruit is sufficient for wild-type rates of pectin solubilization (Smith *et al.*, 1990; Brummell and Labavitch, 1997). However, in another study where endo-PG was anti-sensed, pectin solubilization did not proceed to the same extent as in the wild-type fruit (Carrington *et al.*, 1993). Further work has attempted to refine our understanding of the relationship between the PG isoforms and the processes of pectin solubilization and depolymerization. Expression of the β-subunit has been down-regulated in transgenic tomatoes resulting in elevated levels of EDTA-soluble polyuronide at all stages of ripening and significantly higher degrees of depolymerization late in ripening (Watson *et al.*, 1994). It was concluded that PG2 was responsible for both pectin solubilization and depolymerization *in vivo* and although the β-subunit protein was not necessary for PG2 activity *in vivo*, it played a role in limiting the extent of pectin solubilization and depolymerization that occurred during ripening. A separate study (Chun and Huber, 1997), while confirming the effect of the β-subunit protein on pectin solubilization, was unable to demonstrate that chelator-soluble pectins from β-subunit anti-sensed fruit were of lower molecular mass than polymers from wild-type fruit. In addition, the β-subunit anti-sense fruit showed an increase in the CDTA-soluble pectin from the mature green stage, indicating that the β-subunit can influence pectin solubility independently of PG activity.

The contradictory findings in studies with regard to pectin solubilization and depolymerization following the molecular down-regulation of PG isoforms are likely to be caused by variations in analytical protocols. The use of a chelator-soluble fraction as an indicator of endo-PG-induced pectin solubilization is open to criticism, since the reagent solubilizes a fraction of pectin that is in part insoluble (i.e. calcium complexed pectin). The degree of the PG-induced solubilization early in softening will be obscured by the presence of a pectin fraction that has been solubilized by the chelator and not enzyme action.

3.5.2 *Polygalacturonases in other fruit*

It appears that most fruit contain at least some endo-PG activity, although it may not be ripening regulated. Evidence for endo-PG-independent pectin solubilization had been suggested by a range of fruit that underwent ripening-associated wall breakdown in the apparent absence of endo-PG activity: apple (Knee, 1978), strawberry (Huber, 1984), persimmon (Cutillas *et al.*, 1993), muskmelons (McCollum *et al.*, 1989), cherry (Batisse *et al.*, 1994), banana

(Wade *et al.*, 1992) and pepper (Gross *et al.*, 1986). Subsequently, endo-PG has been shown to be present at very low levels in apple (Wu *et al.*, 1993), strawberry (Nogata *et al.*, 1993), melon (Hadfield *et al.*, 1998) and banana (Neelam *et al.*, 2000). Such levels may approximate those found in endo-PG-anti-sense tomato (Smith *et al.*, 1990; Brummell and Labavitch, 1997) and be sufficient to catalyse pectin breakdown.

Tomato has been accepted by many as a model for research on fruit softening. While many changes in the cell wall polysaccharides in tomato appear similar to other fruit, the genetics of enzyme expression are dramatically different. In ripe tomato, a single endo-PG gene is expressed which can be induced by exogenous ethylene. In kiwifruit, it is likely that endo-PG is encoded by a family of genes none of which has been shown to be fruit specific or strictly ripening regulated (Zhong *et al.*, 2000). The mRNAs for three closely related endo-PG cDNA clones were not expressed in a ripening specific manner. Two clones (PGA and PGB) were closely related to the tomato ripening endo-PG; they were expressed at low levels from harvest onwards, but expression increased during the climacteric in response to endogenous ethylene. The same gene product was expressed in one week old fruit, flower buds and petals. The third PG clone was expressed during fruit development, in senescent petals and root tips. Expression increased significantly in ripe fruit just prior to the production of endogenous ethylene, reaching levels 50 times greater than for PGA and PGB.

Although endo-PG may be necessary for solubilization, there is no correlation between the degree of pectin solubilization and the level of endo-PG activity. A comparison of fruit with varying endo-PG activities showed that pectin solubilization decreased in the order: avocado, kiwifruit, blackberry, plum, persimmon, strawberry and tomato (Redgwell *et al.*, 1997b). However, tomato had high PG activity and strawberry had negligible activity.

It seems that endo-PG depolymerizes pectin late in fruit softening when the substrate (solubilized pectin) is readily accessible to the enzyme. It may be a primary factor in the deterioration of fruits such as tomato late in ripening. Freestone peaches soften appreciably and contain endo-PG activity which increases late in ripening so that soluble depolymerized pectin accumulates (Orr and Brady, 1993; Pressey *et al.*, 1971; Karakurt *et al.*, 2000). The firmer freestone varieties contain only exo-PG and do not soften to a melting texture (Pressey and Avants, 1978). There is increasing evidence that the endo-PG gene corresponds to the melting flesh (M) locus of peach (Lester *et al.*, 1994; Lester *et al.*, 1996). Extensive pectin depolymerization is associated with high endo-PG activity (tomato, avocado) or at least measurable activity (kiwifruit). Where endo-PG activity is either undetectable (cherry, Batisse *et al.*, 1994) or extremely low (strawberry, Matsuhashi and Hatanaka, 1991; Nogata *et al.*, 1993; and apple, Wu *et al.*, 1993) little pectin depolymerization occurs. An exception is persimmon, where pectin depolymerization occurs although endo-PG activity has not been detected (Cutillas *et al.*, 1993; Kang *et al.*, 1998). The high levels of endo-PG in avocado are associated with more exensive depolymerization of the pectic polysaccharides than in tomato (Huber and O'Donoghue, 1993; Redgwell *et al.*, 1997b). The molecular weight decrease in tomato and avocado pectin can be overestimated if endogenous endo-PG is not fully inactivated during isolation (Seymour *et al.*, 1987; Huber, 1992). Avocado contains high levels of fat which may protect endo-PG from inactivation during cell wall preparation. Homogenization of ripe tomato fruit resulted in the formation of polyuronides with a molecular mass distribution nearly indistinguishable from those generated *in vivo* in ripe avocado (Wakabayashi *et al.*, 2000).

3.5.3   Rhamnogalacturonase

Rhamnogalacturonase A (RGase A) was first characterized from *Aspergillus aculeatus* (Schols *et al.*, 1990) but also occurs in apple, grape and tomato (Gross *et al.*, 1995). Unlike endo-PG, RGase A cleaves galacturonosyl-1,2-rhamnosyl linkages and can depolymerize branched pectins that resist attack by endo-PG. There is evidence that highly galactosylated polymers exist in distinct domains within the wall (Roberts, 2001). Their depolymerization by a rhamnogalacturonase or a β-galactosidase may produce quite different changes in wall integrity than those resulting from endo-PG-mediated depolymerization of homogalacturonan pectin zones. RGase A may generate biologically active pectic fragments that stimulate ethylene synthesis (Seymour and Gross, 1996). The RGase A gene has been cloned from *Aspergillus aculeatus* but transgenic experiments with ripe fruit have not been reported (Kofod *et al.*, 1994; Suykerbuyk *et al.*, 1995).

3.5.4   Pectin methylesterase

Pectin methylesterase (PME, EC 3.1.1.11) catalyses the de-esterification of pectin to expose its carboxyl groups and liberate methanol (Rexova-Benkova and Markovic, 1976). The degree of methyl-esterification influences the physico-chemical properties of pectic polymers, relating to charge density of the molecule and gelation properties. Because de-esterification of linear pectic polymers allows more calcium-mediated junction zones it can contribute to the rigidity of the cell wall and fruit tissue integrity (Tieman and Handa, 1994). In tomato, PME activity occurs throughout fruit development and ripening (Hobson, 1963) and arises from the expression of three genes (Tucker and Zhang, 1996). The major isoform PME2 is fruit specific and has a peak of activity at the mature breaker stage. When PME2 expression was down-regulated to 10% of normal the effects on cell wall degradation were not clear. While Tucker and Zhang (1996) reported that the solubility and depolymerization of the pectic polymers was unaffected, Tieman *et al.* (1992) described a reduction in the amount of the EDTA-soluble pectin fraction and higher levels of intermediate sized polyuronides in the transgenic fruit compared to wild-type. In each study the reduction in PME2 isoform levels resulted in higher esterification of pectin at all stages of fruit development. Esterified pectin would be less susceptible to endo-PG-mediated hydrolysis, which would be consistent with the findings of Tieman *et al.* (1992). In Tieman's study the degree of softening during ripening of each genotype was the same but upon storage at room temperature for seven weeks the transgenic fruit lost tissue integrity while the wild-type held their cohesiveness. Thus, reduced pectin depolymerization had a negative effect on shelf-life. It was suggested that the loss of bound calcium in the walls of the transgenic fruit (with more esterified pectin) negated the potential benefits on wall integrity that decreased depolymerization of the pectic polysaccharides may have caused.

Ethylene-treated kiwifruit showed a 10% decrease in the degree of esterification of cell wall pectin (Redgwell *et al.*, 1990), which was accompanied by a 2–3-fold increase in PME activity (Wegrzyn and MacRae, 1992). De-esterification of pectin continued throughout fruit softening while PME activity decreased to undetectable levels. A glycoprotein inhibitor of PME in kiwifruit has been identified (Balestrieri *et al.*, 1990) and purified (Giovane *et al.*, 1995). The inhibitor was present in unripe fruit as an inactive precursor and was transformed into the active protein during the course of ripening. There was no evidence for the presence of the inhibitor during ethylene treatment when PME activity increased in the early stages of softening. The results suggest that transient increases in pectin methylesterase and perhaps pectin de-esterification are a response to ethylene exposure rather than the normal sequence of events during kiwifruit ripening (MacRae and Redgwell, 1992).

Individual walls are de-esterified independently from neighbouring walls in tomato fruits, implying distinct, spatial regulation of PME activity. (Steele *et al.*, 1997). Patterns of de-esterification appear to be dissimilar between fruit types because immunological studies in tomato (Roy *et al.*, 1992) and kiwifruit (Sutherland *et al.*, 1999) using probes for high and low esterified polymers (Knox *et al.*, 1990) showed different patterns of labelling of high and low esterified pectin during ripening.

3.5.5   β-Galactosidase

β-Galactosidase (β-Gal, EC 3.2.1.23) appears to exist as both a constitutive enzyme involved in the turnover of cell-wall associated galactose throughout fruit growth and development and as a ripening regulated catalyst for the degradation of galactose containing polysaccharides during senescence. It hydrolyses terminal non-reducing β-D-galactosyl residues from β-D-galactosides and can be regarded as a pectin degrading enzyme since most of the wall galactose occurs as β-1-4-linked galactan covalently attached to the rhamnogalacturonan backbone. During fruit ripening, up to 70% of the cell wall galactose is lost

(Gross and Sams, 1984; Redgwell *et al.*, 1997a). It is not merely removed from the cell wall along with the pectic polysaccharides which are solubilized during softening, but is cleaved from the galactan side-chains. Existing evidence points to β-Gal acting in an exo-fashion. To date there have been no reports of endo-β-1-4 galactanase in fruit. The action of β-Gal offers an alternative mechanism to that of endo-PG for the depolymerization of pectic polymers during ripening. β-Galactosidases from kiwifruit (Ross *et al.*, 1993), papaya (Ali *et al.*, 1998), Japanese pear (Kitagawa *et al.*, 1995), apple (Ross *et al.*, 1994) and tomato (Carey *et al.*, 1995) have been isolated and shown to remove galactose residues from a range of pectic and hemicellulosic polysaccharides.

For apple (Ross *et al.*, 1994), persimmon (Kang *et al.*, 1994) and tomato (Carey *et al.*, 1995), cDNA clones have been isolated and their sequences are homologous to a carnation petal senescence related cDNA clone PSR12 (Raghothama *et al.*, 1991). Temporal variation in the expression of seven β-Gal genes (*TGB1–TGB7*) in tomato indicates that β-galactosidases with different substrate specificities and functions in the fruit are encoded by a family of genes expressed throughout the life of the plant (Smith and Gross, 2000).

Recently the cDNA clone *TBG1* was used to produce transgenic tomato plants with *TBG1* mRNA reduced to 10% of the levels found in ripening fruit (Carey *et al.*, 2001). No effect on either β-Gal activity, levels of cell wall galactose or phenotype was observed. However, down-regulation of TBG4 corresponding to β-Gal II (Smith *et al.*, 1998) did affect enzyme activity and cell wall galactose levels. The transgenic fruit were 40% firmer than the wild-type even though down-regulation of β-Gal was not complete (K. C. Gross, personal communication). Ripening regulated β-Gal II activity, which correlated with galactose loss in tomatoes, did not appear until after the early stages of softening (Carrington and Pressey, 1996), suggesting that β-Gal, like endo-PG, may catalyse polysaccharide depolymerization in the later stages of softening.

3.5.6 *Xyloglucan endotransglycosylase*

Xyloglucan endotransglycosylase (XET) cleaves a xyloglucan molecule and links the new reducing end to the non-reducing end of another xyloglucan molecule. XET is thought to be cell wall loosening because of its extracellular occurrence (Nishitani and Tominaga, 1992), its mode of action (Smith and Fry, 1991) and the presence of high activity in rapidly growing tissues (Fry *et al.*, 1992).

XET has been implicated in ripening-related changes to the fruit cell wall of kiwifruit (Redgwell and Fry, 1993), tomato (Maclachlan and Brady, 1994) and persimmon (Cutillas *et al.*, 1994). Its activity increased tenfold when softening was promoted by ethylene treatment of kiwifruit. In apple and kiwifruit highest XET activity occurred during the cell division phase in the first weeks after anthesis, but apple did not show the ripening-associated increase in XET activity shown by kiwifruit (Percy *et al.*, 1996).

Schroeder *et al.* (1998) reported that the kiwifruit XET is an *N*-glycosylated protein (34 kDa). Six cDNA clones (*AdXET1–6*) with homology to other reported XETs were isolated and shared 93–99% nucleotide identity. Peptide sequencing indicated that ripe kiwifruit XET was encoded by *AdXET6*. Northern blotting analysis detected the *AdXET1–6* gene products when ethylene production was first detected in kiwifruit. However, low stringency Northern blotting analysis revealed that further XET-like mRNAs were induced early in the softening process, after the application of exogenous ethylene.

In the presence of xyloglucan alone kiwifruit XET acted as a hydrolase. In the presence of xyloglucan-derived oligosaccharides, it catalysed polysaccharide-to-oligosaccharide endotransglycosylation (Schroeder *et al.*, 1998). Both reactions depolymerized xyloglucan but it is not known whether either or both occur *in vivo*. However, xyloglucan-derived oligosaccharides occur in ripe kiwifruit and can act as acceptors for XET in the endo-transglycosylation reaction (Schroeder and Redgwell, unpublished data). In avocado the average molecular weight of xyloglucan did not decrease during ripening but gel-permeation chromatography showed a broadening of the xyloglucan elution profile (O'Donoghue and Huber, 1992) which is symptomatic of endotransglycosylation (Nishitani and Tominaga, 1992). XET is thought to incorporate freshly synthesised XG into a growing cell wall in this way (Thompson *et al.*, 1997). XET activity also increases under stress, such as in flooding in maize plants (Saab and Sachs, 1996). The increase in XET activity during ripening may be a stress response, reflecting an attempt to moderate cell wall deterioration by reinforcing the cellulose-xyloglucan network.

3.5.7   Endo-β-1,4-glucanase

Endo-β-1,4-glucanases (cellulase, EC 3.2.1.4) (EGase) are a class of enzymes which degrade the soluble substrate carboxymethylcellulose (Cx-cellulose). Carboxymethylcellulase (Cx-cellulase) does not degrade native cellulose in plant cell walls and yet the enzyme activity increases during ripening of strawberry (Abeles and Takeda, 1990), pepper (Harpster *et al.*, 1997), peach (Bonghi *et al.*, 1998), tomato (Maclachlan and Brady, 1992), olives (Heredia *et al.*, 1991), raspberry (Sexton *et al.*, 1997) and especially avocado (Awad and Young, 1979). What is the *in vivo* substrate of Cx-cellulases in fruit? Avocado fruit contain one of the highest specific activities reported in plants (Awad and Young, 1979) and activity is associated with softening, suggesting a major role in ripening (Cass *et al.*, 1990). Platt-Aloia *et al.* (1980) suggested that the loss of microfibril structure observed in avocado under the electron microscope was caused by degradation of the cellulose-associated xyloglucan rather than the cellulose itself. However, Cx-cellulase was unable to depolymerize xyloglucan isolated from avocado (O'Donoghue and Huber, 1992). Cellulase-induced fission of individual cellulose molecules in a fibril may be undetectable in terms of hydrolysis products or changes to the amount of cellulose but could affect the architecture and physical appearance of the fibril. O'Donoghue *et al.* (1994) showed that avocado Cx-cellulase was able to decrease cellulose crystallinity and cause cellulose fibrils to lose cohesiveness.

Genes encoding EGases have been isolated and characterized from avocado (Cass *et al.*, 1990), pepper (Harpster *et al.*, 1997) and strawberry (Llop *et al.*, 1999; Trainotti *et al.*, 1999a,b). In tomato, EGases are encoded by a seven-member gene family (Brummell *et al.*, 1999b) and mRNA of one member (*Cel 2*) accumulates during ripening (Lashbrook *et al.*, 1994; Gonzalez *et al.*, 1996). Antisense suppression of *Cel 2* caused no change in the pattern of softening, but the abscission zones of the transgenic fruit were strengthened (Brummell *et al.*, 1999b). It is likely that EGases do contribute in some manner to the overall disruption of the cell wall but the down-regulation of a single enzyme may not impair the cascade of events leading to tissue softening.

3.6   Non-enzymatic mechanisms of change

3.6.1   Expansins

Expansins are small proteins (28 kDa) that catalyse cell wall extension *in vitro* (McQueen-Mason *et al.*, 1992; Cosgrove, 2000). Expansin genes show a limited sequence homology to an EGase cloned from *Trichoderma* and some expansins have EGase activity (Cosgrove, 1999). This activity is not thought to be responsible for the effect of expansins on the extensibility of cell walls. Expansins appear to act as a 'hydrogen bond-ase' to weaken glucan–glucan interactions. Thus they can weaken pure cellulose paper (McQueen-Mason and Cosgrove, 1994) and enhance cellulose breakdown by cellulases (Cosgrove *et al.*, 1999).

An α-expansin gene *Le-EXP1* is specifically expressed in ripening tomato (Rose *et al.*, 1997). cDNAs closely related to *Le-EXP1* have also been identified in ripening melon (Rose *et al.*, 1997) and strawberry (Civello *et al.*, 1999). Expression of *Le-EXP1* occurs late in tomato ripening, and can be induced by ethylene. The expansin gene family in tomato is large and complex. While expression of several expansin genes may contribute to green fruit development, only *EXP1* mRNA is present at high levels during fruit ripening (Brummell *et al.*, 1999a).

Antisense suppression of *EXP1* protein in tomato to 3% of wild-type levels increased firmness between 13% (breaker) and 23% (overripe) (Brummell *et al.*, 1999c). Polyuronides of a CDTA-soluble fraction were less depolymerized late in ripening in the *EXP1* suppressed green fruit, but there was no effect on the degree of depolymerization of the XG fraction. In contrast, overexpression resulted in a 30% increase in compressibility of mature green fruit.

Overexpression of *EXP1* did not increase the degree of polyuronide depolymerization but did induce increased depolymerization of the xyloglucan. This is consistent with the idea that expansins loosen the interaction between xyloglucan and the microfibrils, so that the susceptibility of the xyloglucan to XET, xyloglucanases or EGases is increased. Suppression or overexpression of *EXP1* did not affect solubilization of pectin. Brummell *et al.* (1999c) suggest that *EXP1*-mediated relaxation of the wall could make pectin accessible to endo-PG in the final stages of wall disassembly. This seems unlikely as wall dissolution is so well advanced at this stage that pectin solubilization supplies an easily accessible substrate for endo-PG action. One of the earliest symptoms of fruit softening is a slight wall swelling that occurs before hydrolase-induced changes to the primary structure of matrix polymers are marked. Expansins could promote this swelling with their capacity to weaken glycan–glycan interactions.

3.6.2   Hydroxyl radicals and cell wall lysis

Brennan and Frenkel (1977) hypothesized that H_2O_2-dependent oxidative processes are involved in senescence of pear. Miller (1986) showed that several plant cell wall polysaccharides undergo non-enzymatic scission in the presence of H_2O_2. Fry (1998) found that pectin, xyloglucan and cellulose (carboxymethyl-cellulose, CMC) undergo non-enzymatic scission under physiological conditions in the presence of ascorbate. The active species may be the $\cdot OH^-$ radical which can be formed by the reaction of H_2O_2 and Cu^+. Both these can be produced by the action of ascorbic acid on O_2 and Cu^{2+} and therefore all the necessary components would be readily available in the cell wall. The non-specificity of the reaction means that cellulose and pectin are as vulnerable to $\cdot OH^-$ mediated scission as xyloglucan, but the extent of the decrease in molecular weight of the polysaccharides was not reported. The delayed senescence and maintenance of cellular integrity in a Clipper variety of muskmelon correlated with much lower levels of the $\cdot OH^-$ radical than occurred in the shorter storage Jerac variety, in which there were elevated levels of free radical production during storage (Lacan and Baccou, 1998).

3.6.3   Calcium and the apoplast

Changes in the ionic environment of the apoplast can affect enzyme activity, and the conformation of the pectic polysaccharides that are sensitive to pH and electrolyte composition. The pH of the apoplastic fluid of tomato decreased from 6.7 to 4.4 during ripening (Almeida-Domingos and Huber, 1999). At pH 6.7 endo-PG is almost inactive (Chun and Huber, 1998), providing one explanation for the limited amount of pectin depolymerization in the early stages of tomato softening. Fishman *et al.* (1989) described five macromolecular pectin species in tomato, held together by non-covalent interactions which were dissociated to some extent by dialysis against salt. The pattern of association was affected by the ripening stage and therefore could be susceptible to any change in the electrolyte composition of the apoplast during ripening.

Exogenously applied Ca^{2+} can increase fruit firmness (Knee, 1982) and the tensile strength of tissue (Stow, 1989). Calcium ions can interact ionically and coordinate with the oxygen functions of two adjacent pectin chains, to form the so called 'eggbox structure' and cross-link the chains (Rees *et al.*, 1982). Calcium has many effects on the physiology and biochemistry of fruit (Poovaiah *et al.*, 1988) but it is not clear whether calcium affects the ripening-related changes in wall texture. Electron-energy-loss spectroscopy has been used to map the distribution of Ca^{2+} in the cell walls of Cox's apples grown at two different sites in the UK, one of which consistently gave firmer fruit at harvest and after six months storage (Huxman *et al.*, 1999). Unexpectedly it was found that the firmer fruit had significantly lower levels of cell wall calcium. This contrasts with previous results that reported a strong correlation between apple firmness and calcium concentration in the flesh (Glenn and Poovaiah, 1990).

Ca^{2+} (2 mM) dramatically inhibited PG2-mediated pectin release from tomato cell walls *in vitro* at pH 4.5 (Chun and Huber, 1998). During ripening in tomato Almeida-Domingos and Huber (1999) reported that calcium levels remained relatively constant in the apoplastic fluid, while MacDougall *et al.* (1995) showed a twofold increase in the concentration of free calcium (11 mM to 24 mM). Wall complexed calcium is unlikely to be displaced by the drop in pH of the apoplast during ripening because it is very difficult to remove from the wall (Grignon and Sentenac, 1991). A CDTA-soluble pectin fraction (presumably calcium complexed) is obtained from many fruits (Redgwell *et al.*, 1997b) even after the cell wall has been extracted with acidic solvents. Nevertheless, the pH of the apoplast late in ripening may be sufficiently low to induce pectin disaggregation by a Ca^{2+}-independent mechanism (Jarvis, 1984).

3.7   Wall oligosaccharides as ripening regulators

Particular oligosaccharides can modulate plant growth and act as regulatory signals in fruit ripening (Ryan and Farmer, 1991; Aldington and Fry, 1993; Creelmann and Mullet, 1997). Climacteric fruits have been used as models to test pectin-derived oligosaccharides as ethylene elicitors (Brecht and Huber, 1988; Campbell and Labavitch, 1991). Oligosaccharides produced by acid hydrolysis of citrus pectin not only stimulated ethylene production in tomato discs but some ripening effects (colour development), which were not solely related to ethylene production (Campbell and Labavitch, 1991). Pectin-derived oligosaccharides, which can act as elicitors of ethylene, exist *in vivo* in tomato fruit (Melotto *et al.*, 1994). Pectin oligomers are unlikely to be initiators of ripening if they are products of polygalacturonase (PG) action which occurs after the increase in ethylene production (Brady *et al.*, 1982). However, the monosaccharide composition of the pectic oligomers in the Melotto *et al.* (1994) study suggested that they were derived from the branched sequences of rhamnogalacturonans and were therefore not produced by the action of endo-PG.

Oligosaccharides derived from xyloglucan (XGOs) act as growth regulators that can inhibit auxin-induced growth at nanomolar concentrations and promote growth at micromolar concentrations (Fry, 1993). This together with their participation in XET catalysed reactions, and a report that they can induce ethylene synthesis in persimmon (Cutillas *et al.*, 1998a), suggests that XGOs also have the potential to act as ripening regulators.

While the significance of the *in vivo* role of biologically active cell-wall derived oligosaccharides remains to be established, they are potential biocontrol agents for improving fruit quality (Baldwin and Baker, 1992). They are naturally derived compounds that are active at extremely low concentrations, so that preparation of useful amounts is a practical proposition. Many preharvest factors are involved in postharvest texture problems with fruit (Sams, 1999). It is possible that once the mechanisms of biologically active oligomers are known, they could be used to replace chemicals presently used as preharvest sprays (Child *et al.*, 1984) to control the rate of maturation and ripening.

3.8 Cell wall synthesis during ripening

While ripening related changes to cell wall polysaccharide structure have been attributed to polymer degradation, anabolic as well as catabolic events may dictate alterations to wall architecture during fruit softening. If postharvest cell wall synthesis does occur to a significant degree, then the accommodation of newly synthesized polymers into a changing matrix could contribute to the overall textural changes that accompany ripening.

Tong and Gross (1988) postulated *de novo* synthesis of a mannose-rich hemicellulose late in ripening of tomatoes. However, their results can also be explained by changes in the distribution of polymers between fractions during ripening. Mitcham *et al.* (1989) reported that ^{14}C sucrose injected into the pedicel of ripe tomatoes attached to the plant was incorporated into fractions solubilized from the cell wall. Greve and Labavitch (1991) used ^{13}C glucose to show incorporation into the neutral monosaccharide constituents of chelator-soluble pectin of pericarp discs allowed to ripen in culture after excision from mature green tomato fruit. Later work confirmed these findings and reported that hemicelluloses synthesized during ripening were different in type and/or proportion from those present in developing fruit (Huysamer *et al.*, 1997a,b).

The biosynthetic ability of intact kiwifruit labelled with $^{14}CO_2$ was compared with that of discs excised from the same fruit and allowed to ripen in culture (Redgwell, 1996). Up to 95% of the radioactivity in whole fruit and discs was incorporated into the protein fraction of the wall. There was very low incorporation of label into polysaccharides and no ripening-related enhancement of the level of label into a specific polysaccharide. When discs were excised from unlabelled kiwifruit and treated with ^{14}C-glucose, up to 50% of the total label was incorporated into the polysaccharide fraction, demonstrating that polysaccharide synthesis was favoured by the use of an exogenous precursor. All monosaccharide constituents of the pectic and hemicellulosic polysaccharides were labelled. Certainly, fruit retain polysaccharide synthesizing capacity well into senescence. It is not known whether this occurs in intact fruit, whether it is a ripening regulated process associated with softening, or merely reflects the remnants of constitutive processes, more active earlier in fruit development.

The rate of cell wall polysaccharide turnover (i.e. synthesis versus degradation) has not been measured during fruit ripening, but it is probable some wall components are remetabolized and may be incorporated into new wall polymers. In kiwifruit, although there is a rise in free galactose during ripening, it is insufficient to account for the galactose released, indicating that metabolism of cell-wall derived galactose has occurred (Redgwell et al., 1990). Polyuronide does not seem to be re-metabolized after degradation, as almost all can be accounted for in the form of solubilised polymers (Koch and Nevins, 1989; Redgwell et al., 1992; MacDougall et al., 1995).

3.9   Low temperature disorders

3.9.1   Mealiness and chilling injury

The dryish mouth feel known as 'mealiness' is a symptom of chilling injury in many fruit. Ultrastructural studies show that mealiness in stone fruit results from a separation of the parenchyma cells without a large percentage of the cells fracturing (Luza et al., 1992). This also occurs in non-mealy fruits that ripen to a soft melting texture. The difference is that the cell surfaces are covered in a layer of juice, while in the mealy fruit the layer of free juice was not observed (Harker and Sutherland, 1993). Since there is little apparent difference in the total water content of mealy and non-mealy fruit (Sonego et al., 1995), mealiness may involve some change in the water-holding capacity of the cell wall polysaccharides.

The development of woolly breakdown in peaches stored at $0°C$ was attributed to the enhancement of PME activity and the inhibition of endo-PG activity at low temperature (Ben Arie and Sonego, 1980; Ben Arie et al., 1989). This leads to a build up of high molecular weight, de-esterified pectin as a gel that retains moisture during chewing. Zhou et al. (1999) confirmed a build up of high molecular weight polymers in the juice of chilling injured, mealy nectarines. Mealiness could largely be avoided by storing the fruit at $20°C$ for two days before storage at $0°C$, and this was attributed to a rise in endo-PG activity. Dawson et al. (1992) showed that pectic polysaccharides retained in the wall of mealy fruit contained higher rhamnose levels, suggesting that highly branched pectins were involved.

Similar processes occur in tomato (Jackman *et al.*, 1992). Higher initial PME activity (Marangoni *et al.*, 1995) and a suppression of endo-PG mRNA have been documented in chilling injured fruit (Watkins *et al.*, 1990). In addition, a prestorage heat treatment prevented chilling injury (Lurie *et al.*, 1993). Chilling injury in persimmon is accompanied by mealiness and in more severe cases the formation of a firm gel (MacRae, 1987). Grant *et al.* (1992) showed that gel formation in injured fruit was accompanied by solubilization of high molecular weight pectic polymers containing higher levels of neutral sugars than found in the pectic polysaccharides of normally ripened fruit. During storage at $0°C$ there was an increase in the amount of cell wall polysaccharide implying that cell wall synthesis had occurred at lower temperatures. Woolf *et al.* (1997) showed that heat treatment before cold storage alleviated chilling injury and induced a decrease in the viscosity of both solubilized pectin and the cell wall fraction. But they found no increase in the amount of cell wall material during storage at $0°C$. A similar mechanism based on PG/PME activities could be involved in stone fruits, tomato and persimmon. However, endo-PG has not been detected in persimmon, although solubilization and depolymerization of pectin occur (Cutillas *et al.*, 1993). There could be another pectin depolymerizing enzyme in persimmon that is affected by low temperature, or there are levels of endo-PG which are undetectable by *in vitro* assay.

3.9.2   Low temperature breakdown in kiwifruit

Low temperature breakdown in kiwifruit is a physiological disorder character-ized by a grainy appearance of the outer pericarp, followed by water soaking associated with extreme fruit softening (Lallu, 1997). There was a 30% increase in the amount of cell wall polysaccharides in fruit affected by the disorder compared with unaffected fruit stored under identical conditions (Bauchot *et al.*, 1999). In addition the galactosyl content of the disordered fruit was 70% higher. While it is reasonable to attribute these differences to *de novo* cell wall synthesis, they may be explained by anomalies in cell wall metabolism during cold storage (i.e. lower levels of pectin solubilization and inhibition of β-Gal action).

3.10   Consumer perception

3.10.1   Quality and increased consumption

'Quality is fitness for use' or 'quality is to meet the expectations of the consumer' (Jongen, 2000). Shewfelt (1999) distinguished between product-orientated and consumer-orientated quality of fresh fruit and vegetables and stated that a lack of discrimination between the two may be a limiting factor in improving quality. Postharvest scientists and handlers are product-orientated and attribute fruit quality to specific attributes such as chemical composition, colour or firmness. Consumers and economists on the other hand are consumer-oriented in that quality is described by consumer wants and needs. The fruit industry has concentrated more on quantity than quality and more on disease resistance than consumer appeal. There is a price to be paid for failure to recognize consumer interests in new product development whether it be a processed product (Saguy and Moskowitz, 1999) or fresh fruit. Many products do not succeed because of a failure to interpret the language of the consumer (Hollingsworth, 1996). The connection with consumer requirements can be made by 'preference mapping' which links consumer preference patterns on a hedonic scale to sensory properties perceived by a trained panel (Andani and MacFie, 2000).

3.10.2   Firmness and freshness

Consumers are sensitive to subtle differences in texture and use it as a primary limiting factor for acceptability (Shewfelt, 1999). The important contribution of texture to palatability was reviewed by Szczesniak (1990) who noted that crispness may be the most versatile single texture. Palatable foods are likely to have higher levels of 'dynamic contrast', defined as 'moment-to-moment-sensory contrast from the ever changing properties of foods manipulated in the mouth' (Hyde and Witherly, 1993). It is this property of a crisp apple which contributes significantly to its acceptability.

Sensory science provides the scientific basis for textural differences and reliable tools for texture measurement (Abbott *et al.*, 1997). The texture and appearance of fruit are judged subjectively by the consumer and it seems natural to use a similar judgements in research on fruit quality. In practice sensory assessments require careful planning and preparation and access to trained panellists. Instrumental measurements, which are rapid and require only a single assessor, are widely used in quality control of fruit and vegetables. Firmness, 'a resistance to deformation by applied force' (Jowitt, 1974), is the most widely used attribute to describe fruit texture. Many marketing authorities and organizations use a firmness measurement as an absolute index of acceptable fruit quality. In New Zealand the entire kiwifruit export crop is validated by a puncture test with a 7.9-mm diameter probe (Harker *et al.*, 1997c).

Despite the relevance of a firmness measurement, two apples of similar instrumental firmness may differ considerably in their acceptability to the consumer. This is because they may possess quite different textures in terms of crispness, juiciness and dryness. Recent work has aimed at using knowledge of food–mouth interactions to enable instrumental analysis to be linked directly to consumer responses (Harker *et al.*, 1997a). The research showed that human perception of fruit texture is determined by the way that fruit flesh breaks down during chewing and that tensile strength of the tissue may be a better predictor of texture than other firmness measurements.

Firmness is often equated with freshness and postharvest loss of firmness is a major focus for those trying to improve the shelf-life of fruit. However any fruit that has been harvested at maturity and stored for six months is no longer 'fresh' even if the textural properties are almost identical with that of a tree ripened fruit, picked and eaten immediately. The 'quest for fresh' has been quoted as the second most important food industry trend (Gibson, 1995). Unfortunately for the processed food industry 'fresh' cannot be added to a product and the 'percentage freshness' cannot be defined.

3.10.3   Mealiness

Mealy fruit are a deterrent to the increased consumption of fresh fruit. The purchase of a visually attractive fruit, which after a single bite is unacceptable, is a compelling reason for the 'buyer to beware' and to discourage future purchases. However, mealiness is a weakly defined sensory characteristic. Sensory profiling describes mealiness using words such as softness, dryness, granularity and flouriness. Mealiness appears to affect almost every type of fruit, although it is particularly detrimental for apples, peaches, nectarines and tomatoes that are expected to be juicy when fresh. People can detect differences in degrees of mealiness for apples but the effects on acceptability are variable (Gomez *et al.*, 1998). While mealiness is perceived as unpleasant by most consumers, it is considered as pleasant by some elderly people. European studies showed that although there was a cross-cultural consensus with respect to a mealy texture, perceptions were described differently (Nicolai, 1999; Andani *et al.*, 1998).

The histological (Harker and Sutherland, 1993) and, to some extent, the biochemical basis (Ben Arie *et al.*, 1989) for the property of mealiness are known, but the search for reliable methods for detecting mealiness continues. One approach uses a confined compression procedure in which a sample of fruit is compressed in a cylindrical probe; the breaking force, force/deformation ratio and juice release, can be correlated with sensory scores of mealiness for apple varieties (Barreiro *et al.*, 1998). Non-destructive methods using ultrasonic wave propagation, NMR imaging (Nicolai, 1998) and near infra-red (NIR) measurements (Muresan *et al.*, 1998) have all been tried. The last has given some promising results but the calibration models need further improvement.

3.10.4   Applications of biotechnology

The biochemical elucidation of fruit development will lead to more opportunities for the modification of cell wall metabolism by biotechnology. A number of the hydrolases and other enzymes associated with wall changes have been experimentally manipulated and commercial applications have followed. The first was based on anti-sense suppression of endo-PG in tomato. Following the USA's Food and Drug Administration approval on 18 May 1994 the FLAVR SAVR tomato became the first genetically engineered whole food in the grocery store

(Kramer and Redenbaugh, 1994). However, seven years later the application of biotechnology in agriculture is being intensively discussed and consumer polls, particularly in European states, have revealed a general ambivalence and some hostility in attitudes (Schibeci *et al.*, 1997). The Delphi survey (Menrad, 1998) showed considerable variation in the degree of acceptance from one country to another. In New Zealand (Anon, 2000) the kiwifruit export company Zespri International advised the Royal Commission on Genetic Modification that New Zealand should stay free of genetically altered food. Zespri said it had marketing evidence that adverse consumer opinion caused by the perception of New Zealand as an exporter of genetically engineered foods could jeopardise a significant portion of the kiwifruit industry. Transformation is currently feasible to alter the texture of several fruits, including tomato, kiwifruit, apple and melon. However, the perceived benefits may not be sufficiently dramatic to persuade consumers that the purchase and consumption of genetically modified fruit is worth the risk.

3.11   Conclusions

The 'Albersheim cell wall model' for dicotyledons (Keegstra *et al.*, 1973) has been followed by other versions with various arrangements of the cellulose fibrils, xyloglucan and the pectic polymers (Lamport and Epstein, 1983; Talbott and Ray, 1992; Carpita and Gibeaut, 1993). No one model accommodates all of the evidence on cell wall structure. First, they do not take account of two hemicelluloses that are always present and that also appear to hydrogen bond to some structure in the wall. Second, the diversity in the structural features of the pectic polysaccharides, or the fact that these different forms may have site specificity, is ignored. Some pectic polysaccharides appear to be associated intimately with the cellulose fibrils and may even be part of the fibrillar structure. Finally, the models do not show the nature or extent of all of the interpolymeric connections between polysaccharides.

The complexity of wall structure has made it difficult to understand wall disassembly during ripening. Do the known changes in cell wall chemistry during fruit ripening actually cause softening or are they the consequences of other as yet unknown processes? In many fruits cell wall dissolution is advanced before much depolymerization occurs. The nature of the important early stages of cell wall modification remains to be elucidated.

The idea of pectin solubilization could be misleading. In unripe fruit many pectic polysaccharides are probably entangled with other polysaccharides that were deposited around them during biosynthesis. They are 'insoluble' because they are caught in a molecular sieve. As ripening begins, the cell wall matrix loosens and intermolecular spaces enlarge, allowing increased freedom of movement so that pectic polymers become 'solubilised'. Some transformation within the wall is needed to initiate this process, but 'pectin solubilisation' may not require enzyme-catalysed change in the structure of the pectic polysaccharides themselves. If cell wall loosening involves changes at the secondary and tertiary level of polymer interactions then expansins or expansin-like agents would seem to be ideal candidates to bring this about. Wall loosening may also require the cleavage of covalent cross-links but we do not know what the critical linkages are or how they might be broken. Twelve years ago XET and expansins were unknown, and there is every reason to believe that other undiscovered enzymes and agents for cell wall modification exist in the apoplast.

Since the late 1980s molecular biology has played an increasing role in cell wall research and the potential for improving fruit quality by manipulating monogenic traits is now considerable. Almost all major cell wall hydrolases that are known to be associated with fruit ripening have been cloned. Nevertheless, the benefits of molecular transformation of fruit to manipulate texture may be achieved before we have a clear understanding of the biochemical basis for the acquired advantage. The effects of suppression of enzyme activity have proved to be unpredictable. Since the ground-breaking news in 1989 about the role of endo-PG in tomato softening, molecular biology has not provided a major breakthrough in our understanding of the architecture of the plant cell wall or the mechanisms of fruit softening. The discovery of XET, expansins and the dimeric nature of RG II have resulted from biochemical approaches.

More intensive biochemical investigations are required to identify further targets for genetic manipulation. Unfortunately there have been conflicting data produced for fruit ripening studies even for the same fruit. This is often attributed to cultivar differences, but there is an urgent need for more rigorous validation of some of the procedures used for cell wall preparation and the adoption of more uniform protocols for fruit ripening studies. More refined approaches to cell wall analysis are needed to target specific sites in the fruit cell wall and the use of high resolution electron microscopy in conjunction with specific probes will continue to yield dividends in this area. Because separate cell wall domains have different polymer compositions, the interpretation of changes to cell wall polysaccharides isolated from whole tissue homogenates can be misleading. It is particularly necessary to be able to differentiate between changes in cell-to-cell bonding versus loss of primary wall integrity, to assess the relative contribution of each to texture change during ripening and the development of postharvest disorders.

References

Abbott, J.A., Lu, R., Upchurch, B.L. and Stroshine, R. (1997) Technologies for non-destructive quality evaluation of fruits and vegetables. *Horticultural Reviews*, **20**, 1-120.
Abeles, F.B. and Takeda, F. (1990) Cellulase activity and ethylene in ripening strawberry and apple fruits. *Scientia Horticulturae (Amsterdam)*, **42**, 269-276.
Aldington, S. and Fry, S.C. (1993) Oligosaccharins. *Advances in Botanical Research*, **19**, 1-101.

Ali, Z.M. and Brady, C.J. (1982) Purification and characterization of the polygalacturonases of tomato fruits. *Australian Journal of Plant Physiology*, **9**, 155-169.

Ali, Z.M., Shu, Y.N., Othman, R., Lee, Y.G. and Lazan, H. (1998) Isolation, characterization and significance of papaya beta-galactanases to cell wall modification and fruit softening during ripening. *Physiologia Plantarum*, **104**, 105-115.

Almeida-Domingos, P.F. and Huber, D.J. (1999) Apoplastic pH and inorganic ion levels in tomato fruit: A potential means for regulation of cell wall metabolism during ripening. *Physiologia Plantarum*, **105**, 506-512.

An, J., O'Neill, M.A., Albersheim, P. and Darvill, A.G. (1994) Isolation and structural characterization of beta-D-glucosyluronic acid and 4-*O*-methyl-beta-D-glucosyluronic acid-containing oligosaccharides from the cell wall-pectic polysaccharide, rhamnogalacturonan I. *Carbohydrate Research*, **252**, 235-243.

Andani, Z. and MacFie, H.J.H. (2000) Consumer preference, in *Fruit and Vegetable Quality. An Integrated Approach* (eds Shewfelt, R.L. and Brückner, B.), Technomic Publishing Company, Lancaster, pp. 158-177.

Andani, Z., MacFie, H.J.H. and Wakeling, I. (1998) Mealiness in apples towards a multilingual vocabulary (poster 115789), in *3rd Pangborn Sensory Science Symposium, Sense and Sensibility*, Norway, 9-13 August 1998.

Anon (2000) GE: Debate of our time. GE bad news for exports says kiwifruit marketer. *New Zealand Herald*, **1 December**.

Aspinall, G.O. and Fanous, H.K. (1984) Structural investigations on the non-starchy polysaccharides of apples. *Carbohydrate Polymers*, **4**, 193-214.

Awad, M. and Young, R.E. (1979) Post-harvest variation in cellulase, polygalacturonase and pectin methylesterase in avocado (*Persea americana* Mill, cv. Fuerte) fruits in relation to respiration and ethylene production. *Plant Physiology*, **64**, 306-308.

Bacic, A., Harris, P.J. and Stone, B.A. (1988) Structure and function of plant cell walls, in *The Biochemistry of Plants*, Vol. 14 (ed. J. Preiss), Academic Press, New York, pp. 297-371.

Baldwin, E.A. and Baker, R.A. (1992) Natural enzyme and biocontrol methods for improving fruits and fruit quality, in *Molecular Approaches to Improving Food Quality and Safety* (eds D. Bhatnagar and T.E. Cleveland), Van Nostrand Reinhold, New York, pp. 61-82.

Balestrieri, C., Castaldo, D., Giovane, A., Quagliuolo, L. and Servillo, L. (1990) A glycoprotein inhibitor of pectin methylesterase in kiwi fruit (*Actinidia chinensis*). *European Journal of Biochemistry*, **193**, 183-188.

Barreiro, P., Ortiz, C., Ruiz-Altisent, M., De Smetdt, V., Schotte, S., Andani, Z., Wakeling, I. and Beyts, P.K. (1998) Comparison between sensory and instrumental measurements for mealiness assessment in apples. A collaborative test. *Journal of Texture Studies*, **29**, 509-525.

Batisse, C., Fils, L.B. and Buret, M. (1994) Pectin changes in ripening cherry fruit. *Journal of Food Science*, **59**, 389-393.

Batisse, C., Buret, M., Coulomb, P.J. and Coulomb, C. (1996) Ultrastructure of different textures of Bigarreau Burlat cherries during maturation. *Canadian Journal of Botany*, **74**, 1974-1981.

Bauchot, A.D., Hallett, I.C., Redgwell, R.J. and Lallu, N. (1999) Cell wall properties of kiwifruit affected by low temperature breakdown. *Postharvest Biology and Technology*, **16**, 245-255.

Baydoun, E.-A.H. and Fry, S.C. (1989) *In vivo* degradation and extracellular polymer binding of xyloglucan nonasaccharide, a naturally-occurring anti-auxin. *Journal of Plant Physiology*, **134**, 453-459.

Ben Arie, R. and Sonego, L. (1980) Pectolytic enzyme activity involved in woolly breakdown of stored peaches. *Phytochemistry*, **19**, 2553-2555.

Ben Arie, R., Kislev, N. and Frenkel, C. (1979) Ultrastructural changes in the cell walls of ripening apple and pear fruit. *Plant Physiology*, **64**, 197-202.

Ben Arie, R., Sonego, L., Zeidman, M. and Lurie, S. (1989) Cell wall changes in ripening peaches, in *Cell Separation in Plants: Physiology, Biochemistry and Molecular Biology* (eds D.J. Osbourne and M.B. Jackson), Springer-Verlag, Berlin, pp. 253-262.

Bird, C.R., Smith, C.J.S., Ray, J.A., Moreau, P., Bevan, M.W., Bird, A.S., Hughes, S., Morris, P.C., Grierson, D. and Schuh, W. (1998) The tomato polygalacturonase gene and ripening specific expression in transgenic plants. *Plant Molecular Biology*, **11**, 651-662.

Bonghi, C., Ferrarese, L., Ruperti, B., Tonutti, P. and Ramina, A. (1998) Endo-beta-1,4-glucanases are involved in peach fruit growth and ripening, and regulated by ethylene. *Physiologia Plantarum*, **102**, 346-352.

Bourne, M.C. (1979) Texture of temperate fruit. *Journal of Texture Studies*, **10**, 25-44.

Brady, C.J., MacAlpine, G., McGlasson, W.B. and Ueda, Y. (1982) Polygalacturonase in tomato fruits and the induction of ripening. *Australian Journal of Plant Physiology*, **9**, 171-178.

Brady, C.J., Meldrum, S.K., McGlasson, W.B. and Ali, Z.M. (1983) Differential accumulation of the molecular forms of polygalacturonase in tomato mutants. *Journal of Food Biochemistry*, **7**, 7-14.

Brecht, J.K. and Huber, D.J. (1988) Products released from enzymically active cell wall stimulate ethylene production and ripening in preclimacteric tomato fruit. *Plant Physiology*, **88**, 1037-1041.

Brennan, T. and Frenkel, C. (1977) Involvement of hydrogen peroxide in the regulation of senescence in pear. *Plant Physiology*, **59**, 411-416.

Brown, J.A. and Fry, S.C. (1993) Novel *O*-D-galacturonoyl esters in the pectic polysaccharides of suspension-cultured plant cells. *Plant Physiology*, **103**, 993-999.

Brummell, D.A. and Labavitch, J.M. (1997) Effect of antisense suppression of endopolygalacturonase activity on polyuronide molecular weight in ripening tomato fruit and in fruit homogenates. *Plant Physiology*, **115**, 717-725.

Brummell, D.A., Harpster, M.H. and Dunsmuir, P. (1999a) Differential expression of expansin gene family members during growth and ripening of tomato fruit. *Plant Molecular Biology*, **39**, 161-169.

Brummell, D.A., Hall, B.D. and Bennett, A.B. (1999b) Antisense suppression of tomato endo-1,4-beta-glucanase Cel2 mRNA accumulation increases the force required to break fruit abscission zones but does not affect fruit softening. *Plant Molecular Biology*, **40**, 615-622.

Brummell, D.A., Harpster, M.H., Civello, P.M., Palys, J.M., Bennett, A.B. and Dunsmuir, P. (1999c) Modification of expansin protein abundance in tomato fruit alters softening and cell wall polymer metabolism during ripening. *Plant Cell*, **11**, 2203-2216.

Campbell, A.D. and Labavitch, J.M. (1991) Induction and regulation of ethylene biosynthesis and ripening by pectic oligomers in tomato pericarp discs. *Plant Physiology*, **97**, 706-713.

Carey, A.T., Holt, K., Picard, S., Wilde, R., Tucker, G.A., Bird, C.R., Schuh, W. and Seymour, G.B. (1995) Tomato exo-(1-4)-beta-D-galactanase: isolation, changes during ripening in normal and mutant tomato fruit, and characterization of a related cDNA clone. *Plant Physiology*, **108**, 1099-1107.

Carey, A.T., Smith, D.L., Harrison, E., Bird, C.R., Gross, K.C., Seymour, G.B. and Tucker, G.A. (2001) Down-regulation of a ripening-regulated β-galactosidase gene (TBG1) in transgenic tomato fruits. *Journal of Experimental Botany*, **52**, 663-668.

Carpita, N.C. and Gibeaut, D.M. (1993) Structural models of primary cell walls in flowering plants: consistency of molecular structure in the physical properties of the walls during growth. *Plant Journal*, **3**, 1-30.

Carrington, C.-M.S. and Pressey, R. (1996) Beta-Galactosidase II activity in relation to changes in cell wall galactosyl composition during tomato ripening. *Journal of the American Society for Horticultural Science*, **121**, 132-136.

Carrington, C.M.S., Greve, L.C. and Labavitch, J.M. (1993) Cell wall metabolism in ripening fruit: VI. Effect of the antisense polygalacturonase gene on cell wall changes accompanying ripening in transgenic tomatoes. *Plant Physiology*, **103**, 429-434.

Cartier, N., Chambat, G. and Joseleau, J.P. (1988) Cell wall and extracellular galactoglucomannans from suspension-cultured *Rubus fruticosus* cells. *Phytochemistry*, **27**, 1361-1364.

Cass, L.G., Kirven, K.A. and Christoffersen, R.E. (1990) Isolation and characterization of a cellulase gene family member expressed during avocado fruit ripening. *Molecular and General Genetics*, **223**, 76-86.

Cassab, G.I. (1998) Plant cell wall proteins. *Annual Review of Plant Physiology and Plant Molecular Biology*, **49**, 281-309.

Chapman, H.D., Morris, V.J. Selvendran, R.R. and O'Neill, M.A. (1987) Static and dynamic light-scattering studies of pectic polysaccharides from the middle lamellae and primary cell walls of cider apples. *Carbohydrate Research*, **165**, 53-68.

Child, P.M., Williams, A.A., Hoad, G.V. and Baines, C.R. (1984) The effects of aminoethyoxyvinyl-glycine on maturity and post-harvest changes in Cox's Orange Pippin apples. *Journal of the Science of Food and Agriculture*, **35**, 773-781.

Chun, J.P. and Huber, D.J. (1997) Polygalacturonase isozyme 2 binding and catalysis in cell walls from tomato fruit: pH and beta-subunit effects. *Physiologia Plantarum*, **101**, 283-290.

Chun, J.P. and Huber, D.J. (1998) Polygalacturonase-mediated solubilization and depolymerization of pectic polymers in tomato fruit cell walls. *Plant Physiology*, **117**, 1293-1299.

Civello, P.M., Powell-Ann, L.T., Sabehat, A. and Bennett, A.B. (1999) An expansin gene expressed in ripening strawberry fruit. *Plant Physiology*, **121**, 1273-1279.

Cosgrove, D.J. (1999) Enzymes and other agents that enhance cell wall extensibility. *Annual Review of Plant Physiology and Plant Molecular Biology*, **50**, 391-417.

Cosgrove, D.J. (2000) Loosening of plant cell walls by expansins. *Nature*, **407**, 321-326.

Cosgrove, D.J. (2001) Wall structure and wall loosening. A look backwards and forwards. *Plant Physiology*, **125**, 131-134.

Cousins, S.K. and Brown, R.M. (1995) Cellulose I microfibril assembly: computational molecular mechanics energy analysis favours bonding by van der Waals forces as the initial step in crystallisation. *Polymer*, **36**, 3885-3886.

Creelmann, R.A. and Mullet, J.E. (1997) Oligosaccharins, brassinolides, and jasmonates: nontraditional regulators of plant growth, development and gene expression. *Plant Cell*, **9**, 1211-1223.

Crookes, P.R. and Grierson, D. (1983) Ultrastructure of tomato fruit ripening and the role of polygalacturonase isoenzymes in cell wall degradation. *Plant Physiology*, **72**, 1088-1093.

Cutillas, I.A., Zarra, I. and Lorences, E.P. (1993) Metabolism of cell wall polysaccharides from persimmon fruit: pectin solubilization during fruit ripening occurs in apparent absence of polygalacturonase activity. *Physiologia Plantarum*, **89**, 369-375.

Cutillas, I.A., Zarra, I., Fry, S.C. and Lorences, E.P. (1994) Implication of persimmon fruit hemicellulose metabolism in the softening process. Importance of xyloglucan endotransglycosylase. *Physiologia Plantarum*, **91**, 169-176.

Cutillas, I.A., Fulton, D.C., Fry, S.C. and Lorences, E.P. (1998a) Xyloglucan-derived oligosaccharides induce ethylene synthesis in persimmon (*Diospyros kaki* L.) fruit. *Journal of Experimental Botany*, **49**, 701-706.

Cutillas, I.A., Pena, M.J., Zarra, I. and Lorences, E.P. (1998b) A xyloglucan from persimmon fruit cell walls. *Phytochemistry*, **48**, 607-610.

Daas, P.J.H., Voragen, A.G.J. and Schols, H.A. (2000) Characterization of non-esterified galacturonic acid sequences in pectin with endopolygalacturonase. *Carbohydrate Research*, **326**, 120-129.

Dawson, D.M., Melton, L.D. and Watkins, C.B. (1992) Cell wall changes in nectarines (*Prunus persica*): solubilization and depolymerization of pectic and neutral polymers during ripening and in mealy fruit. *Plant Physiology*, **100**, 1203-1210.

de Veau, E., Gross, K.C., Huber, D.J. and Watada, A.E. (1993) Degradation and solubilization of pectin by beta-galactosidases purified from avocado mesocarp. *Physiologia Plantarum*, **87**, 279-285.

DellaPenna, D., Alexander, D.C. and Bennett, A.B. (1986) Molecular cloning of tomato (*Lycopersicon esculentum* cultivar Castlemart) fruit polygalacturonase: analysis of polygalacturonase messenger RNA levels during ripening. *Proceedings of the National Academy of Sciences of the USA*, **83**, 6420-6424.

DellaPenna, D., Kates, D.S. and Bennett, A.B. (1987) Polygalacturonase gene expression in Rutgers, *rin*, *nor*, and *Nr* tomato fruits. *Plant Physiology*, **85**, 502-507.

Diehl, K.C. and Hamann, D.D. (1980) Relationships between sensory profile parameters and fundamental mechanical parameters for raw potatoes, melons and apples. *Journal of Texture Studies*, **10**, 401-420.

Fischer, R.L. and Bennett, A.B. (1991) Role of cell wall hydrolases in fruit ripening. *Annual Review of Plant Physiology and Plant Molecular Biology*, **42**, 675-704.

Fischer, M., Arrigoni, E. and Amado, R. (1994) Changes in the pectic substance of apples during development and postharvest ripening. Part 2: Analysis of the pectic fractions. *Carbohydrate Polymers*, **25**, 167-175.

Fischer, M., Wegryzn, T.F., Hallett, I.C. and Redgwell, R.J. (1996) Chemical and structural features of kiwifruit cell walls: comparison of fruit and suspension-cultured cells. *Carbohydrate Research*, **295**, 195-208.

Fishman, M.L., Gross, K.C., Gillespie, D.T. and Sondey, S.M. (1989) Macromolecular components of tomato fruit pectin. *Archives of Biochemistry and Biophysics*, **274**, 179-191.

Fry, S.C. (1986) Cross-linking of matrix polymers in the growing cell walls of angiosperms. *Annual Reviews of Plant Physiology*, **37**, 165-186.

Fry, S.C. (1993) Oligosaccharins as plant growth regulators. *Biochemical Society Symposium*, **60**, 5-14.

Fry, S.C. (1998) Oxidative scission of plant cell wall polysaccharides by ascorbate-induced hydroxyl radicals. *Biochemical Journal*, **332**, 507-515.

Fry, S.C., Smith, R.C., Hetherington, P.R. and Potter, I. (1992) Endo-tranglycosylation of xyloglucans in plant cell suspension cultures. *Current Topics in Plant Biochemistry and Physiology*, **11**, 42-62.

Gibson, L.A. (1995) The quest for fresh. Satisfying the customers' desire for just-made taste. *Cereal Food World*, **40**, SR16-SR17.

Gil-Serrano, A., Mateos-Matos, M.I. and Tejero-Mateo, M.P. (1986) Acidic xylan from olive pulp. *Phytochemistry*, **25**, 2563-2564.

Giovane, A., Balestrieri, C., Quagliuolo, L., Castaldo, D. and Servillo, L. (1995) A glycoprotein inhibitor of pectin methylesterase in kiwi fruit. Purification by affinity chromatography and evidence of a ripening-related precursor. *European Journal of Biochemistry*, **233**, 926-929.

Giovannoni, J.J., DellaPenna, D., Bennett, A.B. and Fischer, R.L. (1989) Expression of a chimeric polygalacturonase gene in transgenic rin (ripening inhibitor) tomato fruit results in polyuronide degradation but not fruit softening. *Plant Cell*, **1**, 53-64.

Glenn, G.M. and Poovaiah, B.W. (1990) Calcium-mediated postharvest changes in texture and cell wall structure and composition in 'Golden Delicious' apples. *Journal of the American Society for Horticultural Science*, **115**, 962-968.

Gomez, C., Fiorenza, F., Izquierdo, L. and Costell, E. (1998) Perception of mealiness in apples: a comparison of consumers and trained assessors. *Zeitschrift für Lebensmittel Untersuchung und Forschung A*, **207**, 304-310.

Gonzalez, B.C., Brummell, D.A. and Bennett, A.B. (1996) Differential expression of two endo-1,4-beta-glucanase genes in pericarp and locules of wild-type and mutant tomato fruit. *Plant Physiology*, **111**, 1313-1319.

Grant, T.M., MacRae, E.A. and Redgwell, R.J. (1992) Effect of chilling injury on physicochemical properties of persimmon cell walls. *Phytochemistry*, **31**, 3739-3744.

Greve, L.C. and Labavitch, J.M. (1991) Cell wall metabolism in ripening fruit. V. Analysis of cell wall synthesis in ripening tomato pericarp tissue using a D-(uniformly labeled carbon-13) glucose tracer and gas chromatography-mass spectrometry. *Plant Physiology*, **97**, 1456-1461.

Grignon, C. and Sentenac, H. (1991) pH and ionic conditions of the apoplast. *Annual Review of Plant Physiology and Plant Molecular Biology*, **42**, 103-128.

Gross, K.C. and Sams, C.E. (1984) Changes in cell wall neutral sugar composition during fruit ripening: a species survey. *Phytochemistry*, **23**, 2457-2462.

Gross, K.C., Watada, A.E., Kang, M., Kim, S., Kim, K. and Lee, S. (1986) Biochemical changes associated with the ripening of hot pepper fruit. *Physiologia Plantarum*, **66**, 31-36.

Gross, K.C., Starrett, D.A. and Chen, H.L. (1995) Rhamnogalacturonase, β galactosidase and α-galactosidase: potential role in fruit softening. *Acta Horticulturae*, **398**, 121-130.

Hadfield, K.A. and Bennett, A.B. (1998) Polygalacturonases: many genes in search of a function. *Plant Physiology*, **117**, 337-343.

Hadfield, K.A., Rose, J.K.C., Yaver, D.S., Berka, R.M. and Bennett, A.B. (1998) Polygalacturonase gene expression in ripe melon fruit supports a role for polygalacturonase in ripening-associated pectin disassembly. *Plant Physiology*, **117**, 363-373.

Harker, F.R. and Hallett, I.C. (1994) Physiological and mechanical properties of kiwifruit tissue associated with texture change during cool storage. *Journal of the American Society for Horticultural Science*, **119**, 987-993.

Harker, F.R. and Sutherland, P.W. (1993) Physiological changes associated with fruit ripening and the development of mealy texture during storage of nectarines. *Postharvest Biology and Technology*, **2**, 269-277.

Harker, F.R., Hallett, I.C., Murray, S. and Carter, G. (1997a) Food-mouth interactions: towards a better understanding of food texture. *Food Technologist*, **26**, 136-137.

Harker, F.R., Stec, M.G.H., Hallett, I.C. and Bennett, C.L. (1997b) Texture of parenchymatous plant tissue: a comparison between tensile and other instrumental and sensory measurements of tissue strength and juiciness. *Postharvest Biology and Technology*, **11**, 63-72.

Harker, R., Redgwell, R.J., Hallett, I., Murray, S.H. and Carter, G. (1997c) Texture of fresh fruit. *Horticultural Reviews*, **20**, 121-224.

Harpster, M.H., Lee, K.Y. and Dunsmuir, P. (1997) Isolation and characterization of a gene encoding endo-beta-1,4-glucanase from pepper (*Capsicum annuum* L.). *Plant Molecular Biology*, **33**, 47-59.

Hayashi, T. (1989) Xyloglucans in the primary cell wall. *Annual Review of Plant Physiology and Plant Molecular Biology*, **40**, 139-168.

Hegde, S. and Maness, N.O. (1998) Changes in apparent molecular mass of pectin and hemicellulose extracts during peach softening. *Journal of the American Society for Horticultural Science*, **123**, 445-456.

Heredia, A., Fernandez, B.J. and Guillen, R. (1991) Identification of endoglucanases in olives (*Olea europaea arolensis*). *Zeitschrift für Lebensmittel Untersuchung und Forschung*, **193**, 554-557.

Hobson, G.E. (1963) Pectinesterase in normal and abnormal tomato fruit. *Biochemical Journal*, **86**, 358-365.

Hollingsworth, P. (1996) Sensory testing and the language of the consumer. *Food Technology*, **50**, 65-69.

Huber, D.J. (1983) Polyuronide degradation and hemicellulose modifications in ripening tomato fruit. *Journal of the American Society for Horticultural Science*, **108**, 405-409.

Huber, D.J. (1984) Strawberry (*Fragaria ananassa*) fruit softening: the potential roles of polyuronides and hemicelluloses. *Journal of Food Science*, **49**, 1310-1315.

Huber, D.J. (1992) The inactivation of pectin depolymerase associated with isolated tomato fruit cell wall: implications for the analysis of pectin solubility and molecular weight. *Physiologia Plantarum*, **86**, 25-32.

Huber, D.J. and O'Donoghue, E.M. (1993) Polyuronides in avocado (*Persea americana*) and tomato (*Lycopersicon esculentum*) fruits exhibit markedly different patterns of molecular weight downshifts during ripening. *Plant Physiology*, **102**, 473-480.

Huxman, I.M., Jarvis, M.C., Shakespeare, L., Dover, C.J., Johnson, D., Knox, J.P. and Seymour, G.B. (1999) Electron-energy-loss spectroscopic imaging of calcium and nitrogen in the cell walls of apple fruits. *Planta*, **208**, 438-443.

Huysamer, M., Greve, L.C. and Labavitch, J.M. (1997a) Cell wall metabolism in ripening fruit. IX. Synthesis of pectic and hemicellulosic cell wall polymers in the outer pericarp of mature green tomatoes (cv. XMT-22). *Plant Physiology*, **114**, 1523-1531.

Huysamer, M., Greve, L.C. and Labavitch, J.M. (1997b) Cell wall metabolism in ripening fruit. VIII. Cell wall composition and synthetic capacity of two regions of the outer pericarp of mature green and red ripe cv. Jackpot tomatoes. *Physiologia Plantarum*, **101**, 314-322.

Hyde, R.J. and Witherly, S.A. (1993) Dynamic contrast: a sensory contribution to palatability. *Appetite*, **21**, 1-16.

ISO (1981) *Sensory Analysis—Vocabulary*, International Standard; ISO 5492/4 1981 (E/F), 4 pp.

Jackman, R.L. and Stanley, D.W. (1995) Perspectives in the textural evaluation of plant foods. *Trends in Food Science and Technology*, **6**, 187-194.

Jackman, R.L., Gibson, H.J. and Stanley, D.W. (1992) Effects of chilling on tomato fruit texture. *Physiologia Plantarum*, **86**, 600.

Jarvis, M.C. (1984) Structure and properties of pectin gels in plant cell walls. *Plant Cell and Environment*, **7**, 153-164.

Jarvis, M.C., Hall, M.A., Threlfall, D.R. and Friend, J. (1981) The polysaccharide structure of potato cell walls: chemical fractionation. *Planta*, **152**, 93-100.

John, M.A. and Dey, P.M. (1986) Post harvest changes in fruit cell wall. *Advances in Food Research*, **30**, 139-193.

Jongen, W.M.F. (2000) Food supply chains: from productivity towards quality, in *Fruit and Vegetable Quality. An Integrated View* (eds R.L. Shewfelt and B. Brückner), Technomic Publishing Company, Lancaster, PA, pp. 3-20.

Jowitt, R. (1974) The terminology of food texture. *Journal of Texture Studies*, **5**, 351-358.

Kang, I.K., Suh, S.G., Gross, K.C. and Byun, J.K. (1994) *N*-terminal amino acid sequence of persimmon fruit beta-galactosidase. *Plant Physiology*, **105**, 975-979.

Kang, I.K., Chang, K.H. and Byun, J.K. (1998) Solubilization and depolymerization of pectic and neutral sugar polymers during ripening and softening in persimmon fruits. *Journal of the Korean Society for Horticultural Science*, **39**, 51-54.

Karakurt, Y., Huber, D.J. and Sherman, W.B. (2000) Quality characteristics of melting and non-melting flesh peach genotypes. *Journal of the Science of Food and Agriculture*, **80**, 1848-1853.

Keegstra, K., Talmadge, K.W., Bauer, W.D. and Albersheim, P. (1973) The structure of plant cell walls III. A model of the walls of suspension-cultured sycamore cells based on interconnections of the macromolecular components. *Plant Physiology*, **51**, 188-196.

Kim, J.B. and Carpita, N.C. (1992) Changes in esterification of the uronic acid groups of cell wall polysaccharides during elongation of maize coleoptiles. *Plant Physiology*, **98**, 646-653.

Kitagawa, Y., Kanayama, Y. and Yamaki, S. (1995) Isolation of beta-galactosidase fractions from Japanese pear: activity against native cell wall polysaccharides. *Physiologia Plantarum*, **93**, 545-550.

Knee, M. (1978) Properties of polygalacturonate and cell cohesion in apple fruit cortical tissue *Phytochemistry*, **17**, 1257-1260.

Knee, M. (1982) Fruit softening. III. Requirement for oxygen and pH effects. *Journal of Experimental Botany*, **33**, 1263-1269.

Knox, J.P., Linstead, P.J., King, J., Cooper, C. and Roberts, K. (1990) Pectin esterification is spatially regulated both within cell walls and between developing tissues of root apices. *Planta*, **4**, 512-521.

Koch, J.L. and Nevins, D.J. (1989) Tomato fruit cell wall. I. Use of purified tomato polygalacturonase and pectin methylesterase to identify developmental changes in pectins. *Plant Physiology*, **91**, 816-822.

Kofod, L.V., Kauppinen, S., Christgau, S., Andersen, L.N., Heldt-Hansen, H.P., Dorreich, K. and Dalboge, H. (1994) Cloning and characterization of two structurally and functionally divergent rhamnogalacturonases from *Aspergillus aculeatus. Journal of Biological Chemistry*, **269**, 29182-29189.

Kramer, M.G. and Redenbaugh, K. (1994) Commercialization of a tomato with an antisense polygalacturonase gene: the FLAVR SAVR-TM tomato story. *Euphytica*, **79**, 293-297.

Lacan, D. and Baccou, J.C. (1998) High levels of antioxidant enzymes correlate with delayed senescence in nonnetted muskmelon fruits. *Planta*, **204**, 377-382.

Lallu, N. (1997) Low temperature in kiwifruit. *Acta Horticulturae*, **2**, 579-586.

Lamport, D. and Epstein, L. (1983) A new model for the primary cell wall: concatenated extensin-cellulose network. *Current Topics in Biochemistry and Physiology*, **2**, 73-87.

Lapsley, K.G. (1989) *Texture of fresh apples—evaluation and relationship to structure*, PhD dissertation, Swiss Federal Institute of Technology, Zurich, Switzerland.

Lashbrook, C.C., Gonzalez, B.C. and Bennett, A.B. (1994) Two divergent endo-beta-1,4-glucanase genes exhibit overlapping expression in ripening fruit and abscising flowers. *Plant Cell*, **6**, 1485-1493.

Lester, D.R., Speirs, J., Orr, G. and Brady, C.J. (1994) Peach (*Prunus persica*) endopolygalacturonase cDNA isolation and mRNA analysis in melting and nonmelting peach cultivars. *Plant Physiology*, **105**, 225-231.

Lester, D.R., Sherman, W.B. and Atwell, B.J. (1996) Endopolygalacturonase and the melting flesh (M) locus in peach. *Journal of the American Society for Horticultural Science*, **121**, 231-234.

Llop, T., I, Dominguez, P.E., Palomer, X. and Vendrell, M. (1999) Characterization of two divergent endo-Beta-1,4-glucanase cDNA clones highly expressed in the nonclimacteric strawberry fruit. *Plant Physiology*, **119**, 1415-1421.

Lurie, S., Klein, J.D., Watkins, C., Ross, G., Boss, P. and Ferguson, I.F. (1993) Prestorage heat treatment of tomatoes prevents chilling injury and reversibly inhibits ripening. *Acta Horticulturae*, **343**, 283-285.

Luza, J.G., van Gorsal, R., Polito, V.S. and Kader, A.A. (1992) Chilling injury in peaches: a cytochemical and ultrastructural cell wall study. *Journal of the American Society for Horticultural Science*, **61**, 23-32.

MacDougall, A.J., Parker, R. and Selvendran, R.R. (1995) Nonaqueous fractionation to assess the ionic composition of the apoplast during fruit ripening. *Plant Physiology*, **108**, 1679-1689.

Maclachlan, G. and Brady, C. (1992) Multiple forms of 1,4-beta-glucanase in ripening tomato fruits include a xyloglucanase activatable by xyloglucan oligosaccharides. *Australian Journal of Plant Physiology*, **19**, 137-146.

Maclachlan, G. and Brady, C. (1994) Endo-1,4-beta-glucanase, xyloglucanase, and xyloglucan endo-transglycosylase activities versus potential substrates in ripening tomatoes. *Plant Physiology*, **105**, 965-974.

MacRae, E.A. (1987) Development of chilling injury in New Zealand grown Fuyu persimmon during storage. *New Zealand Journal of Experimental Agriculture*, **15**, 333-344.

MacRae, E. and Redgwell, R.J. (1992) Softening in kiwifruit. *Postharvest News and Information*, **3**, 49N-52N.

Marangoni, A.G., Jackman, R.L. and Stanley, D.W. (1995) Chilling-associated softening of tomato fruit is related to increased pectinmethylesterase activity. *Journal of Food Science*, **60**, 1277-1281.

Matoh, T., Ishigaki, K.I., Ohno, K. and Azuma, J.I. (1993) Isolation and characterization of a boron-polysaccharide complex from radish roots. *Plant and Cell Physiology*, **34**, 639-642.

Matsuhashi, S. and Hatanaka, C. (1991) Polygalacturonase in strawberry fruit. *Nippon Nogeikagaku Kaishi*, **65**, 883-886.

McCollum, T.G., Huber, D.J. and Cantliffe, D.J. (1989) Modification of polyuronides and hemicelluloses during muskmelon fruit softening. *Physiologia Plantarum*, **76**, 303-308.

McQueen-Mason, S.J. and Cosgrove, D.J. (1994) Disruption of hydrogen bonding between plant cell wall polymers by proteins that induce wall extension. *Proceedings of the National Academy of Sciences of the USA*, **91**, 6574-6578.

McQueen-Mason, S.J., Durachko, D.M. and Cosgrove, D.J. (1992) Two endogenous proteins that induce cell wall extension in plants. *Plant Cell*, **4**, 1425-1433.

Melotto, E., Greve, L.C. and Labavitch, J.M. (1994) Cell wall metabolism in ripening fruit. VII. Biologically active pectin oligomers in ripening tomato (*Lycopersicon esculentum* Mill.) fruits. *Plant Physiology*, **106**, 575-581.

Menrad, K. (1998) Consumer attitudes of modern biotechnology in the Agri-Food sector, in *European Research Towards Safer and Better Food*, Proceedings of the 3rd Karlsruhe Nutrition Symposium, Karlsruhe, October 1998, part 1: lectures (eds V. Gaukel and W.E.L. Spiess), Bundesforschungsanstalt für Ernährung, pp. 329-339.

Miller, A.R. (1986) Oxidation of cell wall polysaccharides by hydrogen peroxide: a potential mechanism for cell wall breakdown in plants. *Biochemical and Biophysical Research Communications*, **141**, 238-244.

Mitcham, E.J., Gross, K.C. and Ng, T.J. (1989) Tomato fruit cell wall synthesis during development and senescence: *in vitro* radiolabeling of wall fractions using carbon-14-labelled sucrose. *Plant Physiology*, **89**, 477-481.

Moore, T. and Bennett, A.B. (1994) Tomato fruit polygalacturonase isozyme. 1. Characterization of the beta subunit and its state of assembly *in vivo*. *Plant Physiology*, **106**, 1461-1469.

Moore, P.J., Darvill, A.G., Albersheim, P. and Staehelin, L.A. (1986) Immunogold localization of xyloglucan and rhamnogalacturonan I in the cell walls of suspension-cultured sycamore (*Acer pseudoplatanus*) cells. *Plant Physiology*, **82**, 787-794.

Muresan, S., Ebbenhorst, S.T., Termeer, J., Boeriu, C., Gerritsen, Y., Leguijt, T., Biekman, E. and van Dijk, C. (1998) Mealiness in fruits—consumer perception and means for detection, instrumental assessment of mealiness, in *Interaction of Food Matrix with Small Ligands Influencing Flavour and Texture*, Proceedings of the meeting of the four working groups, Garching, October 1997 (ed. P. Schieberle), pp. 77-80.

Neelam, P., Sanjay, M. and Sanwal, G.G. (2000) Purification and characterization of polygalacturonase from banana fruit. *Phytochemistry*, **54**, 147-152.

Newman, R.H., Ha, M.A. and Melton, L.D. (1994) Solid-state radiolabelled carbon NMR investigation of molecular ordering in the cellulose of apple cell walls. *Journal of Agricultural and Food Chemistry*, **42**, 1402-1406.

Newman, R.H. and Redgwell, R.J. (2001) Cell wall changes in ripening kiwifruit: 13C solid state NMR characterisation of relatively rigid cell wall polymers. *Carbohydrate Polymers*, in press.

Nicolai, B. (1998) Mealiness in fruits—consumer perception and means of detection, in *European Research Towards Safer and Better Food*, Proceedings of the 3rd Karlsruhe Nutrition Symposium, Karlsruhe, October 1998, part 1: lectures (eds V. Gaukel and W.E.L. Spieb), Bundesforschungsanstalt für Ernährung, Karlsruhe, Germany, pp. 320-328.

Nishitani, K. and Tominaga, R. (1992) Endoxyloglucan transferase, a novel class of glycosyltransferase that catalyzes transfer of a segment of xyloglucan molecule to another xyloglucan molecule. *Journal of Biological Chemistry*, **267**, 21058-21064.

Nogata, Y., Ohta, H. and Voragen, A.G.J. (1993) Polygalacturonase in strawberry fruit. *Phytochemistry*, **34**, 617-620.

Nogata, Y., Yoza, K.I., Kusumoto, K.I. and Ohta, H. (1996) Changes in molecular weight and carbohydrate composition of cell wall polyuronide and hemicellulose during ripening in strawberry fruit, in *Progress in Biotechnology* (eds J. Visser and A.G.J. Voragen), Elsevier, Amsterdam, pp. 591-596.

O'Donoghue, E.M. and Huber, D.J. (1992) Modification of matrix polysaccharides during avocado (*Persea americana*) fruit ripening: an assessment of the role of Cx-cellulase. *Physiologia Plantarum*, **86**, 33-42.

O'Donoghue, E.M., Huber, D.J., Timpa, J.D., Erdos, G.W. and Brecht, J.K. (1994) Influence of avocado (*Persea americana*) Cx-cellulase on the structural features of avocado cellulose. *Planta*, **194**, 573-584.

O'Neill, M.A., Albersheim, P. and Darvill, A.G. (1990) The pectic polysaccharides of primary cell walls. *Methods in Plant Biochemistry*, **2**, 415-441.

O'Neill, M.A., Warrenfeltz, D., Kates, K., Pellerin, P., Doco, T., Darvill, A.G. and Albersheim, P. (1996) Rhamnogalacturonan-II, a pectic polysaccharide in the walls of growing plant cell, forms a dimer that is covalently cross-linked by a borate ester. *Journal of Biological Chemistry*, **271**, 22923-22930.

Orr, G. and Brady, C. (1993) Relationship of endopolygalacturonase activity to fruit softening in a freestone peach. *Postharvest Biology and Technology*, **3**, 121-130.

Parr, A.J., Ng, A. and Waldron, K.W. (1997) Ester-linked phenolic components of carrot cell walls. *Journal of Agricultural and Food Chemistry*, **45**, 2468-2471.

Pellerin, P., Doco, T., Vidal, S., Williams, P., Brillouet, J.M. and O'Neill, M.A. (1996) Structural characterization of red wine rhamnogalacturonan II. *Carbohydrate Research*, **290**, 183-197.

Percy, A.E., Brien, I.E.W., Jameson, P.E., Melton, L.D., MacRae, E.A. and Redgwell, R.J. (1996) Xyloglucan endotransglycosylase activity during fruit development and ripening of apple and kiwifruit. *Physiologia Plantarum*, **96**, 43-50.

Pitt, R.E. and Chen, H.L. (1983) Time-dependent aspects of the strength and rheology of vegetative tissue. *Transactions of the American Society for Agricultural Engineering*, **26**, 1275-1780.

Platt-Aloia, K.A., Thomson, W.W. and Young, R.E. (1980) Ultrastructural changes in the cell walls of ripening avocados: transmission, scanning and freeze fracture microscopy. *Botanical Gazette*, **141**, 366-373.

Poovaiah, B.W., Glenn, G.M. and Reddy, A.S.N. (1988) Calcium and fruit softening: physiology and biochemistry. *Horticultural Reviews*, **10**, 107-152.

Pressey, R. and Avants, J.K. (1978) Difference in polygalacturonase composition of clingstone and freestone peaches. *Journal of Food Science*, **43**, 1415-1423.

Pressey, R., Hinton, D.M. and Avants, J.K. (1971) Development of polygalacturonase activity and solubilisation of pectin in peaches during ripening. *Journal of Food Science*, **36**, 1071-1073.

Qi, X., Behrens, B.X., West, P. and Mort, A. (1995) Solubilization and partial characterization of extensin fragments from cell walls of cotton suspension cultures. *Plant Physiology*, **108**, 1691-1701.

Raghothama, K.G., Lawton, K.A., Goldsbrough, P.B. and Woodson, W.R. (1991) Characterization of an ethylene-regulated flower senescence-related gene from carnation. *Plant Molecular Biology*, **17**, 61-72.

Redgwell, R.J. (1991) *Cell wall polysaccharides of kiwifruit* Actinidia deliciosa: *changes during ripening*, PhD dissertation, University of Otago, Dunedin, New Zealand.

Redgwell, R.J. (1996) Cell wall synthesis in kiwifruit following postharvest ethylene treatment. *Phytochemistry*, **41**, 407-413.

Redgwell, R.J. and Fry, S.C. (1993) Xyloglucan endotransglycosylase activity increases during kiwifruit (*Actinidia deliciosa*) ripening. *Plant Physiology*, **103**, 1399-1406.

Redgwell, R.J. and Harker, R. (1995) Softening of kiwifruit discs: effect of inhibition of galactose loss from cell walls. *Phytochemistry*, **39**, 1319-1323.

Redgwell, R.J. and Percy, A.E. (1992) Cell wall changes during on-vine softening of kiwifruit. *New Zealand Journal of Crop and Horticultural Science*, **20**, 453-456.

Redgwell, R.J., Melton, L.D. and Brasch, D.J. (1988) Cell-wall polysaccharides of kiwifruit (*Actinidia deliciosa*): chemical features in different tissue zones of the fruit at harvest. *Carbohydrate Research*, **182**, 241-258.

Redgwell, R.J., Melton, L.D. and Brasch, D.J. (1990) Cell wall changes in kiwifruit following post harvest ethylene treatment. *Phytochemistry*, **29**, 399-408.

Redgwell, R.J., Melton, L.D. and Brasch, D.J. (1991a) Cell-wall polysaccharides of kiwifruit (*Actinidia deliciosa*): effect of ripening on the structural features of cell-wall materials. *Carbohydrate Research*, **209**, 191-202.

Redgwell, R.J., Melton, L.D. and Brasch, D.J. (1991b) Changes to pectic and hemicellulosic polysaccharides of kiwifruit during ripening. *Acta Horticulturae*, **297**, 627-634.

Redgwell, R.J., Melton, L.D. and Brasch, D.J. (1992) Cell wall dissolution in ripening kiwifruit (*Actinidia deliciosa*). Solubilization of the pectic polymers. *Plant Physiology*, **98**, 71-81.

Redgwell, R.J., Fischer, M., Kendal, E. and MacRae, E.A. (1997a) Galactose loss and fruit ripening: high-molecular-weight arabinogalactans in the pectic polysaccharides of fruit cell walls. *Planta*, **203**, 174-181.

Redgwell, R.J., MacRae, E., Hallett, I., Fischer, M., Perry, J. and Harker, R. (1997b) *In vivo* and *in vitro* swelling of cell walls during fruit ripening. *Planta*, **203**, 162-173.

Rees, D.A., Morris, E.R., Thom, D. and Madden, J.K. (1982) Shapes and interactions of carbohydrate chains, in *The Polysaccharides* (ed. G.O. Aspinall), Academic Press, New York, pp. 195-290.

Renard, C.M.G.C., Thibault, J.F., Voragen, A.G.J., Van Den Broek, L.-A.M. and Pilnik, W. (1993) Studies on apple protopectin. VI. Extraction of pectins from apple cell walls with rhamnogalacturonase. *Carbohydrate Polymers*, **22**, 203-210.

Rexova-Benkova, L. and Markovic, O. (1976) Pectic enzymes. *Advances in Carbohydrate Chemistry and Biochemistry*, **33**, 323-385.

Roberts, K. (2001) How the cell wall acquired a cellular context. *Plant Physiology*, **125**, 127-130.

Rose, J.K.C. and Bennett, A.B. (1999) Cooperative disassembly of the cellulose-xyloglucan network of plant cell walls: parallels between cell expansion and fruit ripening. *Trends in Plant Sciences*, **4**, 176-183.

Rose, J.K.C., Lee, H.H. and Bennett, A.B. (1997) Expression of a divergent expansin gene is fruit-specific and ripening-regulated. *Proceedings of the National Academy of Sciences of the USA*, **94**, 5955-5960.

Rose, J.K.C., Hadfield, K.A., Labavitch, J.M. and Bennett, A.B. (1998) Temporal sequence of cell wall disassembly in rapidly ripening melon fruit. *Plant Physiology*, **117**, 345-361.

Ross, G.S., Redgwell, R.J. and MacRae, E.A. (1993) Kiwifruit beta-galactosidase: isolation and activity against specific fruit cell-wall polysaccharides. *Planta*, **189**, 499-506.

Ross, G.S., Wegrzyn, T., MacRae, E.A. and Redgwell, R.J. (1994) Apple beta-galactosidase: activity against cell wall polysaccharides and characterization of a related cDNA clone. *Plant Physiology*, **106**, 521-528.

Roy, S., Vian, B. and Roland, J.C. (1992) Immunocytochemical study of the deesterification patterns during cell wall autolysis in the ripening of cherry tomato. *Plant Physiology and Biochemistry (Paris)*, **30**, 139-146.

Ryan, C.L. and Farmer, E.E. (1991) Oligosaccharide signals in plants: a current assessment. *Annual Review of Plant Physiology and Plant Molecular Biology*, **42**, 651-674.

Saab, I.N. and Sachs, M.M. (1996) A flooding-induced xyloglucan endo-transglycosylase homolog in maize is responsive to ethylene and associated with aerenchyma. *Plant Physiology*, **112**, 385-391.

Saguy, S. and Moskowitz, H.R. (1999) Integrating the consumer into new product development. *Food Technology*, **53**, 68-73.

Sakurai, N. and Nevins, D.J. (1993) Changes in physical properties and cell wall polysaccharides of tomato (*Lycopersicon esculentum*) pericarp tissues. *Physiologia Plantarum*, **89**, 681-686.

Sakurai, N. and Nevins, D.J. (1997) Relationship between fruit softening and wall polysaccharides in avocado (*Persea americana* Mill.) mesocarp tissues. *Plant and Cell Physiology*, **38**, 603-610.

Sams, C.E. (1999) Preharvest factors affecting postharvest texture. *Postharvest Biology and Technology*, **15**, 249-254.

Schibeci, R., Barns, I., Kennealy, S. and Davison, A. (1997) Public attitudes to gene technology: the case of the MacGregor's tomato. *Public Understanding of Science*, **6**, 167-183.

Schols, H.A., Geraeds, C.J.M., Searle-van-Leeuwen, M.F., Kormelink, F.J.M. and Voragen, A.G.J. (1990) Rhamnogalacturonase: a novel enzyme that degrades the hairy regions of pectins. *Carbohydrate Research*, **206**, 105-115.

Schols, H.A., Bakx, E.J., Schipper, D. and Voragen, A.G.J. (1995a) A xylogalacturonan subunit present in the modified hairy regions of apple pectin. *Carbohydrate Research*, **279**, 265-279.

Schols, H.A., Vierhuis, E., Bakx, E.J. and Voragen, A.-G.J. (1995b) Different populations of pectic hairy regions occur in apple cell walls. *Carbohydrate Research*, **275**, 343-360.

Schroeder, R., Atkinson, R.G., Langenkaemper, G. and Redgwell, R.J. (1998) Biochemical and molecular characterisation of xyloglucan endotransglycosylase from ripe kiwifruit. *Planta*, **204**, 242-251.

Schroeder, R., Nicolas, P., Vincent, S.J.F., Fischer, M., Reymond, S. and Redgwell, R.J. (2001) Purification and characterisation of a galactoglucomannan from kiwifruit (*Actinidia deliciosa*). *Carbohydrate Research*, **331**, 291-306.

Sexton, R., Palmer, J.M., Whyte, N.A. and Littlejohns, S. (1997) Cellulase, fruit softening and abscission in red raspberry *Rubus idaeus* L. cv. Glen Clova. *Annals of Botany*, **80**, 371-376.

Seymour, G.B. and Gross, K.C. (1996) Cell wall disassembly and fruit softening. *Postharvest News and Information*, **7**, 45N-52N.

Seymour, G.B., Harding, S.E., Taylor, A.J., Hobson, G.E. and Tucker, G.A. (1987) Polyuronide solubilization during ripening of normal and mutant tomato fruit. *Phytochemistry*, **26**, 1871-1875.

Seymour, G.B., Colquhoun, I.J., DuPont, M.S., Parsley, K.R. and Selvendran, R.R. (1990) Composition and structural features of cell wall polysaccharides from tomato fruits. *Phytochemistry*, **29**, 725-731.

Shewfelt, R.L. (1999) What is quality? *Postharvest Biology and Technology*, **15**, 197-200.

Sims, I.M., Craik, D.J. and Bacic, A. (1997) Structural characterisation of galactoglucomannan secreted by suspension-cultured cells of *Nicotiana plumbaginifolia*. *Carbohydrate Research*, **303**, 79-92.

Smith, R.C. and Fry, S.C. (1991) Endotransglycosylation of xyloglucans in plant cell suspension cultures. *Biochemical Journal*, **279**, 529-536.

Smith, D.L. and Gross, K.C. (2000) A family of at least seven beta-galactosidase genes is expressed during tomato fruit development. *Plant Physiology*, **123**, 1173-1183.

Smith, C.J.S., Watson, C.F., Morris, P.C., Bird, C.R., Seymour, G.B., Gray, J.E., Arnold, C., Tucker, G.A., Schuh, W., Harding, S. and Grierson, D. (1990) Inheritance and effect on ripening of antisense polygalacturonase genes in transgenic tomatoes. *Plant Molecular Biology*, **14**, 369-379.

Smith, D.L., Starrett, D.A. and Gross, K.C. (1998) A gene coding for tomato fruit Beta-galactosidase II is expressed during fruit ripening. Cloning, characterization, and expression pattern. *Plant Physiology*, **117**, 417-423.

Sonego, L., Ben Arie, R., Raynal, J. and Pech, J.C. (1995) Biochemical and physical evaluation of textural characteristics of nectarines exhibiting woolly breakdown: NMR imaging, X-ray computed tomography and pectin composition. *Postharvest Biology and Technology*, **5**, 187-198.

Steele, N.M., McCann, M.C. and Roberts, K. (1997) Pectin modification in cell walls of ripening tomatoes occurs in distinct domains. *Plant Physiology*, **114**, 373-381.

Stow, J. (1989) The involvement of calcium ions in maintenance of apple fruit tissue structure. *Journal of Experimental Botany*, **40**, 1053-1057.

Sutherland, P., Hallett, I., Redgwell, R., Benhamou, N. and MacRae, E. (1999) Localization of cell wall polysaccharides during kiwifruit (*Actinidia deliciosa*) ripening. *International Journal of Plant Sciences*, **160**, 1099-1109.

Suykerbuyk, M.E.G., Schaap, P.J., Stam, H., Musters, W. and Visser, J. (1995) Cloning, sequence and expression of the gene coding for rhamnogalacturonase of *Aspergillus aculeatus*; a novel pectinolytic enzyme. *Applied Microbiology and Biotechnology*, **43**, 861-870.

Szczesniak, A.S. (1990) Texture: is it still an overlooked food attribute? *Food Technology*, September, 1990, pp. 86-95.

Talbott, L.F. and Ray, P.M. (1992) Molecular size and separability features of pea cell wall polysaccharides: implications for models of primary wall structure. *Plant Physiology*, **98**, 357-368.

Thompson, J.E., Smith, R.C. and Fry, S.C. (1997) Xyloglucan undergoes interpolymeric transglycosylation during binding to the plant cell wall *in vivo*: evidence from $^{13}C/^{3}H$ dual labelling and isopycnic centrifugation in cesium trifluoroacetate. *Biochemical Journal*, **327**, 699-708.

Tieman, D.M. and Handa, A.K. (1994) Reduction in pectin methylesterase activity modifies tissue integrity and action levels in ripening tomato (*Lycopersicon esculentum* Mill.) fruits. *Plant Physiology*, **106**, 429-436.

Tieman, D.M., Harriman, R.W., Ramamohan, G. and Handa, A.K. (1992) An antisense pectin methylesterase gene alters pectin chemistry and soluble solids in tomato fruit. *Plant Cell*, **4**, 667-679.

Tong, C.B.S. and Gross, K.C. (1988) Glycosyl-linkage composition of tomato fruit cell wall hemicellulosic fractions during ripening. *Physiologia Plantarum*, **74**, 365-370.

Trainotti, L., Ferrarese, L., Dalla, V.F., Rascio, N. and Casadoro, G. (1999a) Two different endo-beta-1,4-glucanases contribute to the softening of the strawberry fruits. *Journal of Plant Physiology*, **154**, 355-362.

Trainotti, L., Spolaore, S., Pavanello, A., Baldan, B. and Casadoro, G. (1999b) A novel E-type endo-beta-1,4-glucanases with a putative cellulose-binding domain is highly expressed in ripening strawberry fruits. *Plant Molecular Biology*, **40**, 323-332.

Tucker, G. and Zhang, J. (1996) Expression of polygalacturonase and pectinesterase in normal and transgenic tomatoes, in *Progress in Biotechnology 14, Pectins and pectinases* (eds J. Visser and A.G.J. Voragen), Elsevier, Amsterdam, pp. 347-353.

Tucker, G.A., Robertson, N.G. and Grierson, D. (1980) Changes in polygalacturonase isoenzymes during the 'ripening' of normal and mutant tomato fruit. *European Journal of Biochemistry*, **112**, 119-124.

Tucker, G.A., Robertson, N.G. and Grierson, D. (1981) The conversion of tomato fruit polygalacturonase isoenzymes 2 into isoenzyme 1 *in vitro*. *European Journal of Biochemistry*, **33**, 396-400.

Vidal, S., Doco, T., Williams, P., Pellerin, P., York, W.S., O'Neill, M.A., Glushka, J., Darvill, A.G. and Albersheim, P. (2000) Structural characterization of the pectic polysaccharide rhamnogalacturonan II: evidence for the backbone location of the aceric acid-containing oligosyl side chain. *Carbohydrate Research*, **326**, 277-294.

Wade, N.L., Kavanagh, E.E., Hockley, D.G. and Brady, C.J. (1992) Relationship between softening and the polyuronides in ripening banana fruit. *Journal of the Science of Food and Agriculture*, **60**, 61-68.

Wakabayashi, K. (2000) Changes in cell wall polysaccharides during fruit ripening. *Journal of Plant Research*, **113**, 231-237.

Wakabayashi, K., Chun, J.P. and Huber, D.J. (2000) Extensive solubilization and depolymerization of cell wall polysaccharides during avocado (*Persea americana*) ripening involves concerted action of polygalacturonase and pectinmethylesterase. *Physiologia Plantarum*, **108**, 345-352.

Waldron, K.W. (1998) Vegetables on the test stand. On the track of the Chinese water chestnut's secret. *Lebensmitteltechnik*, **30**, 50-51.

Waldron, K.W., Smith, A.C., Parr, A.J., Ng, A. and Parker, M.L. (1997) New approaches to understanding and controlling cell separation in relation to fruit and vegetable texture. *Trends in Food Science and Technology*, **8**, 213-221.

Watkins, C.B., Picton, S. and Grierson, D. (1990) Stimulation and inhibition of expression of ripening-related messenger RNA in tomatoes as influenced by chilling temperatures. *Journal of Plant Physiology*, **136**, 318-323.

Watson, C.F., Zheng, L. and DellaPenna, D. (1994) Reduction of tomato polygalacturonase beta subunit expression affects pectin solubilization and degradation during fruit ripening. *Plant Cell*, **6**, 1623-1634.

Wegrzyn, T.F. and MacRae, E.A. (1992) Pectinesterase, polygalacturonase, and beta-galactosidase during softening of ethylene-treated kiwifruit. *Hortscience*, **27**, 900-902.

Whitney, S.E.C., Brigham, J.E., Darke, A.H., Reid, J.-S.G. and Gidley, M.J. (1998) Structural aspects of the interaction of mannan-based polysaccharides with bacterial cellulose. *Carbohydrate Research*, **307**, 299-309.

Wojcik, P., Mika, A. and Cieslinski, G. (1999) Effect of boron fertilization on the storage ability of apples (*Malus domestica*), in *Effect of Preharvest and Postharvest Factors on Storage of Fruit*, Proceedings of an international symposium, Warsaw, August 1997 (ed L. Michalczuk), ISHS (International Society for Horticultural Science), Leuven, pp. 393-397.

Woolf, A.B., MacRae, E.A., Spooner, K.J. and Redgwell, R.J. (1997) Changes to physical properties of the cell wall and polyuronides in response to heat treatment of 'Fuyu' persimmon that alleviate chilling injury. *Journal of the American Society for Horticultural Science*, **122**, 698-702.

Wu, Q., Szakcs, D.M., Hemmat, M. and Hrazdina, G. (1993) Endopolygalacturonase in apples (*Malus domestica*) and its expression during fruit ripening. *Plant Physiology*, **102**, 219-225.

Yu, L. and Mort, A.J. (1996) Partial characterisation of xylogalacturonans from cell walls of ripe watermelon: inhibition of endopolygalacturonase activity by xylosylation, in *Progress in Biotechnology 14, Pectins and Pectinases* (eds J. Visser and A.G.J. Voragen), Elsevier, Amsterdam, pp. 79-88.

Zheng, L.S., Heupel, R.C. and DellaPenna, D. (1992) The beta-subunit of tomato fruit polygalacturonase isoenzyme-1. Isolation, characterisation and identification of unique structural features. *Plant Cell*, **4**, 1147-1156.

Zhong, Y.W., MacRae, E.A., Wright, M.A., Bolitho, K.M., Ross, G.S. and Atkinson, R.G. (2000) Polygalacturonase gene expression in kiwifruit: relationship to fruit softening and ethylene production. *Plant Molecular Biology*, **42**, 317-328.

Zhou, H.W., Sonego, L., Ben Arie, R. and Lurie, S. (1999) Analysis of cell wall components in juice of 'Flavortop' nectarines during normal ripening and woolliness development. *Journal of the American Society for Horticultural Science*, **124**, 424-429.

4 Fruit flavor, volatile metabolism and consumer perceptions

Elizabeth A. Baldwin

4.1 Introduction

There has been growing dissatisfaction among consumers about the flavor of fruits and vegetables. First-time purchases are often based on appearance and firmness. Repeat buys, however, are determined by internal quality traits such as mouth-feel and flavor. Fruit breeders select lines based on color, size, disease resistance, yield or other horticultural characteristics, thereby inadvertently breeding out flavor by not selecting for it. For this reason, the flavor of many fruits has deteriorated into mediocrity or worse, which has affected repeat sales resulting in industry concern.

4.2 Flavor is an elusive trait

Flavor is a complex trait, determined by genetics (Baldwin *et al.*, 1991b, 1992; Cunningham *et al.*, 1985), followed by environmental (Romani *et al.*, 1983; Baldwin *et al.*, 1995a) and cultural influences (Wright and Harris, 1985); it is further affected by harvest maturity (Fellman *et al.*, 1993; Maul *et al.*, 1998; Baldwin *et al.*, 1999a) and postharvest handling (Mattheis *et al.*, 1991, 1995; Fellman *et al.*, 1993; Maul *et al.*, 1998, 1999; Baldwin *et al.*, 1999a,b) (table 4.1). Breeders often do not select for flavor because they lack the necessary information. Simplistically, flavor is composed of sweetness, sourness, bitterness, saltiness and aroma. Conducting elaborate sensory panels to measure these traits, or quantifying the chemical components that correspond to them, is usually not possible for breeders because the equipment is expensive, and they lack expertise in sensory science. Even when equipment and expertise is available to breeders, there is much that we do not understand in terms of fruit flavor components, their biosynthetic pathways and how they are perceived by humans.

Table 4.1 Examples of genetic, environmental, cultural and harvest maturity effects on postharvest fruit flavor

Fruit	Effect	Reference
	Genetic effects	
Apple	Aroma volatiles differ in cultivars	Cunningham *et al.*, 1985
Strawberry		Pérez *et al.*, 1999a;
		Larsen *et al.*, 1992;
		Morton and MacLeod, 1990
Tomato		Baldwin *et al.*, 1991a,b, 1992, 2000;
		Kopeliovitch *et al.*, 1982
	Environmental effects	
Tomato	Heavy rain before harvest reduced aroma levels (dilution from excess water)	Baldwin *et al.*, 1995a
	Cultural effects	
Pears	Preharvest AVG inhibited volatile production	Romani *et al.*, 1983
Tomato	Nitrogen and Potassium fertilizers increased several volatiles	Wright and Harris, 1985
	Lower levels of volatiles in greenhouse than in field	Dalal *et al.*, 1967
Strawberry	Mite control in the field increased flavor relative to untreated	Podoski *et al.*, 1997
	Harvest maturity	
Apple	Ester formation affected by harvest maturity	Fellman *et al.*, 1993
	More 'fruity' if harvested later	Cliff *et al.*, 1998
Tomato	Higher volatile levels and better sensory scores when harvested more mature	Maul *et al.*, 1998a
	More fruity/floral aroma if harvested more mature	Watada and Aulenbach, 1979
	More 'tomato-like' flavor when harvested more mature	Kader *et al.*, 1977
Mango	Aroma ratings for various descriptors higher when fruit harvested more mature	Baldwin *et al.*, 1999a
	Postharvest handling	
Tomato	Chilling reduced aroma volatiles	Buttery *et al.*, 1987;
		Maul *et al.*, 2000
Apple	Bruising altered aroma profiles	Moretti *et al.*, 1997
	Heat-treatment altered aroma profiles	McDonald *et al.*, 1996
	Heat treatment decreased ester volatiles	Fallik *et al.*, 1997
	CA altered flavor and reduced volatile emission; some recovery of volatiles when moved from CA to air	Fellman *et al.*, 1993; Mattheis *et al.*, 1995; Plotto *et al.*, 1999
	1-MCP and methyl jasmonate reduced volatiles	Fan and Mattheis, 1999
	Pressure treatment with calcium chloride caused transient reduction in volatiles	Saftner *et al.*, 1999
	Edible coatings affected aroma profiles	Saftner, 1999; Saftner *et al.*, 1999
	Exposure of apples to hypoxia induced changes in aroma volatile concentrations	Dixon and Hewitt, 2001

Table 4.1 (continued)

Fruit	Effect	Reference
Banana	1-MCP suppressed volatile production	Golding *et al.*, 1999
Strawberry	Ozone for decay control decreased volatile esters	Pérez *et al.*, 1999b
Tomato	CA storage increased certain volatiles compared to air-stored fruit	Crouzet *et al.*, 1986
	Ethylene treatment altered volatile profiles	McDonald *et al.*, 1996
Citrus	Edible coatings altered aroma profiles	Baldwin *et al.*, 1995b; Cohen *et al.*, 1990
Mango	Edible coating altered aroma profiles	Baldwin *et al.*, 1999b

AVG = aminoethoxyvinyl glycine; CA = controlled atmosphere; 1-MCP = 1-methylcyclopropene.

4.3   Flavor components in fruits

Sweetness in fruits is related mainly to sugars which include: sucrose, as in citrus (Baldwin, 1993); glucose and/or fructose (Wills *et al.*, 1981), as in tomato (Baldwin *et al.*, 1991a,b); and even sorbitol, which persists in ripe apple along with other sugars (Knee, 1993). Of these, fructose is perceived as more sweet than sucrose, which is, in turn, perceived as sweeter than glucose. Thus a weighting of these sugars in relation to sucrose and subsequent combining can give a single value 'sucrose equivalent' (Koehler and Kays, 1991). Soluble solids are often thought to be synonymous with sugars and are easily measured. Thus, breeders will select for higher solids in an attempt to increase sweetness. In fruits like orange, solids appear to relate to sweetness while in others, like tomato and mango, the relationship is not clear (Baldwin *et al.*, 1998, 1999a; Malundo *et al.*, 2001a).

Acidity is mainly contributed by organic acids such as citric acid in lemon (Baldwin, 1993), malic acid in apples (Knee, 1993), quinic acid in pears (Knee, 1993), tartaric acid in grapes (Kanellis and Roubelakis-Angelakis, 1993) and oxalic acid in bananas, which gives some astringency to complement the citric and malic acids (Seymour, 1993).

Other components of flavor include: bitterness, which can depend on terpenoid lactones such as limonin in orange, or flavonoid glycosides, such as naringin in grapefruit (Baldwin, 1993; Maier, 1969); saltiness contributed by various natural salts; and astringency derived from flavonoids, alkaloids (Zitnak and Filadelfi, 1985; DeRovira, 1997), tannins (Taylor, 1993) and other factors. Often compounds interact either chemically or in terms of perception to affect intensity of flavor descriptors. For example, recent studies have shown that sourness, or acid levels, and aroma compounds affect sweetness perception (Malundo *et al.*, 1995). Aroma, in general, has gained increasing attention for its part in fresh produce flavor quality. In addition to contributing to fruit odor, aroma compounds can affect perception of sweetness and sourness as was found for tomato (Baldwin *et al.*, 1998).

Aroma is derived from the volatile components in fruits. Those that are present in concentrations that can be perceived by the human nose are assumed to contribute to flavor or aroma. Odor thresholds are determined by placing a compound in a background similar to a food medium and testing to determine the level at which it can be detected by smell as described by the 'Ascending Method of Limits' of the American Society for Testing and Materials (ASTM, 1991; Meilgaard *et al.*, 1991). Log odor units can then be calculated from the ratio of the concentration of a component in a food to its odor threshold. It is assumed that compounds with positive odor units contribute to the flavor of a food (Buttery *et al.*, 1989). Contributions to tomato flavor were evaluated by Buttery (1993) for volatiles present at levels of 1 ppb or more (about 30 of the more than 400 identified compounds). The aroma perception of volatile compounds can be affected by the medium of evaluation, however. In one study, both the thresholds and descriptors of some volatile compounds in tomato were different in different media (water versus aqueous alcohol, versus tomato homogenate) (Tandon *et al.*, 2000).

In tomato (Buttery, 1993), citrus (Shaw and Wilson, 1980; Shaw, 1991) and apple (Cunningham *et al.*, 1985) 15 to 40 compounds contribute to flavor, with no one compound being responsible for the characteristic flavor of the fruit. Banana aroma, on the other hand, can be approximated quite well by 3-methylbutylacetate (Berger, 1991), peach aroma by γ-undecalactone, known as the 'peach aldehyde' (Crouzet *et al.*, 1990), raspberry by 4-(4-hydroxyphenyl)-butan-2-one, known as the 'raspberry ketone' (Larsen and Poll, 1990) and grapefruit by 1-*p*-menthene-8-thiol (Demole *et al.*, 1982). These compounds that approximate the aroma of a fruit are considered 'character impact' compounds.

4.4   Measurement of flavor components

Soluble solids can be easily be measured by refractometry, while measurement of individual sugars requires gas chromatography (GC) or high performance liquid chromatography (HPLC). Acids can be measured individually by HPLC (Baldwin *et al.*, 1991a,b), by titration (TA) with sodium hydroxide (Jones and Scott, 1984) or by pH (Baldwin *et al.*, 1998). Sometimes measurement of solids, the ratio of solids/TA, or pH relate better to sourness than TA itself (Baldwin *et al.*, 1998, Malundo *et al.*, 2001a).

Initially flavor isolation procedures of steam distillation and solvent extraction were commonly used to extract and even quantify aroma compounds (Teranishi and Kint, 1993). Unfortunately these techniques can modify the flavor profile of a sample both qualitatively and quantitatively (Schamp and Dirinck, 1982). They are also time consuming and, therefore, difficult to apply to large sample sets. Internal standards that cover the boiling points of volatiles of interest must be incorporated to determine recovery. The resulting concentration of material, however, allows identification of compounds by gas chromatography–mass spectrometry (GC/MS). Currently, purge and trap headspace sampling methods are more popular. These involve trapping and concentrating volatile components on a solid support. Volatiles are later released from the trap using heat and analyzed on GC or GC/MS (Teranishi and Kint, 1993; Schamp and Dirinck, 1982). Static headspace methods better reflect the true flavor profile, but compounds present at low levels may not be detected or easily quantified. Cryofocusing (using a cold trap) of static headspace volatiles (Teranishi and Kint, 1993) can help overcome this problem since samples can be concentrated and recovered without heating while avoiding the possibility of adulteration. Moshonas and Shaw (1997) used this method for quantification of orange juice volatiles. Solid phase microextraction (SPME) is a rapid sampling technique where volatiles interact with a fiber-coated probe inserted into the sample headspace. The probe is then transferred to a GC injection port where the volatiles are desorbed. This technique has been applied to apples, tomatoes (Song *et al.*, 1997, 1998) and strawberries (Golaszewski *et al.*, 1998; Song *et al.*, 1997).

The 'electronic nose' incorporates a sensor array with a broad range of selectivity. The interaction of volatile components with the various sensors allows for discrimination between samples. A particular sample or flavor component mixture can be identified by pattern recognition. This instrument does not, however, identify or quantify individual compounds. Four basic sensor technologies have been commercialized to date that use two classes of sensors. The four types of sensors are metal oxide semiconductors (MOS), metal oxide semiconductor field effect transistors (MOSFET), conducting organic polymers (CP), piezoelectric crystals (bulk acoustic wave, BAW) or quartz crystal microbalance. Such sensors are divided into two classes since they either operate 'hot' (MOS, MOSFET) or cold (CP, BAW). The 'hot' sensors are less sensitive to moisture and have less carry over from one measurement to the next. The sensor output can be combined with mass data from a modified mass spectrometer. The next generation of electronic noses may employ fiberoptic, electrochemical and bimetal sensors that are currently in the developmental stage (Schaller *et al.*, 1998).

4.5 Biochemical pathways that produce flavor components in fruits

The two major pathways in plants are photosynthesis and respiration. Sugars are derived from photosynthesis while acids are generated in reactions of the tricarboxylic acid cycle via glycolytic precursors (Wills *et al.*, 1981). Aroma compounds, on the other hand, come from several different pathways in fatty acid, amino acid, phenolic and terpenoid metabolism (figure 4.1). Some

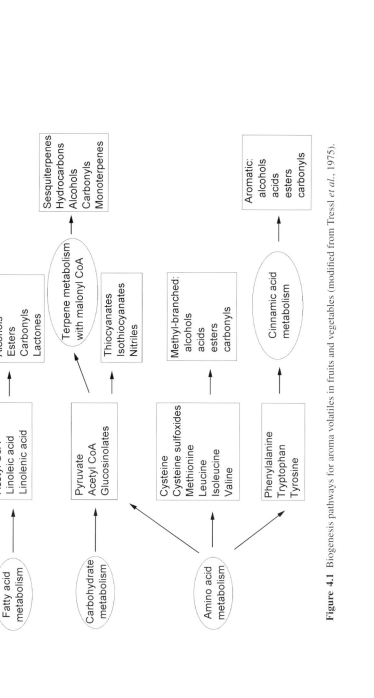

Figure 4.1 Biogenesis pathways for aroma volatiles in fruits and vegetables (modified from Tressl *et al.*, 1975).

reviews on flavor genesis cover tomatoes (Buttery and Ling, 1993a,b; Buttery, 1993), apples (Fellman *et al.*, 1993, 2000), olives (Olías *et al.*, 1993), melons (Yabumoto *et al.*, 1977; Wyllie *et al.*, 1995), cucumbers (Hatanaka *et al.*, 1975) and fruits and vegetables in general (Sanz *et al.*, 1997; Wyllie *et al.*, 1996b).

As fruits ripen, characteristic aroma compounds are produced. This is often coupled with ethylene synthesis in climacteric fruits such as tomato (Baldwin *et al.*, 1991a), apples (Fan and Mattheis, 1999) and melons (Wyllie *et al.*, 1996a). Four or five classes of chemical compounds account for most important flavor volatiles in fruits: aldehydes, esters, ketones, terpenoids and sulfur-containing compounds. Often aroma compounds are only released upon cell disruption, when enzymes and substrates, that were formerly compartmentalized, come in contact (Buttery, 1993). Some volatiles are glycosylated and only exert a sensory effect when released from the sugar. For example, grapes show an increase in free and glycosylated aroma compounds at the end of the ripening period, after sugar increase has slowed (Coombe and McCarthy, 1997). The volatiles in wine grapes are particularly complex and are classified into five groups, of which the first four have glycosylated forms: monoterpene (abundant in 'floral' grapes), norisoprenoid, benzenoid, aliphatic and methoxypyrazine.

Aldehydes are formed from linoleic and linolenic acids via the lipoxygenase pathway and are important to the flavor of tomato, cucumber, peppers and other vegetables. They also play a less significant but important role in fruit flavors. Precursors and enzymes in flavor volatile pathways are often deduced in extracts with added C^{14} radio-labeled precursors (by observing which subsequent products are radioactive) and, in some cases, added enzymes (Galliard *et al.*, 1977; Myers *et al.*, 1970; Stone *et al.*, 1975; Tressl and Drawert, 1973). The enzyme lipoxygenase (LOX), along with hydroperoxide lyase (HPL) and a hydroperoxy cleavage enzyme, convert linoleic (18:2) and linolenic (18:3) acids to hexanal and *cis*-3-hexenal, respectively, via 9- and 13-hydroperoxy-C18:2 and -C18:3 intermediates. Hexanal and *cis*-3-hexenal can be reduced to hexanol and *cis*-3-hexenol, respectively, by a reductase enzyme such as alcohol dehydrogenase (ADH). Further isomerization of *cis*-3-hexenal to *trans*-2-hexenal can occur, either enzymatically or non-enzymatically (Galliard *et al.*, 1977; Jadhav *et al.*, 1972; Riley *et al.*, 1996; Stone *et al.*, 1975). The C6 aldehydes contribute green and grassy flavor notes. Hexyl acetate synthesis in crushed olives is associated with lipid degrading enzymes (Oliás *et al.*, 1993).

Fruits produce C1 to C20 fatty acids which are converted to alcohols as described for the LOX pathway. These alcohols are then converted to esters using acetyl CoA and alcohol acyltransferase (AAT) as demonstrated for apple (Fellman and Mattheis, 1995; Fellman *et al.*, 1993), banana (Ueda *et al.*, 1992; Harada *et al.*, 1985), melons (Ueda *et al.*, 1997) and strawberry (Ke *et al.*, 1994; Pérez *et al.*, 1993). Lactones are important for peach flavor and are also formed by LOX (Crouzet *et al.*, 1990).

Amino acids such as alanine, leucine, isoleucine, valine and phenylalanine are involved in volatile synthesis. Leucine, isoleucine or valine were converted into branched chain alcohols and esters in muskmelon (Yabumoto *et al.*, 1977) and banana (Myers *et al.*, 1970; Tressl and Drawert, 1973). In tomatoes, valine is a reported precursor for 1-*N*-2-methylpropane, 3-methylbutanal, 3-methylbutanol, 3-methylbutylnitrile, 1-*N*-3-methylbutane, 3-methylbutyric acid and 2-isobutylthiazole (Buttery and Ling, 1993b); isoleucine is a precursor of 2-methylbutanol and 2-methylbutyric acid; and phenylalanine is a precursor of phenylacetaldehyde, 2-phenylethanol, 1-*N*-2-phenylethane and phenylacetonitrile. Isoleucine was converted into esters in apple (Hansen and Poll, 1993). Alanine was converted into esters in strawberries (Pérez *et al.*, 1992; Drawert, 1981). Conversion of an amino acid to an ester involves deamination, decarboxylation, several reductions and esterification (Pérez *et al.*, 1992) via α-aminotransferase, α-ketoacid decarboxylase, α-ketoacid dehdrogenase, ADH and AAT (Wyllie *et al.*, 1996b). Phenylalanine was determined to be the precursor for phenylethanol, phenethylacetate, phenethylbutanoate and phenolic ethers in banana (Tressl and Drawert, 1973). Conversion of phenylalanine to phenolic ethers such as eugenol, eugenol methylester and elimicin was catalyzed by phenylalanine ammonia lyase (PAL), cinnamic acid-4-hydrolase, phenolase and methyltransferase (Tressl and Drawert, 1973).

Terpenes are formed from the mevalonic pathway along with carotenoids and other secondary metabolites. Aromatic terpenes include limonene, valencene, γ-terpinene and 3-carene, α-copaene, α- and β-pinene and myrcene that are important in citrus fruits (Shaw *et al.*, 1977; Wilson and Shaw, 1981) and mango (Malundo *et al.*, 1996). Oxygenated terpenoids such as linalool, neral and geranial are important in orange (Baldwin, 1993) and other fruits. Orange fruit contain enzymes for conversion of mevalonic acid to isopentenyl pyrophosphate and dimethylallyl pyrophosphate as well as for the synthesis of linalool and its cyclization to 2,8-menthadiene-1-ol, α-terpineol and D-limonene. Linalool may be a key intermediate in terpenoid metabolism in citrus (Bruemmer, 1975).

Some important volatile compounds in tomato are thought to be breakdown products of pigments such as lycopene and other carotenoids (Buttery and Ling, 1993b). These compounds include β-ionone, cyclocitral and β-damascenone from cyclic carotenoids; and geranylacetone, 6-methyl-5-hepten-2-one, 6-methyl-5-hepten-2-ol, and pseudoionone from open chain carotenoids (Buttery and Ling, 1993b).

4.6   Perception of flavor components

As with flavor chemistry, the science of flavor perception is also complicated. Tongue receptors detect non-volatile compounds in parts per hundred for sugars or polyalcohols, hydronium ions, sodium ions, glucosides and alkaloids, etc. These receptors are responsible for the perception of sweet, sour, salty and bitter tastes in food. The olfactory nerve endings at the back of the nose detect aromas in parts per billion (DeRovira, 1997) and, therefore, are much more sensitive to flavor compounds than the tongue is sensitive to taste compounds.

Smelling an aromatic food through the front of the nose may produce a different experience than when the aroma is perceived during chewing of food (Voirol and Daget, 1987). Temperature, viscosity and polarity of the food can affect relative vapor pressure and aroma release such as occurs in the mouth (Land, 1994; Taylor and Linforth, 1994; Voirol and Daget, 1987). The extent of aroma release, in turn, alters the concentration in the headspace of the mouth of volatile compounds that rise through the back of the nose to bind olfactory receptors (Land, 1994; Taylor and Linforth, 1994). In addition, odor and taste can interact to give an integrated perception (Voirol and Daget, 1987). Texture can also play a role. For example a crisp, juicy apple is likely to have more flavor than a mealy one (see Chapter 3). This is thought to depend on the texture and the state of the cell wall (in particular, the condition of the middle lamella), which affect the mechanism of tissue disruption. Fruit cells may break across cell walls, releasing cellular components, or between cells (middle lamellae) without release of cell contents, as in mealy fruit. Both juiciness and flavor impact are affected (Vickers, 1977).

To further complicate things, there are primary and secondary odorants. Primary odorants bind one receptor and define one odor individually and can then, in combination, define another aroma, such as methyl salicylate, which is an important volatile in tomato. Alone this is responsible for the aroma 'wintergreen'. Secondary odorants bind more than one olfactory receptor (Amoore, 1952). An example would be safrole, which is thought to bind four receptors, and is described as anise, with a wintergreen character, vanilla background and camphoraceous overtone (DeRovira, 1997). The aroma of safrole can be duplicated by a combination of the primary odorants anise, methyl salicylate, vanillin and camphor, which bind the same four olfactory receptors. Furthermore, there are top note and background note flavors. Top notes are generally compounds of relatively low molecular weight and high volatility that are heat labile and polar (DeRovira, 1997). They are usually easy to recognize in a food item. Background notes, on the other hand, are generally of low molecular weight, heat stable, non-polar and have a more subtle impact on flavor compared with top notes. However, they can affect perception of top notes.

4.7 Measurement of flavor perception

Sensory science attempts to measure human perception of flavor. Consumer preference and acceptance testing, investigating the likes and dislikes of consumers, typically requires large numbers (50–100) of panelists and its results can vary depending on socioeconomic, ethnic and geographical background. This necessitates the segmenting of subpopulations for a particular study (O'Mahony, 1995) and generally a traditional nine-point hedonic scale is used. A very simple three-point scale (outstanding, acceptable and unacceptable) was also shown to be effective for tomato fruit evaluation (Baldwin et al., 1995a). Adaptation of logistic regression from medical science can also be used, where a zero or one (a two-point scale) is designated for a consumer decision to purchase or not purchase an item. For example, consumers were asked to base their purchase decision on flavor, which was then related to chemical constituents (Malundo et al., 2001b). Difference testing can be used to measure slight differences between foods (usually arising from one particular aspect of flavor) and descriptive analysis measures intensities of a set of sensory attributes (O'Mahony, 1995). For descriptive analysis, panelists are trained to detect a range of flavor attributes and score their intensity, generally on a 150 mm unstructured line scale. Sensory studies for fruits can be used to identify optimal harvest maturity, evaluate flavor quality in breeding material, determine optimal storage and handling conditions, assess effects of disinfestation or preconditioning techniques on flavor quality, and measure flavor quality over the postharvest life of the product.

To understand flavor, it is necessary to relate sensory and chemical data in order to determine important flavor compounds as well as how to measure best those that correspond to the sensory experience. Chemical analysis of flavor compounds is informative, but alone does not give information on flavor from the consumer experience. Likewise sensory data alone does not provide information on how flavor components, combinations of components and relative proportions of components result in desirable or undesirable flavor quality. Sniff ports (olfactometry detectors) can be used with GCs, allowing a person to determine if odors are detectable, as well as their relative intensity and characteristics as they are separated by the GC column. This technique has been used successfully on apples (Cunningham et al., 1985; Rizzolo et al., 1989; Young et al., 1996). Descriptive terms can also be assigned to the respective peaks on the GC chromatogram that have odor activity (Acree, 1993). The drawback to this method is that the interactive effects of volatile compounds with each other and with sugars and acids, both chemically and in terms of human perception, are eliminated.

Statistics can be used to relate sensory attributes, preferences, and intensity to chemical components in foods (Martens et al., 1994; Bett 1993). Correlation of physical measurements with sensory analysis can give meaning to instrumental data as was shown with apple and tomato (Guadagni et al., 1966; Baldwin et al., 1998; Sinesio et al., 2000). Correlation, linear and multiple regression (stepwise forwards or backwards) can be used (Baldwin et al., 1998). For example, linear regression was used to establish relationships between levels of sesquiterpene lactones and bitterness in chicory (Peters and Van Amerongen, 1998). Multivariate methods require large data sets. These regression techniques, such as principle component or discriminant analysis, however, yield useful results for discriminating between fruit or fruit juice samples based on data for flavor volatiles or for other flavor compounds as was demonstrated for citrus (Shaw *et al.*, 1993; Moshonas and Shaw, 1997; Shaw *et al.*, 1998), strawberry (Shamaila *et al.*, 1992) and tomato (Resurreccion and Shewfelt, 1985; Maul *et al.*, 1998, 1999).

4.8   Conclusions

Flavor in fruits is complex, both in terms of chemical and sensory measurements and in terms of interfacing the two approaches. Flavor is also an important aspect of fruit quality for consumers and, consequently, for industry. Flavor quality of fresh fruits is at maximum at harvest (except for climacteric fruits that continue to ripen after harvest) and can only be maintained, at best, during storage, shipping and marketing. This is because the bottom line for fruit flavor is genetic. Breeders need more information and analytical tools in order to select for flavor quality, such as molecular markers that relate to flavor. These markers could help identify important enzymes in flavor pathways. Flavor life is usually shorter than shelf-life based on appearance. The effect of harvest maturity on flavor quality needs to be determined for each commodity. In an ever more competitive and global market flavor quality is increasingly important to consumers. Maintenance of flavor quality in fruits after harvest is the challenge, while marketing distances increase as a result of new storage, handling and transport technologies. Postharvest physiologists, sensory scientists, flavor chemists, breeders and molecular biologists, armed with new technologies, are taking an integrated approach to flavor quality to confront this challenge.

References

Acree, T.E. (1993) Bioassays for flavor, in *Flavor Science: Sensible Principles and Techniques* (eds T.E. Acree and R. Teranishi), ACS Books, Washington, DC, pp. 1-20.
Amoore, J.E. (1952) The stereochemical specificities of human olfactory receptors. *Perfumer Essential Oils*, **43**, 321-323.

ASTM (1991) *Standard Practice E679. Determination of Odor and Taste Thresholds by a Force Choice Ascending Concentration Series Methods of Limits, 1–5,* American Society for Testing and Materials, Philadelphia, PA.

Baldwin, E.A. (1993) Citrus, in *Biochemistry of Fruit Ripening* (eds G.B. Seymour, J.E. Taylor and G.A. Tucker), Chapman & Hall, New York, pp. 107-149.

Baldwin, E.A., Nisperos, M.O. and Moshonas, M.G. (1991a) Quantitative analysis of flavor and other volatiles and for other constituents of two tomato varieties during ripening. *Journal of the American Society for Horticultural Science,* **116,** 265-269.

Baldwin, E.A., Nisperos, M.O. and Moshonas, M.G. (1991b) Quantitative analysis of flavor parameters in six Florida tomato varieties (*Lycopersicon esculentum* Mill). *Journal of Agricultural and Food Chemistry,* **39,** 1135-1140.

Baldwin, E.A., Nisperos-Carriedo, M.O. and Scott, J.W. (1992) Levels of flavor volatiles in a normal cultivar, ripening inhibitor and their hybrid. *Proceedings of the Florida State Horticultural Society,* **104,** 86-89.

Baldwin, E.A., Scott, J.W. and Shewfelt, R.L. (1995a) Quality of ripened mutant and transgenic tomato cultigens. *Proceedings of the Tomato Quality Workshop,* **503,** 47-57.

Baldwin, E.A., Nisperos-Carriedo, M.O., Shaw, P.E. and Burns, J.K. (1995b) Effect of coatings and prolonged storage conditions on fresh orange flavor volatiles, degrees Brix, and ascorbic acid levels. *Journal of Agricultural and Food Chemistry,* **43,** 1321-1331.

Baldwin, E.A., Scott, J.W., Einstein, M.A., Malundo, T.M.M., Carr, B.T., Shewfelt, R.L. and Tandon, K.S. (1998) Relationship between sensory and instrumental analysis for tomato flavor. *Journal of the American Society for Horticultural Science,* **12,** 906-915.

Baldwin, E.A., Malundo, T.M.M., Bender, R. and Brecht, J.K. (1999a) Interactive effects of harvest maturity, controlled atmosphere and surface coatings on mango (*Mangifera indica* L.) flavor quality. *HortScience,* **34,** 514.

Baldwin, E.A., Burns, J.K., Kazokas, W., Brecht, J.K., Hagenmaier, R.D., Bender, R.J. and Pesis, E. (1999b) Effect of two edible coatings with different permeability characteristics on mango (*Mangifera indica* L.) ripening during storage. *Postharvest Biology and Technology,* **17,** 215-226.

Baldwin, E.A., Scott, J.W., Shewmaker, C.K. and Schuch, W. (2000) Flavor trivia and tomato aroma: biochemistry and possible mechanisms for control of important aroma components. *HortScience,* **35,** 1013-122.

Berger, R.G. (1991) Fruits I, in *Volatile Compounds in Foods and Beverages* (ed. H. Maarse), Marcel Dekker, Inc., New York, pp. 283-204.

Bett, K. (1993) Measuring sensory properties of meat in the laboratory. *Food Technology,* **47(11),** 121-125.

Bruemmer, J.H. (1975) Aroma substances of citrus fruits and their biogenesis, in *Symposium on Fragrances and Flavor substances* (ed. F. Drawert), Verlag Hans Carl, Nurenburg, pp. 167-176.

Buttery, R.G. (1993) Quantitative and sensory aspects of flavor of tomato and other vegetables and fruits, in *Flavor Science: Sensible Principles and Techniques* (eds T.E. Acree and R. Teranishi), American Chemical Society, Washington, DC, pp. 259-286.

Buttery, R.G. and Ling, L.C. (1993a) Enzymatic production of volatiles in tomatoes, in *Flavor Precursors* (eds P. Schreier and P. Winterhalter), Allured Publishing, Wheaton, IL, pp. 137-146.

Buttery, R.G. and Ling, L.C. (1993b) Volatiles of tomato fruit and plant parts: relationship and biogenesis, in *Bioactive Volatile Compounds From Plants* (eds R. Teranishi, R. Buttery and H. Sugisawa), ACS Books, Washington, DC, pp. 23-34.

Buttery, R.G., Teranishi, R. and Ling, L.C. (1987) Fresh tomato aroma volatiles: a quantitative study. *Journal of Agricultural and Food Chemistry,* **35,** 540-544.

Buttery, R.G., Teranishi, R., Flath, R.A. and Ling, L.C. (1989) Fresh tomato volatiles: composition and sensory studies, in *Flavor Chemistry: Trends and Developments* (eds R. Teranishi, R.G. Buttery and F. Shahidi), American Chemical Society, Washington, DC, pp. 213-222.

Cliff, M.A., Lau, O.L. and King, M.C. (1998) Sensory characteristics of controlled atmosphere- and air-stored 'Gala' apples. *Journal of Food Quality*, **21**, 239-249.

Cohen, E., Shalom, Y. and Rosenberger, I. (1990) Postharvest ethanol buildup and off-flavor in Murcott tangerine fruits. *Journal of the American Society for Horticultural Science*, **115**, 775-778.

Coombe, B.G. and McCarthy, M.G. (1997) Identification and naming of the inception of aroma development in ripening grape berries. *Australian. Journal of Grape and Wine Research*, **3**, 18-20.

Crouzet, J., Signoret, A., Coulibaly, J. and Roudsari, M.H. (1986) Influence of controlled atmosphere storage on tomato volatile components, in *The Shelf Life of Foods and Beverages* (ed. G. Charalambous), Elsevier Science Publishers, Amsterdam, pp. 355-367.

Crouzet, J., Etievant, P. and Bayonove, C. (1990) Stoned fruit: apricot, plum, peach, cherry, in *Food Flavors Part C: The Flavor of Fruits* (eds I.D. Morton and A.J. MacLeod), Elsevier, New York, pp. 43-91.

Cunningham, D.G., Acree, T.E., Barnard, J., Butts, R.M. and Braell, P.A. (1985) Charm analysis of apple volatiles. *Food Chemistry*, **19**, 137-147.

Dalal, K.B., Olson, L.E., Yu, M.H. and Salunkhe, D.K. (1967) Gas chromatography of the field-glass-greenhouse-grown, and artificially ripened tomatoes. *Phytochemistry*, **6**, 155-157.

Dan, K., Todoriki, S., Nagata, M. and Yamashita, I. (1997) Formation of volatile sulfur compounds in broccoli stored under anaerobic condition. *Journal of the Japanese Society for Horticultural Science*, **65**, 867-875.

Demole, E., Enggist, P. and Ohloff, G. (1982) 1-*p*-menthene-8-thiol: a powerful flavor impact constituent of grapefruit juice (*Citrus paradisi* MacFayden). *Helvetica-Chimica-Acta*, **65**, 1785-1794.

DeRovira, D. (1997) *Flavor Nomenclature Workshop: An Odor Description and Sensory Evaluation Workshop*, Flavor Dynamics Manual, Inc., Somerset, NJ.

Dixon, J. and Hewett, E.W. (2001) Temperature of hypoxic treatment alters volatile composition of juice from 'Fugi' and 'Royal Gala' apples. *Postharvest Biology and Technology*, **21**, 71-83.

Drawert, F. (1981) Possibilities of the biotechnological production of aroma substances by plant tissue cultures, in *Food Flavorings* (ed. P. Schreier), Walter de Gruyter, New York, pp. 509-527.

Fallik, E., Archbold, D.D., Hamilton-Kemp, T.R., Loughrin, J.H. and Collins, R.W. (1997) Heat treatment temporarily inhibits aroma volatile compound emission from Golden Delicious apples. *Journal of Agricultural and Food Chemistry*, **45**, 4038-4041.

Fan, X. and Mattheis, J.P. (1999) Impact of 1-methylcyclopropene and methyl jasmonate on apple volatile production. *Journal of Agricultural and Food Chemistry*, **47**, 2847-2853.

Fellman, J.K. and Mattheis, J.P. (1995) Ester biosynthesis in relation to harvest maturity and controlled-atmosphere storage of apples, in *Fruit Flavors: Biogenesis, Characterization, and Authentication* (eds R.L. Rouseff and M.M. Leahy), American Chemical Society, Washington, DC, pp. 149-162.

Fellman, J.K., Mattheis, J.P., Patterson, M.E., Mattinson, D.S. and Bostick, B.C. (1993) Study of ester biosynthesis in relation to harvest maturity and controlled-atmosphere storage of apples (*Malus domestica* Borkh.), in *Proceedings of the 6th International Controlled Atmosphere Research Conference*, Cornell University, Ithaca, NY, 15-17 June 1993.

Fellman, J.F., Miller, T.W., Mattison, D.S. and Mattheis, J.P. (2000) Factors that influence biosynthesis of volatile flavor compounds in apple fruit. *HortScience*, **35**, 1026-1033.

Galliard, T., Matthew, J.A., Wright, A.J. and Fishwick, M.J. (1977) The enzymic breakdown of lipids to volatile and non-volatile carbonyl fragments in disrupted tomato fruits. *Journal of the Science of Food and Agriculture*, **28**, 863-868.

Golaszewski, R., Sims, C.A., O'Keefe, S.F., Braddock, R.J. and Littell, E.C. (1998) Sensory attributes and volatile components of stored strawberry juice. *Journal of Food Science*, **63**, 734-738.

Golding, J.B., Shearer, D., McGlasson, W.B. and Wyllie, S.G. (1999) Relationships between respiration, ethylene, and aroma production in ripening banana. *Journal of Agricultural and Food Chemistry*, **47**, 1646-1651.

Guadagni, D.G., Okano, S., Buttery, R.G. and Burr, H.K. (1966) Correlation of sensory and gas-liquid chromatographic measurements of apple volatiles. *Food Technology*, **20**(**4**), 166-169.

Hansen, K. and Poll, L. (1993) Conversion of L-isoleucine into 2-methylbut-2-enyl esters in apples. *Lebensmittel Wissenschaft und Technologie*, **26**, 178-180.

Harada, M., Ueda, Y. and Iwata, T. (1985) Purification and some properties of alcohol acetyltransferase from banana fruit. *Plant and Cell Physiology*, **26**, 1067-1074.

Hatanaka, A., Kajiwara, T. and Harada, T. (1975) Biosynthetic pathway of cucumber alcohol: *trans*-2, *cis*-6-nonadienol via *cis*-3, *cis*-6-nonadienal. *Phytochemistry*, **14**, 2589-2592.

Jadhav, S., Singh, B. and Salunkhe, D.K. (1972) Metabolism of unsaturated fatty acids in tomato fruit: linoleic and linolenic acid as precursors of hexanal. *Plant and Cell Physiology*, **13**, 449-459.

Jones, R.A. and Scott, S.J. (1984) Genetic potential to improve tomato flavor in commercial F_1 hybrids. *Journal of the American Society for Horticultural Science*, **109**, 318-321.

Kader, A.A., Stevens, M.A., Albight-Holton, M., Morris, L.L. and Algazi, M. (1977) Effect of fruit ripeness when picked on flavor and composition in fresh market tomatoes. *Journal of the American Society for Horticultural Science*, **102**, 724-731.

Kanellis, A.K. and Roubelakis-Angelakis, K.A. (1993) Grape, in *Biochemistry of Fruit Ripening* (eds G.B. Seymour, J.E. Taylor and G.A. Tucker), Chapman & Hall, New York, pp. 189-234.

Ke, D., Zhou, L. and Kader, A.A. (1994) Mode of oxygen and carbon dioxide action on strawberry ester biosynthesis. *Journal of the American Society for Horticultural Science*, **119**, 971-975.

Knee, M. (1993) Pome fruits, in *Biochemistry of Fruit Ripening* (eds G.B. Seymour, J.E. Taylor and G.A. Tucker), Chapman & Hall, New York, pp. 325-346.

Koehler, P.E. and Kays, S.J. (1991) Sweet potato flavor: quantitative and qualitative assessment of optimum sweetness. *Journal of Food Quality*, **14**, 241-249.

Kopeliovitch, E., Mizrahi, Y., Rabinowitch, H.D. and Kedar, N. (1982) Effect of the fruit-ripening mutant genes *rin* and *nor* on the flavor of tomato fruit. *Journal of the American Society for Horticultural Science*, **107**, 361-364.

Land, D.G. (1994) Savory flavors—an overview, in *Understanding Natural Flavors* (eds J.R. Piggott and A. Paterson), Blackie Academic & Professional/Chapman & Hall, New York, pp. 298-306.

Larsen, M. and Poll, L. (1990) Odour thresholds of some important aroma compounds in raspberries. *Zeitschrift für Lebensmittel Untersuchung und Forschung*, **191**, 129-131.

Larsen, M., Poll, L. and Olsen, C.E. (1992) Evaluation of the aroma composition of some strawberry (*Fragaria ananassa* Duch) cultivars by use of odour threshold values. *Zeitschrift für Lebensmittel Untersuchung und Forschung*, **195**, 536-539.

McDonald, R.E., McCollum, T.G. and Baldwin, E.A. (1996) Prestorage heat treatments influence free sterols and flavor volatiles of tomatoes stored at chilling temperature. *Journal of the American Society for Horticultural Science*, **12**, 531-536.

Maier, V.P. (1969) Compositional studies of citrus: significance in processing, identification, and flavor. *Proceedings of the First International Citrus Symposium*, **1**, 235-243.

Malundo, T.M.M., Shewfelt, R.L. and Scott, J.W. (1995) Flavor quality of fresh tomato (*Lycopersicon esculentum* Mill.) as affected by sugar and acid levels. *Postharvest Biology and Technology*, **6**, 103-110.

Malundo, T.M.M., Baldwin, E.A., Ware, G.O. and Shewfelt, R.L. (1996) Volatile composition and interaction influence flavor properties of mango (*Mangifera indica* L.). *Proceedings of the Florida State Horticultural Society*, **109**, 264-268.

Malundo, T.M.M., Shewfelt, R.L., Ware, G.O. and Baldwin, E.A. (2001a) Sugars and acids influence flavor properties of mango. *Journal of the American Society for Horticultural Science*, **26**, 115-121.

Malundo, T.M.M., Shewfelt, R.L., Ware, G.O. and Baldwin, E.A. (2001b) Alternative methods for consumer testing and data analysis used to identify critical flavor properties of mango (*Mangifera indica* L.). *Journal of Sensory Studies*, **16**, 119-214.

Martens, M., Risvik, E. and Martens, H. (1994) Matching sensory and instrumental analyses, in *Understanding Natural Flavors* (eds J.R. Piggott and A. Paterson), Blackie Academic and Professional/Chapman & Hall, New York, pp. 60-76.

Mattheis, J.P., Buchanan, D.A. and Fellman, J.K. (1991) Change in apple fruit volatiles after storage in atmospheres inducing anaerobic metabolism. *Journal of Agricultural and Food Chemistry*, **39**, 1602-1605.

Mattheis, J.P., Buchanan, D.A. and Fellman, J.K. (1995) Volatile compound production by Bisbee Delicious apples after sequential atmosphere storage. *Journal of Agricultural and Food Chemistry*, **43**, 194-199.

Maul, F., Sargent, S.A., Balaban, M.O., Baldwin, E.A., Huber, D.J. and Sims, C.A. (1998) Aroma volatile profiles from ripe tomatoes are influenced by physiological maturity at harvest: an application for electronic nose technology. *Journal of the American Society for Horticultural Science*, **123**, 1094-1101.

Maul, F., Sargent, S.A., Huber, D.J., Balaban, M.O., Sims, C.A. and Baldwin, E.A. (1999) Harvest maturity and storage temperature affect volatile profiles of ripe tomato fruits: electronic nose and gas chromatographic analyses, in *Electronic Noses and Sensor Array Based Systems: Design and Applications. Proceedings of the 5th International Symposium on Olfaction and the Electronic Nose* (ed W.J. Hurst), Technomic Pub. Co., Lancaster, PA. pp. 1-13.

Maul, F., Sargent, S.A., Sims, C.A., Baldwin, E.A., Balaban, M.O. and Huber, D.J. (2000) Recommended storage temperatures affect tomato flavor and aroma quality. *Journal of Food Science*, **65**, 1228-1237.

Meilgaard, M.C., Civille, G.V. and Carr, B.T. (1991) *Sensory Evaluation Techniques*, 2nd edn. CRC Press, Inc., Boca Raton, FL.

Moretti, C.L., Sargent, S.A., Baldwin, E.A., Huber, D.J. and Puschmann, R. (1997) Pericarp, locule and placental tissue volatile profiles are altered in tomato fruits with internal bruising, in *6th Brazilian Congress of Plant Physiology*, 10-15 August, 1997, Belem, Brazil.

Morton, I.D. and MacLeod, A.J. (1990) *Food Flavors: Part C. The Flavor of Fruits*. Elsevier Science, New York.

Moshonas, M.G. and Shaw, P.E. (1997) Dynamic headspace gas chromatography combined with multivariate analysis to classify fresh and processed orange juices. *Journal of Essential Oil Research*, **9**, 133-139.

Myers, M.J., Issenberg, P. and Wick, E.L. (1970) L-Leucine as a precursor of *iso*amyl alcohol and *iso*amyl acetate, volatile aroma constituents of banana fruit discs. *Phytochemistry*, **9**, 1693-1700.

Olías, J.M., Pérez, A.G., Rios, J.J. and Sanz, L.C. (1993) Aroma of virgin olive oil: biogenesis of the 'Green' odor notes. *Journal of Agricultural and Food Chemistry*, **41**, 2368-2373.

O'Mahony, M. (1995) Sensory measurement in food science: fitting methods to goals. *Food Technology*, **49(4)**, 72-82.

Pérez, A.G., Rios, J.J., Sanz, C. and Olías, J.M. (1992) Aroma components and free amino acids in the strawberry variety Chandler during ripening. *Journal of Agricultural and Food Chemistry*, **40**, 2232-2235.

Pérez, A.G., Sanz, L.C. and Olías, J.M. (1993) Partial purification and some properties of alcohol acyltransferase from strawberry fruits. *Journal of Agricultural and Food Chemistry*, **41**, 1462-1466.

Pérez, A.G., Sanz, C., Olías, R. and Olías, J.M. (1999a) Lipoxygenase and hydroperoxide lyase activities in ripening strawberry fruits. *Journal of Agricultural and Food Chemistry*, **47**, 249-253.

Pérez, A.G., Sanz, C., Rios, J.J., Olías, R. and Olías, J.M. (1999b) Effects of ozone treatment on postharvest strawberry quality. *Journal of Agricultural and Food Chemistry*, **47**, 1652-1656.

Peters, A.M. and Van Amerongen, A. (1998) Relationship between levels of sesquiterpene lactones in chicory and sensory evaluation. *Journal of the American Society for Horticultural Science*, **123**, 326-329.

Plotto, A., McDaniel, M.R. and Mattheis, J.P. (1999) Characterization of 'Gala' apple aroma and flavor: differences between controlled atmosphere and air storage. *Journal of the American Society for Horticultural Science*, **124**, 416-423.

Podoski, B.W., Sims, C.A., Sargent, S.A., Price, J.F., Chandler, C.K. and O'Keefe, S.F. (1998) Effects of cultivar, modified atmosphere, and pre-harvest conditions on strawberry quality. *Proceedings of the Florida State Horticultural Society*, **110**, 246-252.

Resurreccion, A.V.A. and Shewfelt, R.L. (1985) Relationship between sensory attributes and objective measurements of postharvest quality of tomatoes. *Journal of Food Science*, **50**, 1242-1245.

Riley, J.C.M., Willemot, C. and Thompson, J.E. (1996) Lipoxygenase and hydroperoxide lyase activities in ripening tomato fruit. *Postharvest Biology and Technology*, **7**, 97-107.

Rizzolo, A., Polesello, A. and Teleky-Vamossy, G. (1989) CGC/sensory analysis of volatile compounds developed from ripening apple fruit. *Journal of High Resolution Chromatography*, **12**, 824-827.

Romani, R., Labavitch, J., Yamashita, T., Hess, B. and Rae, H. (1983) Preharvest AVG treatment of 'Bartlett' pear fruits: effects on ripening, color change, and volatiles. *Journal of the American Society for Horticultural Science*, **108**, 1046-1049.

Saftner, R.A. (1999) The potential of fruit coating and film treatments for improving the storage and shelf-life qualities of 'Gala and 'Golden Delicious' apples. *Journal of the American Society for Horticultural Science*, **124**, 682-689.

Saftner, R.A., Conway, W.S. and Sams, C.E. (1999) Postharvest calcium infiltration alone and combined with surface coating treatments influence volatile levels, respiration, ethylene production, and internal atmospheres of 'Golden Delicious' apples. *Journal of the American Society for Horticultural Science*, **124**, 553-558.

Sanz, C., Olías, J.M. and Pérez, A.G. (1997) Aroma biochemistry of fruits and vegetables. *Phytochemistry of Fruit and Vegetables*, Proceedings of the Phytochemical Society of Europe, 41, Clarendon Press, Oxford and Oxford University Press, New York, pp. 125-155.

Schaller, E., Bosset, J.O. and Escher, F. (1998) 'Electronic noses' and their application to food. *Lebensmittell Wissenschaft und Technologie*, **31**, 305-316.

Schamp, N. and Dirinck, P. (1982) The use of headspace concentration on Tenax for objective evaluation of fresh fruits, in *Chemistry of Foods and Beverages* (ed. G. Charalambous), Academic Press, New York, pp. 25-47.

Seymour, G.B. (1993) Banana, in *Biochemistry of Fruit Ripening* (eds G.B. Seymour, J.E. Taylor and G.A. Tucker), Chapman & Hall, New York, pp. 83-106.

Shamaila, M., Powrie, W.D. and Skura, B.J. (1992) Analysis of volatile compounds from strawberry fruit stored under modified atmosphere packaging (MAP). *Journal of Food Science*, **57**, 1173-1176.

Shaw, P.E. (1991) Fruits II, in *Volatile Compounds in Foods and Beverages* (ed. H. Maarse), Marcel Dekker, Inc., New York, pp. 305-328.

Shaw, P.E. and Wilson, C.W. (1980) Importance of selected volatile components to natural orange, grapefruit, tangerine, and mandarin flavors, in *Citrus Nutrition and Quality* (eds S. Nagy and J.A. Attaway), ACS Symposium Series no. 143, American Chemical Society, Washington, DC, pp. 167-190.

Shaw, P.E., Ahmed, E.M. and Dennison, R.A. (1977) Orange juice flavor: contribution of certain volatile components as evaluated by sensory panels. *Proceedings of the International Society for Citriculture*, **3**, 804-807.

Shaw, P.E., Buslig, B.S. and Moshonas, M.G. (1993) Classification of commercial orange juice types by pattern recognition involving volatile constituents quantified by gas chromatography. *Journal of Agricultural and Food Chemistry*, **41**, 809-813.

Shaw, P.E., Buslig, B.S. and Goodner, K.L. (1998) Monitoring changes in citrus juice quality during processing. *Proceedings of the Florida State Horticultural Society*, **111**, 280-282.

Sinesio, F., Di Natale, C., Quaglia, G.B., Bucarelli, F.M., Moneta, E., Macagnano, A., Paolesse, R. and D'Amico, A. (2000) Use of electronic nose and trained sensory panel in the evaluation of tomato quality. *Journal of the Science of Food and Agriculture*, **80**, 63-71.

Song, J., Gardner, B.D., Holland, J.F. and Beaudry, R.M. (1997) Rapid analysis of volatile flavor compounds in apple fruit using SPME and GC/time of flight spectrometry. *Journal of Agricultural and Food Chemistry*, **45**, 1801-1807.

Song, J., Fan, L. and Beaudry, R.M. (1998) Application of solid phase microextraction and gas chromatography/time-of-flight mass spectrometry for rapid analysis of flavor volatiles in tomato and strawberry fruits. *Journal of Agricultural and Food Chemistry*, **46**, 3721-3726.

Stone, E.J., Hall, R.M. and Kazeniac, S.J. (1975) Formation of aldehydes and alcohols in tomato fruit from U^{14} C-labeled linolenic and linoleic acids. *Journal of Food Science*, **40**, 1138-1141.

Tandon, K.S., Baldwin, E.A. and Shewfelt, R.L. (2000) Aroma perception of individual volatile compounds in fresh tomatoes (*Lycopersicon esculentum* Mill.) As affected by the medium of evaluation. *Postharvest Biology and Technology*, **17**, 215-226.

Taylor, J.E. (1993) Exotics, in *Biochemistry of Fruit Ripening* (eds G.B. Seymour, J.E. Taylor, and T.A. Tucker), Chapman & Hall, London, pp. 151-187.

Taylor, A.J. and Linforth, R.S.T. (1994) Methodology for measuring volatile profiles in the mouth and nose during eating, in *Trends in Flavor Research* (eds J. Maarse and D.G. van der Heij), Elsevier Science, New York, pp. 3-14.

Teranishi, R. and Kint, S. (1993) Sample preparation, in *Flavor Science: Sensible Principles and Techniques* (eds T. Acree and R. Teranshi), ACS Books, Washington, DC, pp. 137-167.

Tressl, R. and Drawert, F. (1973) Biogenesis of banana volatiles. *Journal of Agricultural and Food Chemistry*, **21**, 560-565.

Tressl, R., Holzer, M. and Apetz, M. (1975) Biogenesis of volatiles in Fruit and Vegetables, *in Aroma Research* (eds H. Maarse and P.J. Groenen), Proceedings of the International Symposium on Aroma Research at Central Institute for Nutrition and Food Research TNO, Zeist, the Netherlands. Centre for Agricultural Publishing and Documentation, Wageningen, the Netherlands, pp. 41-62.

Ueda, Y., Tsuda, A., Bai, J.H., Fujishita, N. and Hachin, K. (1992) Characteristic pattern of aroma ester formation from banana, melon, and strawberry with reference to the substrate specificity of ester synthetase and alcohol contents in pulp. *Journal of the Japanese Society for Food Science and Technology*, **39**, 183-187.

Ueda, Y., Fujishita, N. and Hachin, K. (1997) Presence of alcohol acetyltransferase in melons (*Cucumis melo* L.). *Postharvest Biology and Technology*, **10**, 121-126.

Vickers, A. (1977) Structural and mechanical indicators of flavor quality, in *Flavor Quality: Objective Measurement* (ed. R. Scanlan), ACS Symposium Series 51, American Chemical Society, Washington, DC, pp. 45-50.

Voirol, E. and Daget, N. (1987) Nasal and retronasal olfactory perception of a meat aroma, in *Flavor Science and Technology* (eds M. Martens, G.A. Dalen and H. Russwurm), Wiley, New York, pp. 309-316.

Watada, A.E. and Aulenbach, B.B. (1979) Chemical and sensory qualities of fresh market tomatoes. *Journal of Food Science*, **44**, 1013-1016.

Wills, R.H.H., Lee, T.H., Graham, D., McGlasson, W.B. and Hall, E.G. (1981) *Postharvest: An Introduction to the Physiology and Handling of Fruit and Vegetables*, AVI Publishing Co., Inc., Westport, CT.

Wilson, C.W. and Shaw, P.E. (1981) Importance of thymol, methyl-*N*-methylanthranilate, and monoterpene hydrocarbons to the aroma and flavor of mandarin cold-pressed oils. *Journal of Agricultural and Food Chemistry*, **29**, 494-496.

Wright, D.H. and Harris, N.D. (1985) Effect of nitrogen and potassium fertilization on tomato flavor. *Journal of Agricultural and Food Chemistry*, **33**, 355-358.

Wyllie, S.G., Leach, D.N., Wang, Y. and Shewfelt, R.L. (1995) Key aroma compounds in melons: Their development and cultivar dependence, in *Fruit Flavors: Biogenesis, Characterization, and Authentication* (eds R.L. Rouseff and M.M. Leahy), American Chemical Society, Washington, DC, pp. 248-257.

Wyllie, S.G., Leach, D.N. and Wang, Y. (1996a) Development of flavor attributes in the fruit of *C. melo* during ripening and storage, In *Biotechnology for Improved Foods and Flavors* (eds G.R. Takeoka, R. Teranishi, P.J. Williams and A. Kobayashi), American Chemical Society, Washington, DC, pp. 228-239.

Wyllie, S.G., Leach, D.N., Nonhebel, H.N. and Lusunzi, I. (1996b) Biochemical pathways for the formation of esters in ripening fruit, in *Flavor Science: Recent Developments* (eds A.J. Taylor and D.S. Mottram), Royal Society of Chemistry, Cambridge, pp. 52-57.

Yabumoto, K., Yamaguchi, M. and Jennings, W.G. (1977) Biosynthesis of some volatile constituents of muskmelon, *Cucumis melo. Chemie Mikrobiologie Technologie der Lebensmittel*, **5**, 53-56.

Young, H., Gilbert, J.M., Murray, S.H. and Ball, R.D. (1996) Causal effects of aroma compounds on Royal Gala apple flavors. *Journal of the Science of Food and Agriculture*, **71**, 329-326.

Zitnak, A. and Filadelfi, M.A. (1985) Estimation of taste thresholds of three potato glycoalkaloids. *Canadian Institute of Food Science and Technology Journal*, **18**, 337-339.

5 Temperature management
Susan Lurie

5.1 Preharvest temperatures

5.1.1 Fruit growth and development

Temperatures during the growing season will have a direct effect on the final size and shape of the fruit. Many fruits, particularly those with seeds or pits, show a double sigmoidal growth curve (figure 5.1). Early development results from cell division; then there is a period of slow growth while the pit hardens, and growth resumes with cell expansion. Cool spring weather will extend the period of cell division, leading to greater fruit size at the end of the season when these cells expand. Temperate fruits grown in hot climates tend to be smaller than the same cultivars grown in a cooler climate for this reason. Apples are smaller and rounder when grown in subtropical areas but larger and more elongated when grown in temperate climates.

Temperatures close to harvest also influence fruit development. High temperatures will advance fruit ripening and lead to early harvest. Extended temperatures above 30°C can also delay fruit color development. The carotenoid lycopene, responsible for red color development in tomatoes, accumulates more slowly when temperatures are high. Anthocyanin accumulation is also decreased by high temperatures. The lack of red color development on apples grown in hot climates is due primarily to high night temperatures (Blankenship, 1987).

5.1.2 Susceptibility to storage disorders

There are few postharvest disorders of fruits that are completely independent of preharvest factors. Even incidence of disorders induced specifically by storage conditions, such as low temperature, can be modified by preharvest environmental conditions. Long-term exposure of fruit to high temperatures and direct sunlight can have effects on the morphology of the skin, pigmentation, carbohydrate metabolism and water relations of the fruit, with consequences in ripening and response to low temperature. Water core in apples is a disorder associated with dysfunction in carbohydrate metabolism. Low temperatures during fruit maturation can exacerbate water core (Yamada *et al.*, 1994). However, water core can also be caused by exposure of fruit to high temperatures on the tree before ripening (Ferguson *et al.*, 1999). The difference between these high and low temperature responses relates to long-term effects where low temperatures

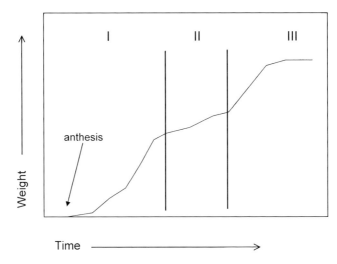

Figure 5.1 Stages of growth of a fruit. Stage I is cell division, stage II is seed hardening and stage III cell expansion.

are influential, and to short-term effects with high localized temperatures in the exposed fruit.

High temperatures will hasten ripening. However, within a tree there will be a gradation in fruit maturity due to variation in temperature, light intensity, nutrient transport and water relations. A study of shaded compared with sun-exposed apples showed that there were differences in both maturity at harvest and ripening behavior postharvest (Klein *et al.*, 2001). Exposed apples were firmer and had higher soluble solids than shaded apples, but also produced more ethylene and had higher starch levels than shaded apples at harvest. The firmness difference between exposed and shaded apples was maintained during ripening. Exposed avocados were also found to be firmer and to ripen more slowly than those in the interior of the tree (Woolf *et al.*, 2000).

Exposure to high temperatures on the tree, particularly close to harvest, can induce tolerance to low temperatures in postharvest storage. Many fruits, not only tropical and subtropical but also those growing under temperate conditions with relatively low air temperatures, can experience high flesh temperatures when exposed to sunlight. Avocados exposed to high field temperatures before harvest developed less chilling injury when held at low temperatures (Woolf *et al.*, 1999, 2000). In contrast, citrus fruits harvested from the exterior of the tree were more sensitive to chilling injury than interior harvested fruit (Nordby and McDonald, 1995; McDonald *et al.*, 1993).

Preharvest factors predispose apples to the storage disorder 'superficial scald' (Emonger *et al.*, 1994; Bramlage and Weis, 1997). Shaded parts of the fruit tend to develop scald more readily, suggesting a link with light intensity or temperature. Bramlage and Weis (1997) concluded, from analysis of results collected over several years from four countries, that low preharvest temperatures reduced susceptibility to scald.

Some fruits have been shown to withstand low temperatures postharvest better when conditioned by preharvest exposure to temperatures that are low but above those which induce injury. Examples include grapefruit and bell pepper (Harding *et al.*, 1957; Wang, 1990; Wheaton and Morris, 1967). These findings suggest some similarity between the effects of preharvest high and low temperature conditioning on tolerance to low storage temperatures.

5.2   Removing field heat

It is an axiom that the faster field heat can be removed and the fruit cooled to storage temperature, the longer will be its storage life. This is generally but not universally correct, as will be discussed below.

Once a fruit is harvested it is more sensitive to water loss and temperature changes than when attached to the tree. Respiration releases heat and the fruit temperature increases. In small fruits, such as cherries or berries, rapid cooling is essential to maintain quality. Cherry fruits harvested and held in the shade increase in temperature by $1°C$ an hour, and when held in sun the temperature increase is much faster (Kupferman, 1995). More quality is lost in one hour at $20°C$ than 24 h at $0°C$.

The major reasons for cooling fruits immediately after harvest include reducing the heat of respiration, minimizing water loss from the fruit and retarding development of decay caused by pathogens. Two main methods are used for rapid removal of field heat from fruits: hydrocooling and forced air cooling.

5.2.1   *Hydrocooling*

This method is used for many types of tree fruits as well as melons. Grapes and berries are generally not hydrocooled because the surface water after cooling encourages decay. Hydrocooling does not remove water from the produce and may revive slightly wilted commodities. Effective hydrocooling requires cooling water to move over and be in contact with all the surface of the produce, to remain cold and to be free of decay causing organisms.

There are two main types of hydrocoolers—a shower type and an immersion type. In shower type hydrocoolers, water is pumped above the produce and rains down past it. The fruit can either move slowly but continuously through the shower, or can be loaded as a batch. Immersion coolers are mainly used for bulk produce. They are best suited for fruits such as cherries that are denser than water and therefore stay completely submerged.

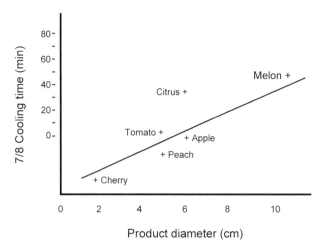

Figure 5.2 Effect of fruit diameter on product cooling time.

Hydrocooling time depends on produce minimum diameter and water flow rate. As shown in Figure 5.2, the cooling time increases with increasing minimum produce diameter. Thus, small diameter produce such as asparagus and radishes cool in less than 10 min, while a large cantaloupe can take 1 h to cool. Water temperature should be kept at 0–0.5°C for most produce. Even chilling sensitive produce cooled to temperatures above 5°C can be cooled with 0°C water if cooling time is limited. Chilling sensitive produce is not damaged if the flesh temperature does not drop below the damage threshold (Thompson, 1997).

5.2.2   Forced air cooling

This is a widely used method of cooling fruits, vegetables and cut flowers. Cooling is achieved by forcing cold air through containers and past individual fruits. Nearly all fresh market commodities can be cooled with forced air, but it is commonly used for grapes, berries and some tree fruits. It does not require a water resistant packaging container as does hydrocooling. Disadvantages of forced air are that it is usually slower than hydrocooling and it may cause excessive water loss in some fruits (Thompson *et al.*, 1998).

The most common design for forced air cooling is the tunnel cooler (figure 5.3). Pallet loads of fruit are placed in two lanes on either side of an open channel. A tarpaulin is placed over the containers, covering the open channel, and a fan removes air from the channel, forcing air through the packed fruit. The warmed air is directed to evaporator coils, recooled and returned to the room. Fruit in bins can also be cooled by this method if the bins are ventilated. This system cools large amounts of fruit in a single batch without specifically managing the temperature of individual boxes or pallet loads.

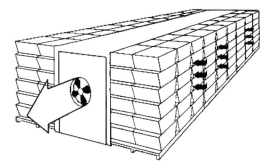

Figure 5.3 Diagrammatic view of a forced air cooling set-up. Palletized boxes are placed to form a tunnel from which air is exhausted. The negative pressure causes the cold air to pass through the air passages between the staked boxes.

The rate of cooling is related to the temperature difference between the fruit and the cold air. The curve of cooling time is rapid at the beginning of cooling and slows as the fruit nears the final temperature. Often fruits are cooled to a few degrees above their storage temperature to save on cooling time. Grapes are cooled to 4°C with forced air cooling and the pallets are then wrapped and moved into store rooms held at 0°C (Ben-Arie *et al.*, 1987). Total cooling time depends on airflow rate and product diameter. As in hydrocooling a larger fruit takes longer to cool than smaller fruit. Causing the air to move faster around the fruit will cool it faster.

If the relative humidity of the cooling air is above 80% it has a negligible effect on moisture loss. Most of the vapor pressure difference between the cooling air and the fruit is a result of the warm fruit, not the humidity of the cooling air. An advantage of high humidity during cooling is that it will add moisture to the boxes and therefore decrease the amount of moisture removed from the fruit during subsequent handling.

5.2.3 Delayed cooling

While most fruits benefit from rapid cooling, there are a few exceptions where the storage quality of the fruit will be enhanced by delayed cooling. Two examples are kiwifruit and nectarine. Both these fruits benefit from remaining at ambient temperature for a period of time after harvest, for very different reasons.

Kiwifruit develops decay in storage due to *Botrytis cinerea*, which penetrates into the fruit at the stem end where the fruit is picked from the vine. If the fruit is held for 24 to 72 h after harvest at ambient temperatures and then cooled to its storage temperature of 0°C, the cut end develops a physical barrier to the fungus, which reduces the level of storage decay (Pennycook and Manning, 1992). After the period of curing, suberin and phenol compounds were found in the cell walls of the cells of the stem scar (Ipollito *et al.*, 1997). The best temperature for this delayed storage is between 10 and 20°C (Bautista-Banos *et al.*, 1997; Retamales *et al.*, 1997).

Late ripening nectarines are sensitive to low temperature storage and develop mealiness or woolliness after a few weeks at 0°C. Fruits affected by this non-pathogenic disorder have a good appearance when removed from storage but do not ripen normally (see Chapter 3). The fruits soften but instead of becoming juicy have dry, woolly flesh. The disorder is caused by abnormal solubilization of cell wall pectins (Lurie *et al.*, 1994), resulting from an imbalance in the activities of the enzymes polygalacturonase and pectin esterase (Ben-Arie and Sonego, 1980). With relatively high pectin esterase and low polygalacturonase activities in chilling injured fruit, the pectin matrix is de-esterified, without subsequent depolymerization, leading to large pectin molecules with low esterification. This kind of pectin can form a gel which will bind free water and cause woolliness symptoms (Zhou *et al.*, 2000a).

The problem of mealiness or woolliness has been recognized for a long time in peaches and nectarines and one strategy to prevent it is delayed storage (Lill *et al.*, 1989; Zhou *et al.*, 2000b). The fruits are held at ambient temperature for 36 to 48 h after harvest before being cooled to 0°C. The fruits may soften a little more in storage than immediately cooled fruit, but they retain their ability to soften and become juicy following storage (Zhou *et al.*, 2000b). It has recently been shown that the time at ambient temperature allows the fruit to retain the ability to produce ethylene after storage and initiate normal ripening, while the immediately chilled fruits lose their ability to produce ethylene after a few weeks in storage (Zhou *et al.*, 2001a).

In the same way that low temperatures before harvest can condition a fruit to be less sensitive to chilling injury in storage, fruit can be treated after harvest with the same effect. There are a number of commodities for which step-down cooling will allow storage at a temperature that would normally be detrimental to fruit quality. Gradual reduction of temperature is a treatment that has been effective on subtropical fruits, including citrus fruits, avocados and tomatoes. The commercial storage practice of avocado storage in South Africa has been based on a step-down temperature procedure (Vorster *et al.*, 1990). Using step-down temperature storage for citrus allows them to withstand the disinfestations treatment against fruit fly which involves holding the fruit at 1°C, a temperature which normally causes chilling injury (Cohen *et al.*, 1983). Mature green tomatoes can also be stored at 2°C without chilling injury if the temperature is gradually reduced over three days (Lurie and Sabehat, 1997).

5.3   Storage temperatures

Temperature management is the most effective tool for extending the storage life of fresh horticultural commodities. The major reason that postharvest life of a fruit is extended by cooling is that metabolism is slowed by low temperatures. Respiration depletes the food reserves that were accumulated in the fruit while on the tree and reduces eating quality. Each 10°C decrease in temperature will reduce respiratory activity by a factor of 2 to 4 (Mitchell, 1992).

Another benefit of lowering temperature is that ethylene production is reduced. The ethylene synthesizing enzymes, 1-aminocyclopropane carboxylic acid (ACC) oxidase and ACC synthase, are sensitive to low temperatures and, as the temperature is lowered, less ethylene will be produced (Larrigaudiere et al., 1997). Fruits are also less sensitive to ethylene at low than at ambient temperatures (Zhou et al., 2001a). This is important for climacteric fruits which continue to ripen after harvest. During ripening, sugars increase and volatile constituents (flavors and odors) develop. Flesh softening accompanies ripening, and these changes should be minimized until the fruit is marketed.

Once harvested, fruit constantly lose water to the environment. Since this water cannot be replaced by the tree, weight loss occurs. Many commodities show visible signs of wilting or shriveling after losing 3 to 5% of their initial weight (Mitchell, 1992). The rate of water loss is controlled by the vapor-pressure difference between the fruit and the surrounding air, which is governed by temperature and relative humidity. Because warm air can hold much more water vapor than cold air, high humidity and low weight loss are easier to achieve at low temperatures than at high temperatures.

Temperature also affects the rate of growth and spread of pathogens and decay in the same way that it affects the fruit; the lower the temperature the slower the metabolism. Certain fungi that can cause severe losses do not grow at low temperatures. *Rhizopus stolonifera* ceases growth below 5°C and germinating spores can be killed at 0°C (Dennis and Cohen, 1976; Sommer, 1985). *Botrytis cinerea* can survive at 0°C but develops very slowly. At 2°C or below, germinating spores do not penetrate into fruit and mycelia do not spread from a decayed fruit to a healthy neighbor (Sommer, 1985). Thus good temperature management is important for reducing decay on harvested fruit.

The temperature in a storage room should be kept within 1°C of the desired temperature for the fruit being stored. For storage close to or below 0°C a narrower range may be needed. Temperatures below the optimal range for a fruit can cause freezing or chilling injury; temperatures above it shorten storage life. In addition, wide temperature fluctuations can result in water condensing on the stored product and more rapid water loss.

Temperate climate crops are best stored near or at 0°C (table 5.1). These include the pome fruits, stone fruits, persimmons and kiwifruit. Subtropical fruits are stored between 6 and 13°C, depending on the fruit and the specific cultivar. At these temperatures ripening processes and decay are slowed but not stopped and storage time is limited. Below a threshold temperature, which may differ from fruit to fruit but hovers around 10°C, chilling injury will develop. To prevent, or delay, the onset of chilling injury, and to extend the maximum

Table 5.1 Storage temperature and duration of fruits

Fruit	Storage temperature (°C)	Duration (weeks)
Apricot	0	2–3
Apple	0–3	16–20
Avocado	5–10	4–6
Banana	12–14	4–6
Cherry	0	2–4
Grape	0	4–12
Grapefruit	10–13	4–8
Guava	8–12	2
Kiwifruit	0	16–20
Lemon	12–14	4–16
Loquat	0–2	3
Lychee	0–2	4
Mandarin	5–8	2–6
Nectarine	0	2–4
Orange	4–6	4–8
Papaya	12–14	2
Peach	0	2–4
Pear	0	16–20
Persimmon	0–1	12–16
Plum	0	3–6
Pomegranate	6–10	4–12
Pummelo (pomelo)	8–10	8–12
Tangerine	5–8	2–6

storage period without compromising fruit quality, a number of temperature manipulations have been devised for different commodities.

5.3.1 Prestorage heat treatments

Delayed storage at ambient temperature, or at temperatures close to but above the storage temperature, has been discussed in section 5.2.3. However, the use of high temperatures (38°C or higher) to induce resistance to low temperatures has also been found to be effective (Lurie, 1998). In practice, a prestorage high temperature treatment is usually used for insect disinfestations or for fungal control, but it can also affect the response of the fruit to post-treatment storage temperatures. It can also affect the progression of fruit ripening and therefore maintenance of fruit quality.

There are three methods in use to heat commodities: hot water, vapor heat and hot air. Hot water was originally used for fungal control but has been extended to disinfestations of insects. Vapor heat was developed specifically for insect control and hot air has been used for both fungal and insect control. Humidity and air circulation rate can be varied to control heat transfer rate, the time the fruit will need to be in the treatment and the effects on fruit physiology.

A recent extension of hot water treatment has been the development of a hot water spray machine (Fallik *et al.*, 1999). Fruit on a sorting line is moved by brush rollers through a pressurized spray of hot water. The fruit is exposed to temperatures from 50 to 65°C for periods of 10 to 20 s. This machine is in use both to clean and reduce pathogen presence on a number of fruits and vegetables, including mangoes (Prusky *et al.*, 1999), peppers (Fallik *et al.*, 1999), melon (Fallik *et al.*, 2000) and citrus fruits (Porat *et al.*, 2000a). Even though the exposure to high temperatures is very short, this treatment induces responses in the fruit similar to longer exposures at lower temperatures (Porat *et al.*, 2000b).

Exposure to high temperatures is known to induce thermotolerance, i.e. ability to withstand an even higher temperature which is normally lethal (Vierling, 1991). However, in harvested commodities it also induced resistance to low temperature (Lurie and Klein, 1991) and can be used to reduce chilling injury in subtropical fruits. A high temperature prestorage treatment was found to inhibit chilling injury in avocado (Woolf *et al.*, 1995), citrus fruits (Wild, 1993; Rodov *et al.*, 1995; Schirra and Mulas, 1995), mango (McCollum *et al.*, 1995) and papaya (Paull, 1994). The heat treatment in these studies varied from a few minutes in hot water to hours in hot air. A study comparing various treatments in citrus found that chilling injury could be prevented by treatments ranging from 15 s at 59°C in a hot water brush machine to 3 days of curing at 36°C in moist hot air (Porat *et al.*, 2000b). Therefore, the method to be chosen depends on the other purposes (fungal or insect control) for which the treatment is given.

Prestorage heat treatments have also been found to benefit temperate fruits that are normally stored at low temperatures. Chilling injury in persimmon is reduced by both hot water and high temperature treatments (Burmeister *et al.*, 1997; Lay-Yee *et al.*, 1997; Woolf *et al.*, 1997). Apples can be protected from developing superficial scald by a prestorage hot air treatment (Lurie *et al.*, 1990). However, prestorage heat treatments do not diminish the development of mealiness or woolliness in peaches and nectarines (Oberland and Carroll, 2000).

The means by which a prestorage heat treatment confers protection are not fully known. It was found that the high temperature induces heat shock proteins that are involved in thermotolerance (Picton and Grierson, 1988; Lurie and Klein, 1990; Paull and Chen, 1990). These proteins remain present for extended times when the fruit is placed in cold storage and may play a role in protecting tissue from chilling injury (Sabehat *et al.*, 1996; 1998). However, other mechanisms are probably involved, including membrane adaptation (Lurie *et al.*, 1995; Whitaker *et al.*, 1997). Tomatoes treated prior to 2°C storage for 48 h at 38°C or immersed in 46 to 48°C water for 2 to 3 min showed similar changes in lipid composition, an indication that even short exposure to heat can instigate processes leading to tissue adaptation to low temperature (Lurie *et al.*, 1997).

There is, of course, the danger that a high temperature prestorage treatment will cause heat damage to the fruit. Damage can be both external and internal. External damage is generally peel browning, pitting or yellowing (Lurie, 1998 and references therein). Tissue damage will also lead to increased decay development (Jacobi and Wong, 1992; Jacobi et al., 1993; Lay-Yee and Rose, 1994; Shellie and Mangan, 1998). Symptoms of internal damage on mango and papaya include poor color development, abnormal softening, lack of starch breakdown and development of internal cavities (An and Paull, 1990; Jacobi and Wong, 1992; Mitcham and McDonald, 1993; Paull, 1995). Symptoms on other fruits include flesh darkening on lychee and nectarines (Jacobi et al., 1993; Lay-Yee and Rose, 1994). If the fruit is stored at low temperature after the heat treatment some symptoms can be confused with chilling injury.

All these responses may be more or less severe depending on the fruit and cultivar being treated as well as the preharvest heat exposure history. In order to minimize damage many disinfestation treatments utilize a step-up treatment of more than one temperature, or gradual heating to the final disinfestation temperature. These methods induce thermotolerance and minimize fruit damage. Papaya can be pretreated for 1 h at 42°C before being heated to 47°C, or can be exposed to a slow heating rate of 8 h to reach the final temperature (Paull and Chen, 2000). These methods reduce the heat injury shown as failure to ripen.

Prestorage heat treatments can increase storage life and improve the flavor of the fruit (Liu, 1978; Lurie, 1998; Shellie and Mangan, 1994, 1998). Ethylene production ceases at high temperatures (Biggs et al., 1988; Klein, 1989) and fruit will not respond to exogenous ethylene (Seymour et al., 1987; Yang et al., 1990). Ripening processes initiated by ethylene are therefore inhibited by high temperatures. Fruits often soften more slowly after heat treatment, although mangoes and papaya may show faster softening (Ketsa et al., 2000). Flavor characteristics can also be affected by a prestorage heat treatment. Titratable acidity, soluble solids and volatiles may be affected, depending on treatment and type of fruit (Lurie, 1998; Paull and Chen, 2000).

5.3.2   Intermittent warming, step-down cooling and dual temperature regimes

In an effort to delay or prevent the chilling injury, researchers have tested the effects of interrupting cold storage, with short intervals of warming. Ben-Arie et al. (1970) reported substantial control of woolliness in peaches if the fruit was warmed for two days every two weeks during a six-week storage at 0°C. Lill (1985) showed that intermittent warming was effective at temperatures as low as 12°C but at this temperature the length of the warming period had to be longer than if warming was done at 20°C. Either intermittent warming or delayed storage can be more effective in reducing woolliness in particular cultivars of stone fruits. Scott et al. (1969) found intermittent warming to be more effective than delayed storage, while von Mollendorff and de Villiers (1988) had success with delayed storage.

These procedures appeared to be effective because they resulted in more normal ethylene production by the fruit after storage (Zhou et al., 2001a,b).

Chilling injured fruit had reduced ethylene production and altered cell wall hydrolytic enzyme activities. Thus softening in woolly fruit seemed to depend on endoglucanase, while in healthy fruit polygalacturonase was the major enzyme (Zhou et al., 2000b). In the absence of polygalacturonase activity the pectin molecules bind the extracellular juice and produce woolly fruit.

A number of subtropical fruits can be held at low temperatures if they are stored first at a permissible temperature and then the temperature is gradually lowered. Lemons can be stored at 6 to 8°C if the temperature is gradually lowered over a week to that temperature (Cohen et al., 1983). Alternatively, intermittent warming for five days at 15°C before lowering the temperature again will also prevent chilling injury and extend storage (Cohen et al., 1983). Avocados can also be stored below 10°C when they are given a step-down temperature procedure (Vorster et al., 1990). These regimes allow the fruit to acclimate and resist the low temperature stress.

Another method, in use for plums, is a dual-temperature regime. The fruit is rapidly cooled and stored at 0 to −1°C for 10 to 14 days; the temperature is then raised to 7°C for another 10 to 14 days until marketing (Taylor et al., 1994). This prevents the storage disorders that would develop at 0°C, and 7°C is low enough so that fruit do not overripen during the time spent at that temperature. The opposite technique can also be employed to produce the proper color on yellow plums. Depending on the stage of color development at harvest, Songold plums are held 8 to 14 days at 7°C and then transferred to 0°C for another two weeks (Nerya et al., 2000). This allows the yellow color to develop but maintains fruit firmness adequate for marketing.

5.4   Conclusions

Effective temperature management requires an understanding of the fruit metabolism and physiology as well as the length of time needed for storage and marketing. If a fruit is to be marketed quickly it may be enough to cool it rapidly and maintain the cold chain during marketing. If the fruit needs to be held for an extended time before marketing, specific temperature manipulations then become more important in order to minimize damage and maintain quality. There is still a great deal to be learned about fruit physiology and greater knowledge will lead to more accurate techniques to maintain postharvest quality.

References

An, J.F. and Paull, R.E. (1990) Storage temperature and ethylene influence on ripening of papaya fruit. *Journal of the American Society of Horticultural Science*, **115**, 949-953.

Bautista-Banos, S., Long, P.G. and Ganesh, S. (1997) Curing of kiwifruit for control of postharvest infection by *Botrytis cinerea. Postharvest Biology and Technology*, **12**, 137-145.

Ben-Arie, R. and Sonego, L. (1980) Pectolytic enzyme activity involved in woolly breakdown of stored peaches. *Phytochemistry*, **10**, 531-538.

Ben-Arie, R., Lavee, S. and Guelfat-Reich, S. (1970) Control of woolly breakdown of Elberta peaches in cold storage by intermittent exposure to room temperature. *Journal of the American Society of Horticultural Science*, **95**, 801-802.

Ben-Arie, R., Zuthi, Y. and Zeidman, M. (1987) Export of Thompson grapes after extended storage. *HaSadeh*, **67**, 1150-1157.

Biggs, M.S., Woodson, W.R. and Handa, A.K. (1988) Biochemical basis of high temperature inhibition of ethylene biosynthesis in ripening tomato fruits. *Physiologia Plantarum*, **72**, 572-578.

Blankenship, S.M. (1987) Night-temperature effects on rate of apple fruit maturation and fruit quality. *Scientia Horticulturae*, **33**, 205-212.

Bramlage, W.J. and Weis, S.A. (1997) Effects of temperature, light, and rainfall on superficial scald susceptibility in apples. *HortScience*, **32**, 808-811.

Burmeister, D., Ball, S., Green, S. and Woolf, A.B. (1997) Interaction of hot water treatments and controlled atmosphere storage on quality of 'Fuyu' persimmons. *Postharvest Biology and Technology*, **12**, 71-82.

Cohen, E., Shuali, M. and Shalom, Y. (1983) Effect of temperature manipulation on the reduction of chilling injury of Villa Franka lemon fruits stored at cold temperature. *Journal of Horticultural Science*, **58**, 593-598.

Dennis, C. and Cohen, E. (1976) The effect of temperature on strains of soft fruit spoilage fungi. *Annals of Applied Biology*, **79**, 141-147.

Emonger, V.E., Murr, D.P. and Lougheed, E.C. (1994) Preharvest factors that predispose apples to superficial scald. *Postharvest Biology and Technology*, **4**, 289-300.

Fallik, E., Grinberg, S., Alkalai, S., Yekutieli, O., Wiseblum, A., Regev, R., Beres, H. and Bar-Lev, E. (1999) A unique rapid hot water treatment to improve storage quality of sweet pepper. *Postharvest Biology and Technology*, **15**, 25-32.

Fallik, E., Aharoni, Y., Copel, A., Rodov, V., Tuvai-Alkalai, S., Horev, B., Yekutielli, O., Wiseblum, A. and Regev, R. (2000) Reduction of postharvest losses of Galia melon by a short hot-water rinse. *Plant Pathology*, **49**, 333-338.

Ferguson, I., Volz, R. and Woolf, A. (1999) Preharvest factors affecting physiological disorders of fruit. *Postharvest Biology and Technology*, **15**, 255-262.

Harding, P.L., Soule, M.J. and Sunday, M.B. (1957) *Storage Studies on Marsh Grapefruit*. US Department of Agriculture Research Service Marketing Research Report, AMS-202.

Ippolito, A., Nigro, F., Lima, G., Castellano, M.A., Salerno, M. Lattanzio, V. and Di Venera, D. (1997) Mechanisms of resistance to *Botrytis cinerea* in wounds of cured kiwifruits. *Acta Horticulturae*, **444**, 719-724.

Jacobi, K.K. and Wong, L.S. (1992) Quality of 'Kensington' mango (*Mangifera indica* L.) following hot water and vapour-heat treatments. *Postharvest Biology and Technology*, **1**, 349-359.

Jacobi, K.K., Wong, L.S. and Giles, J.E. (1993) Lychee (*Lichi chinesis* Sonn.) fruit quality following vapour heat treatment and cool storage. *Postharvest Biology and Technology*, **3**, 111-119.

Ketsa, S., Chidragoo., S. and Lurie, S. (2000) Effect of prestorage heat treatment on poststorage quality of mango fruits. *HortScience*, **35**, 247-249.

Klein, J.D. (1989) Ethylene biosynthesis in heat treated apples, in *Biochemical and Physiological Aspects of Ethylene Production in Lower and Higher Plants* (eds H. Clijster, M. de Proft, R. Marcelle and M. van Pouche), Kluwer, Dordrecht, pp. 184-190.

Klein, J.D., Dong, L., Zhou, H.W. and Lurie, S. (2001) Ripeness of shade and sun grown apples (*Malus domestica*). *Acta Horticulturae*, **401**, 135-138.

Kupferman, E.M. (1995) Cherry temperature management. *Tree Fruit Postharvest Journal*, **6**, 3-6.

Larrigaudiere, C., Graell, J., Salas, J. and Vendrell, M. (1997) Cultivar differences in the influence of a short period of cold storage on ethylene biosynthesis in apples. *Postharvest Biology and Technology*, **10**, 21-27.

Lay-Yee, M. and Rose, K.J. (1994) Quality of 'Fantasia' nectarines following forced air heat treatments for insect disinfestations. *HortScience*, **29**, 663-666.

Lay-Yee, M., Ball, S., Forbes, S.K. and Woolf, A.B. (1997) Hot water treatment for insect disinfestations and reduction of chilling sensitivity of 'Fuyu' persimmon. *Postharvest Biology and Technology*, **10**, 81-87.

Lill, R.E. (1985) Alleviation of internal breakdown of nectarines during cold storage by intermittent warming. *Scientia Horticulturae*, **25**, 241-246.

Lill, R.E., O'Donoghue, E.M. and King, G.A. (1989) Postharvest physiology of peaches and nectarines. *Horticultural Review*, **11**, 413-452.

Liu, F.W. (1978) Modification of apple quality by high temperature. *Journal of the American Society of Horticultural Science*, **103**, 730-732.

Lurie, S. (1998) Postharvest heat treatments. *Postharvest Biology and Technology*, **14**, 257-269.

Lurie, S. and Klein, J.D. (1990) Heat treatment of ripening apples: differential effects on physiology and biochemistry. *Physiologia Plantarum*, **78**, 181-186.

Lurie, S. and Klein, J.D. (1991) Acquisition of low temperature tolerance in tomatoes by exposure to high temperature stress. *Journal of the American Society of Horticultural Science*, **116**, 1007-1012.

Lurie, S. and Sabehat, A. (1997) Prestorage temperature manipulations to reduce chilling injury in tomatoes. *Postharvest Biology and Technology*, **11**, 57-62.

Lurie, S., Klein, J.D. and Ben-Arie, R. (1990) Postharvest heat treatments as a possible means of reducing superficial scald of apples. *Journal of Horticultural Science*, **65**, 503-509.

Lurie, S., Levin, A., Greve, L.C. and Labavitch, J.M. (1994) Pectic polymer changes in nectarines during normal and abnormal ripening. *Phytochemistry*, **36**, 11-17.

Lurie, S., Othman, S. and Borochov, A. (1995) Effects of heat treatment on plasma membrane of apple fruit. *Postharvest Biology and Technology*, **5**, 29-38.

Lurie, S., Laamim, M., Lapsker, Z. and Fallik, E. (1997) Heat treatments to decrease chilling injury in tomato fruit. Effects on lipids, pericarp lesions and fungal growth. *Physiologia Plantarum*, **100**, 297-302.

McCollum, T.G., Doostdar, H., Mayer, R.T. and McDonald, R.E. (1995) Immersion of cucumber fruit in heated water alters chilling-induced physiological changes. *Postharvest Biology and Technology*, **6**, 55-64.

McDonald, R.E., Nordby, H.E. and McCollum, T.G. (1993) Epicuticular wax morphology and composition are related to grapefruit chilling injury. *HortScience*, **28**, 311-312.

Mitcham, E.J. and McDonald, R.E. (1993) Respiration rate, internal atmosphere, and ethanol and acetaldehyde accumulation in heat treated mango fruit. *Postharvest Biology and Technology*, **3**, 77-86.

Mitchell, F.G. (1992) Cooling horticultural commodities, in *Postharvest Technology of Horticultural Crops* (ed. A.A. Kader), University of California Press, CA, pp. 53-78.

Nerya, O., Gizis, A., Zvilling, A. *et al.* (2000) Storage regimes to improve the color and quality of 'Songold' plums for export. *Alon HaNotea*, **54**, 404-407.

Nordby, H.E. and McDonald, R.E. (1995) Variations in chilling injury and epicuticular wax composition of white grapefruit with canopy position and fruit development during the season. *Journal of Agricultural and Food Chemistry*, **43**, 1823-1833.

Oberland, D.M. and Carroll, T.R. (2000) Mealiness and pectolytic activity in peaches and nectarines in response to heat treatment and cold storage. *Journal of the American Society of Horticultural Science*, **125**, 723-738.

Paull, R.E. (1994) Response of tropical horticultural commodities to insect disinfestations treatments. *HortScience*, **29**, 988-996.

Paull, R.E. (1995) Preharvest factors and the heat sensitivity of field grown ripening papaya fruit. *Postharvest Biology and Technology*, **6**, 167-175.

Paull, R.E. and Chen, N.J. (1990) Heat shock response in field grown ripening papaya fruit. *Journal of the American Society of Horticultural Science*, **115**, 623-631.

Paull, R.E. and Chen, N.J. (2000) Heat treatment and fruit ripening. *Postharvest Biology and Technology*, **21**, 21-37.

Pennycook, S.R. and Manning, M.A. (1992) Picking wound curing to reduce botrytis rot of kiwifruit. *New Zealand Journal of Crop and Horticultural Science*, **20**, 357-360.

Picton, S. and Grierson, D. (1988) Inhibition of expression of tomato ripening genes at high temperature. *Plant Cell and Environment*, **11**, 265-272.

Porat, R., Daus, A., Weiss, B., Cohen, L, Fallik, E. and Droby, S. (2000a) Reduction of postharvest decay in organic citrus fruit by a short water brushing treatment. *Postharvest Biology and Technology*, **18**, 151-157.

Porat, R., Pavoncello, D., Peretz, J., Ben-Yehoshua, S. and Lurie, S. (2000b) Effects of various heat treatments on the induction of cold tolerance and on the postharvest qualities of 'Star Ruby' grapefruit. *Postharvest Biology and Technology*, **18**, 159-166.

Prusky, D., Fuchs, Y., Kobiler, I. *et al.* (1999) The effect of hot water brushing, prochloraz treatment and waxing on the incidence of black spot decay caused by *Alternaria alternata. Postharvest Biology and Technology*, **15**, 165-174.

Retamales, J., Cooper, T. and Montealegre, J. (1997) Effects of curing and cooling regime on ethylene production and storage behavior of kiwifruit. *Acta Horticulturae*, **444**, 567-571.

Rodov, V., Ben-Yehoshua, S., Albagli, R. and Fang, D.Q. (1995) Reducing chilling injury and decay of stored citrus fruit by hot water dips. *Postharvest Biology and Technology*, **5**, 119-127.

Sabehat, A., Weiss, D. and Lurie, S. (1996) The correlation between heat-shock protein accumulation and persistence and chilling tolerance in tomato fruit. *Plant Physiology*, **110**, 531-537.

Sabehat, A., Lurie, S. and Weiss, D. (1998) Expression of small heat-shock proteins at low temperatures: a possible role in protecting against chilling injuries. *Plant Physiology*, **117**, 651-656.

Schirra, M. and Mulas, M. (1995) Improving storability of 'Tarocco' oranges by postharvest hot-dip fungicide treatments. *Postharvest Biology and Technology*, **6**, 129-138.

Scott, K.J., Wills, R.B.H. and Roberts, E.A. (1969) Low temperature injury of peaches in relation to weight loss during storage. *Australian Journal of Experimental Agriculture*, **9**, 364-366.

Seymour, G.B., Joh, P. and Thompson, A.K. (1987) Inhibition of degreening in the peel of bananas ripened at tropical temperatures. II. Role of ethylene, oxygen and carbon dioxide. *Annals of Applied Biology*, **110**, 153-161.

Shellie, K.C. and Mangan, R.L. (1994) Postharvest quality of 'Valencia' orange after exposure to hot, moist, forced air for fruit fly disinfestations. *HortScience*, **29**, 1524-1527.

Shellie, K.C. and Mangan, R.L. (1998) Navel orange tolerance to heat treatments for disinfesting Mexican fruit fly. *Journal of the American Society of Horticultural Sciences*, **123**, 288-293.

Sommer, N.F. (1985) Role of controlled environments in suppression of postharvest diseases. *Canadian Journal of Plant Pathology*, **7**, 331-339.

Taylor, M.A., Rabe, E., Dodd, M.C. and Jacobs, G. (1994) Effect of storage regimes on pectolytic enzymes, pectic substrates, internal conductivity and gel breakdown in cold stored Songold plums. *Journal of Horticultural Science*, **69**, 527-534.

Thompson, J.F. (1997) Hydrocooling fresh market commodities. *Tree Fruit Postharvest Journal*, **8**, 3-11.

Thompson, J.F., Rumsey, T.R. and Mitchell, G.F. (1998) Forced-air cooling. *Tree Fruit Postharvest Journal*, **9**, 3-15.

Vierling, E. (1991) The roles of heat shock proteins in plants. *Annual Review of Plant Physiology and Plant Molecular Biology*, **42**, 579-620.

von Mollendorff, L.J. and de Villiers, O.T. (1988) Physiological changes associated with the development of woolliness in 'Peregrine' peaches during low temperature storage. *Journal of Horticultural Science*, **63**, 47-51.

Vorster, L., Toerien, J. and Bezuidenhout, J. (1990) Temperature management of avocados—an integrated approach. *South African Avocado Growers Association Yearbook*, **13**, 43-46.

Wang, C.Y. (1990) Alleviation of chilling injury of horticultural crops, in *Chilling Injury of Horticultural Crops* (ed. C.Y. Wang), CRC Press, Boca Raton, FL, pp. 281-301.

Wheaton, T.A. and Morris, L.L. (1967) Modification of chilling sensitivity by temperature conditioning. *Proceedings of the American Society of Horticultural Science*, **91**, 529-533.

Whittaker, B.D., Klein, J.D., Conway, W.S. and Sams, C.E. (1997) Influence of prestorage heat and calcium treatments on lipid metabolism in 'Golden Delicious' apples. *Phytochemistry*, **45**, 465-472.

Wild, B.L. (1993) Reduction of chilling injury in grapefruit and oranges stored at 1°C by prestorage hot dip treatments, curing, and wax application. *Australian Journal of Experimental Agriculture*, **33**, 495-498.

Woolf, A.B., Watkins, C.B., Bowen, J.H., Lay-Yee, M., Maindonald, J.H. and Ferguson, I.B. (1995) Reducing external chilling injury in stored 'Hass' avocados with dry heat treatments. *Journal of the American Society of Horticultural Science*, **120**, 1050-1056.

Woolf, A.B., Ball, S., Spooner, K.J., Lay-Yee, M., Ferguson, I.B., Watkins, C.B., Gunson, A. and Forbes, S.K. (1997) Reduction of chilling injury in the sweet persimmon 'Fuyu' during storage by dry air heat treatments. *Postharvest Biology and Technology*, **11**, 155-164.

Woolf, A.B., Bowen, J.H. and Ferguson, I.B. (1999) Preharvest exposure to the sun influences postharvest responses of 'Hass' avocado fruit. *Postharvest Biology and Technology*, **15**, 143-153.

Woolf, A., Weksler, A., Prusky, D., Kobiler, E. and Lurie, S. (2000) Direct sunlight influences postharvest temperature responses and ripening of five avocado cultivars. *Journal of the American Society of Horticultural Science*, **125**, 370-376.

Yamada, H., Ohmura, H., Ara, C. and Terui, M. (1994) Effect of preharvest fruit temperature on ripening, sugars, and occurrence of watercore in apples. *Journal of the American Society of Horticultural Science*, **119**, 1208-1214.

Yang, R.F., Cheng, T.S. and Shewfelt, R.L. (1990) The effect of high temperature and ethylene treatment on the ripening of tomatoes. *Journal of Plant Physiology*, **136**, 368-372.

Zhou, H.W., Ben-Arie, R. and Lurie, S. (2000a) Pectin esterase, polygalacturonase and gel formation in peach pectin fractions. *Phytochemistry*, **55**, 191-195.

Zhou, H.W., Lurie, S., Lers, A., Khatchitski, A., Sonego, L. and Ben-Arie, R. (2000b) Delayed storage and controlled atmosphere storage of nectarines: two strategies to prevent woolliness. *Postharvest Biology and Technology*, **18**, 133-141.

Zhou, H.W., Dong, L., Ben-Arie, R. and Lurie, S. (2001a) The role of ethylene in the prevention of chilling injury in nectarines. *Journal of Plant Physiology*, **158**, 55-61.

Zhou, H.W., Lurie, S., Ben-Arie, R. *et al.* (2001b) Intermittent warming of peaches reduces chilling injury by enhancing ethylene production and enzymes mediated by ethylene. *Journal of Hortical Science and Biotechnology*, **76**, 321-328.

6 Atmosphere control using oxygen and carbon dioxide

Nazir Mir and Randolph Beaudry

6.1 Introduction

The maintenance of physical and chemical attributes that confer quality to harvested fruits depends, in part, upon our ability to impose conditions that minimize change in these attributes. Atmospheric modification is one tool at our disposal to achieve this goal. In general, the capacity for tissues to respond positively to modified atmospheres reflects the capacity of the atmosphere to alter metabolic activity favorably without inducing negative quality attributes such as off-flavors or tissue discoloration. Modified atmospheres improve storability and shelf-life of some fruits, but most experience little or no benefit and, in response to extreme atmosphere modification, fermentation and quality loss result.

Responses to atmosphere modification integrate a variety of processes affected by low O_2 and elevated CO_2. It is widely recognized that the partial pressure of O_2 can be reduced and that of CO_2 increased to reduce the rate of respiratory metabolism. Oxygen and CO_2 are also active participants in many chemical reactions and can also have a marked impact on many of the biochemical processes in plant cells, including the metabolism of pigments, phenolics, cell wall constituents, volatiles, starch, sugars, organic acids and phytonutrients. Importantly, low O_2 and elevated CO_2 impact ethylene production and perception. For those fruit for which ethylene forms an integral part of the ripening process, the impact of modified atmospheres on suppressing ethylene responses probably outweighs the effect of reducing metabolic activity via respiratory suppression.

In addition to the effects of modified atmospheres on the fruit tissue, its effects on decay organisms and insect pests are also important in quality preservation. Elevated CO_2 in the range of 8 to 20% is effective at suppressing many fungal decay organisms (Brown, 1922; Ceponis and Cappellini, 1985). Low O_2, however, has little impact on the growth of fungi (Brown, 1922). Of growing importance is the possibility of using extremes of O_2 and CO_2 for short durations for disinfestation of perishables (Mitcham *et al.*, 1997). While it is recognized that the response of fungal, bacterial and insect pests to modified atmospheres comprises a significant component of some quality preservation systems, it is the response of the plant material to modified atmospheres that ultimately determines the potential for the success of a particular atmosphere.

Early applications of modified atmospheres to preserve the quality of crops predates knowledge of the atmosphere's constituents and is presumed to date back several thousand years in the Middle East (Kays, 1997). It was not until the early 1800s, however, that Jacques Berard (1821) found that fruits utilize O_2 and produce CO_2 and that excluding O_2 from the storage environment prevented ripening. From the early work of Kidd and West (1927) on apple, the effectiveness of modified atmospheres was quickly recognized to be of extremely high value. Although a storage facility using the basic principles of controlled atmosphere (CA) storage was built in the 1860s, the first successful commercial CA storage in England was built only two years after the Kidd and West (1927) publication (Kays, 1997). Modified atmospheres are currently used for the preservation of pome fruit, kiwifruit, banana fruit and berry crops.

6.2 Respiratory metabolism

Of the reactions influenced by modified atmospheres, the reactions of the respiratory chain have the most global influence on cellular metabolism. Respiratory metabolism recaptures stored energy released from carbohydrates and other energetically rich organic compounds and generates carbon skeletons for reactions needed for the maintenance and development of the harvested fruit. The rates of these processes can be reduced by restricting the availability of O_2, which is a substrate in the terminal step of the respiratory pathway, and, to a more limited extent, by elevating the concentration of the product of respiration, CO_2. Restricting the rate of respiration results in shifts in primary metabolism that have the potential to improve or impair quality.

6.2.1 Biochemical responses of the respiratory chain to O_2 and CO_2

The amount of energy liberated during respiration is, by convention, often expressed as function of the complete oxidation of a mole (6.02×10^{23}) of hexose molecules (Brownleader, 1997; Plaxton, 1996; Taiz and Zeiger, 1998). One mole of hexose molecules has a weight of 180 g and releases 2880 kJ (686 kcal) of energy during respiration. Each molecule of glucose is oxidized to six molecules of CO_2, utilizing 32 molecules of adenosine diphosphate (ADP) and inorganic phosphate (P_i) and six molecules of O_2, which are reduced to form 12 molecules of water (equation 6.1). Under normal physiological conditions, approximately 50 to 60% of this energy is captured chemically in the form of approximately 32 adenosine triphosphate (ATP) molecules, which form a useful 'currency' in metabolic processes for powering cellular reactions.

$$C_6H_{12}O_6 + 6O_2 + 32ADP + 32P_i \rightarrow 6CO_2 + 12H_2O + 32ATP + heat$$

$$(6.1)$$

Central to the concept of limiting respiratory metabolism by low O_2 partial pressures is the reduction in chemical energy production. In concept, this reduction in energy availability should reduce the global rate of metabolism and thereby minimize some of the undesirable chemical reactions occurring during postharvest storage. When respiration is excessively limited by reduced O_2 availability, there can be a shortage of energy for maintenance reactions required for normal cellular activity, leading to undesirable reactions. The degree to which respiration can be suppressed by low O_2, therefore, is limited in most tissues and is a function of the degree to which the tissue can adapt to the stress without inducing negative quality attributes.

Three processes collectively comprise the respiratory pathway (Taiz and Zeiger, 1998). These include glycolysis, the tricarboxylic acid (TCA) cycle (which involves CO_2 release) and the electron transport pathway (which involves O_2 uptake). Each of these processes has been relatively well characterized and their potential for altering quality attributes related to sugar metabolism can be largely anticipated. To understand how these processes impact quality, an examination of each at the biochemical level is required.

6.2.1.1   Glycolysis

Glycolysis is the series of enzymatic reactions that oxidize 6-carbon sugar molecules such as glucose and fructose to the three-carbon molecule pyruvate (figure 6.1). In addition, glycolytic reactions include the pathways of fermentation in which pyruvate is converted to lactate and/or ethanol. The enzymes of glycolysis are located in the cytosol of the plant cell, but apart from the enzymes of the fermentation pathways, most, if not all, of these enzymes are also present in plastids (Givan, 1999). The physiological implications of this parallel plastidic pathway are not yet fully appreciated.

A single molecule of glucose can be converted to two molecules of pyruvate, with the potential production of two ATP molecules and two molecules of reduced nicotinamide adenine dinucleotide (NADH), which allows the cell to store reducing power. These numbers can be a little misleading. The breakdown products of glucose and other sugars do not always proceed cleanly through the glycolytic pathway; many of the intermediates in the pathway are utilized for reactions catalyzing the synthesis of numerous other cellular constituents such as cellulose, nucleotides and nucleic acids, alkaloids, flavonoids and lignins (Taiz and Zeiger, 1998). Thus, the amount of energy realized by the plant cell may not be stoichiometrically related to the number of glucose equivalents consumed. Oxygen is not required for glycolysis, but under conditions of sufficient O_2, pyruvate is the primary end product. When O_2 levels become limiting to respiration and prevent the further oxidation of pyruvate, fermentation results. If all the carbon flux occurs through the fermentation pathway, two NADH molecules are lost (reoxidized) and the total ATP yield is limited to a maximum of two per glucose, rather than the 32 noted in equation 6.1.

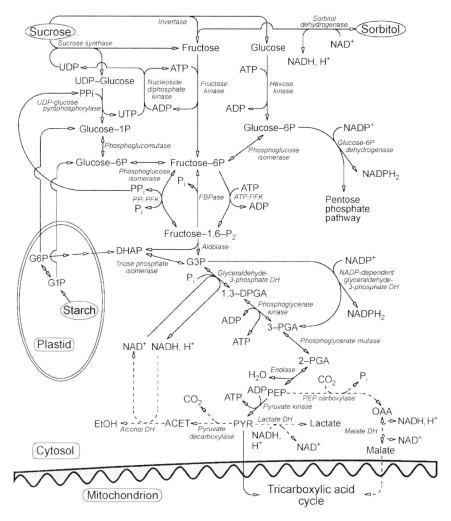

Figure 6.1 The reactions of plant glycolysis. A complete glycolytic pathway probably exists in the plastid. A bypass of pyruvate kinase by the vacuolar enzyme phosphoenolpyruvate phosphatase is not shown. Sucrose and sorbitol can be derived from intracellular (vacuole) and extracellular (apoplast) sources. Arrows with single heads represent essentially irreversible reactions; those with two heads represent freely reversible reactions. Dashed lines indicate pathways probably used under conditions of plant stress. On a per hexose basis, there is the potential to produce two molecules of glyceraldehyde-3-phosphate and, hence, pyruvate. Abbreviations: UDP, uridine diphosphate; UTP, uridine triphosphate; ADP, adenine diphosphate; ATP, adenine triphosphate; PP_i, pyrophosphate; P_i, phosphate; NAD^+, oxidized nicotinamide adenine dinucleotide; NADH, reduced nicotinamide adenine dinucleotide; $NADP^+$, oxidized nicotinamide adenine dinucleotide phosphate; $NADPH_2$, reduced nicotinamide adenine dinucleotide phosphate; G1P and glucose-1P, glucose-1-phosphate; G6P and glucose-6P, glucose-6-phosphate; fructose-6P, fructose-6-phosphate; fructose-1,6-P_2, fructose-1,6-bisphosphate; G3P, glyceraldehyde-3-phosphate; 1,3-DPGA, 1,3-diphosphoglycerate; 3-PGA, 3-phosphoglycerate; 2-PGA, 2-phosphoglycerate; PEP, phosphoenolpyruvate; PYR, pyruvate; ACET, acetaldehyde; EtOH, ethanol; and OAA, oxaloacetate.

The source of carbon in the glycolytic pathway is typically sucrose, a disaccharide composed of a glucose and fructose molecule. Sucrose is a non-reducing sugar and therefore a less reactive molecule than the reducing sugars fructose and glucose. Very often it is sucrose that is imported into cells, sequestered in the vacuole and subsequently used to support respiration; however, while not commonly recognized, other sugars are also translocated. Some horticulturally important plant species translocate significant amounts of sugar alcohols such as mannitol and sorbitol, both non-reducing sugars. Sorbitol is produced in fruit tree species such as cherry, plum, apricot and apple. Sorbitol accumulation in the fruit can be significant, especially in cherry, and it enters the glycolytic pathway as fructose with the concomitant production of NADH following the action of the enzyme sorbitol dehydrogenase (Loescher and Everard, 1996).

Sucrose enters the glycolytic pathway via two routes: 1. hydrolysis in an irreversible reaction to glucose and fructose by invertase; and 2. combining with uridine diphosphate (UDP) via the action of sucrose synthase (SS) in a reversible reaction to form UDP-glucose and fructose. Invertase exists in two classes based on the pH optima of the enzyme. Acid invertase has a pH optimum of approximately 5.0 and resides in the vacuole and the cell wall, both acid environments. Neutral invertase has a pH optimum of approximately 7.0 and is found in the cytoplasm. Invertases are sensitive to inhibition by high glucose and fructose levels. Sucrose synthase vies with invertase for the degradation of sucrose and is considered to play a major role in the entry of carbon into anabolic reactions in active sink tissues such as growing tips and storage organs (Winter and Huber, 2000). Sucrose synthase activity is enhanced at the gene level by hypoxic and anaerobic conditions in some tissues (McCarty et al., 1986; Zeng et al., 1998), suggesting the possibility of atmospheric effects on sucrose degradation. Sucrose synthase is also stimulated by high concentrations of sucrose (Koch et al., 1992). At this point in time, there are no known activators of the enzyme, but protein phosphorylation is likely. In addition to the effect of sucrose on sucrose synthase, sucrose and other sugars affect the expression of a number of genes (Avigad and Dey, 1997; Smeekens, 2000). In potato, sucrose promotes the expression of a gene responsible for production of the storage protein patatin (Wenzler et al., 1989); in sweet potato, sucrose promotes expression of the gene for β-amylase (Hattori and Nakamura, 1988), a protein that affects starch degradation; in petunia, it promotes the chalcone synthase gene, which is involved in anthocyanin biosynthesis (Tsukaya et al., 1991). Thus, factors that affect sucrose metabolism, such as low O_2, have the potential to influence a number of quality-related reactions.

The degradation of sucrose or other respiratory substrate as the first step in glycolysis has important implications in terms of quality maintenance of harvested products. Sucrose is a sweet molecule and its flavor contributes significantly to the perception of sweetness in many tissues. Nevertheless, its sweetness is only roughly half the combined sweetness of its constituents, glucose and fructose. A molar sucrose solution has a sweetness rating of 1.0 whereas of glucose has a rating of 0.6 and fructose, 1.6. Breakdown of sucrose to individual sugars can, therefore, potentially enhance sweetness, especially if the degradation is by invertase. Hypoxic conditions are likely to promote the depletion of sucrose in some plant tissues through the activation of sucrose synthase. Most fruits have large carbohydrate reserves, so sucrose depletion may not impose a serious physiological limitation and a shift in sweetness of the tissue may not be detectable. For tissues that contain little reserves, however, sucrose loss may have serious negative implications, altering the expression of genes that may be related to quality (King $et\ al.$, 1995). In asparagus, extremely low O_2 concentrations were associated with a marked loss in the concentration of sucrose and glucose, an increase in fructose concentration and a decline in spear quality (Silva, 1998).

Following sucrose breakdown, its products are phosphorylated at the expense of ATP or the energetically rich pyrophosphate (PP_i) to generate six-carbon sugar phosphates, which are freely interconvertible. The energy requirement for sucrose metabolism via sucrose synthase is potentially less than that of invertase. In the sucrose synthase reaction, UDP-glucose is phosphorylated with the use of PP_i and the production of glucose-1-P_i (G1P) and an ATP equivalent; the G1P is eventually converted to fructose-1,6-bisphosphate (F1,6P) with the use of an ATP molecule. If F1,6P is dephosphorylated via PP_i-dependent phospho-fructokinase, PP_i can be reformed, providing substrates for the UDP-glucose reaction noted; another ATP is required to again form F1,6P from F6P. The net ATP use for the glucose portion of the molecule would therefore be 1 ATP. The fructose portion of the molecule would require 2 ATP equivalents for a total of 3 ATP per sucrose. Four ATP are required for the formation of two F1,6P molecules via invertase.

Fructose-6-P_i(F6P) is converted to F1,6P at a step in the pathway known to be an important regulatory point. Three enzymes catalyze the interconversion between these two compounds: 1. ATP-dependent phosphofructokinase (ATP-PFK), which is irreversible in the glycolytic direction; 2. PP_i-dependent phosphofructokinase (PP_i-PFK), which operates in a freely reversible reaction; and 3. fructose-1,6-bisphosphatase (FBPase), which acts irreversibly in the sugar-forming (gluconeogenic) direction. Importantly, the reaction catalyzed by ATP:PFK is inhibited by ATP, one of the reaction substrates and also by one of the down-stream components of the glycolytic pathway, phosphoenolpyruvate (PEP). Furthermore, P_i, a product of the PP_i-PFK reaction, can alleviate this inhibition. This step is markedly affected by factors that alter global respiratory metabolism such as transferring plant tissues between aerobic and anaerobic atmospheres (ap Rees $et\ al.$, 1985). The loss in ATP-forming capacity associated with low O_2 stress and the increase in the P_i pool (Silva, 1998) may be important in enhancing glycolysis when O_2 is limiting. For carrot tissue, a drop in O_2 from

2 kPa to 0.5 kPa resulted in an increase in the pool of F1,6P relative to F6P, which is indicative of a promotion of this step (Kato-Noguchi and Watada, 1996a), by stimulation of PP_i-PFK or inhibition of FBPase via enhanced levels of the regulator fructose-2,6-bisphosphate (Kato-Noguchi and Watada, 1996b). Similar responses have been seen for a number of tissues. Elevated CO_2 has also been associated with the suppression of carbon flux through this step in the pathway. In pears held at room temperature, 10 kPa CO_2 reduced the activity of PP_i-PFK as well as ATP-PFK; analyses of glycolytic intermediates indicated that the conversion of F6P to F1,6P was inhibited *in vivo* (Kerbel *et al.*, 1988). Similar responses have been found in asparagus where CO_2 partial pressures between 5 and 10 kPa enhanced this step, but a CO_2 level of 20 kPa was inhibitory (Silva, 1998).

After several more steps in glycolysis, another likely control point step is encountered in the conversion of PEP to pyruvate, the final step in the pathway prior to the TCA cycle. In this reaction catalyzed by pyruvate kinase (PK), the glycolytic intermediates shed their last phosphate molecule with the production of an ATP molecule. Pyruvate kinase is inhibited by a number of metabolites including the product of its reaction, ATP, and the TCA cycle intermediates, citrate, malate and α-ketoglutarate. Conditions that favor ATP depletion such as low O_2 stress, would be expected to enhance the flux of carbon through this step by reducing ATP inhibition of the reaction. In O_2-stressed tissues, ATP levels decline within minutes of application of the stress (Givan, 1968), PEP drops rapidly and glycolysis often accelerates in a process known as the Pasteur effect (Barker *et al.*, 1967). The drop in ATP and accumulation of ADP is suggested to enhance the activity of PK sufficiently to reduce the level of PEP and removing the inhibitory influence of this intermediate on the previously noted control point involving ATP-PFK at the interconversion of F6P and F1,6P (Givan, 1999). Hatzfeld and Stitt (1991) found that when starved cells were suddenly provided with sugar to promote glycolysis, the ratio of F6P to F1,6P changed slightly later than the drop in PEP, suggesting that pyruvate kinase is the first control point enzyme to respond. This possibility has yet to be established, however. It is interesting to note that the nitrogen-containing compounds glutamate and glutamine also inhibit this enzyme (Podestá and Plaxton, 1994). The importance of this latter finding is not entirely clear, but it may link energy production to nitrogen metabolism and protein biosynthesis. There is an additional possible route for PEP to undergo conversion to pyruvate via the action of PEP phosphatase in the vacuole, but conclusive evidence of its operation in living tissue is not available. Alternatively, PEP can be converted to oxaloacetate by the action of PEP carboxylase; oxaloacetate can then be reduced to malate by malate dehydrogenase with the consumption of an NADH molecule. Malate can enter the mitochondrion and enter into the TCA cycle. In this way, PEP can enter the TCA cycle bypassing pyruvate kinase. Under normal, non-stressed circumstances, however, it has been estimated that little carbon enters the mitochondria via PEP carboxylase (Edwards *et al.*, 1998).

To this point in the discussion, oxygen has the potential to impact the glycolytic pathway at three points: sucrose synthase, ATP-PFK and pyruvate kinase. In each case, as far as is known, the effect is largely indirect through alteration of gene expression by an as yet unknown mechanism(s) or via the limitation of O_2 uptake by the oxidases of the mitochondrion and the concomitant limitation of ATP formation.

In response to O_2 limitation, the plant tissue attempts to adapt/survive by diverting its carbon to ethanol and/or lactate. Lactate production is a one-step process catalyzed by the enzyme lactate dehydrogenase (LDH). Ethanol production from pyruvate is a two-step process. Pyruvate decarboxylase (PDC) generates CO_2 and acetaldehyde from pyruvate in the first step and alcohol dehydrogenase (ADH) generates ethanol from acetaldehyde in the second. Lactate and ethanolic fermentation reactions result in the regeneration of NAD^+ from NADH, thus providing substrate for the reaction catalyzed by glyceraldehyde-3-P dehydrogenase earlier in glycolysis and maintaining glycolytic carbon flux.

The process of fermentation is important to plant tissue survival under low O_2 atmospheres (Schwartz, 1969). Fermentation fulfills two functions: it re-oxidizes NADH to NAD^+ so that glycolysis can proceed as previously noted; and it provides for the diversion of carbon to a relatively non-toxic end-product (ethanol) so that some energy production can occur. The energy yield per glucose molecule is significantly lower than that of aerobic metabolism, yielding 2 ATP per glucose molecule and no net NADH; an efficiency of only 4%. Under conditions of an energy shortfall, the preferential metabolism of sucrose via the more energy efficient sucrose synthase pathway, as opposed to the invertase pathway, may be advantageous to the plant tissue. In asparagus, O_2 levels resulting in fermentation caused a marked decline in steady-state ATP levels, which were reduced approximately tenfold relative to spears held in air (Silva, 1998). In maize roots, ADH activity was responsible for improved ATP levels of tissues in anoxia and improved tissue tolerance to anoxia (Johnson et al., 1994). Maize roots lacking ADH-1, one of the two forms of ADH in that tissue, were killed by as little as 6 h of anoxia (Johnson et al., 1994). Interestingly, preconditioning of plant tissues with low levels of O_2 boosted ADH levels, improved tolerance to subsequent anoxia and improved energy status, probably through enhanced fermentative capacity (Hole et al., 1992; Johnson et al., 1994).

Low oxygen induces the expression of a number of genes, including those for PDC and ADH in the ethanolic fermentation pathway as well as some (e.g. sucrose synthase, enolase and aldolase) that are involved in glycolysis (Sachs et al., 1980, 1996; Andrews et al., 1993). The patterns for induction of these genes differ from each other for reasons that are not clear at this time (Andrews et al., 1994b). The proteins associated with these genes have also been shown to accumulate in amount (Chang et al., 2000) as well as activity (Andrews et al., 1994a). The diversion of carbon to ethanol production, however, may initially require a mechanism not dependent on these changes in gene expression and protein synthesis. There is a growing body of evidence using maize, pea and rice systems that suggests that a drop in cytoplasmic pH may play a significant role in fermentative metabolism.

The prevailing theory for the involvement of cytoplasmic pH, based on data for root tissues, is that the O_2 stress causes an upset in pH regulation in the tissue. The source of this pH shift has been hypothesized to result from a transient increase in lactate concentration (Davies *et al.*, 1974). The concept is that O_2 restriction initially causes carbon to be diverted to lactate, much the same as in animal tissues. This reaction is expected to be favored since the pH of the cytosol is between 7.0 and 7.5, which is near the optimum for LDH. The pH optimum of PDC, however, is well below 7.0 and the enzyme is essentially inactive above pH 7.0. The formation of lactate causes a drop in the pH of approximately 0.5 units, thereby favoring PDC and the ethanolic fermentation pathway and ceasing lactate accumulation (Roberts *et al.*, 1984a,b). In mutants deficient in ADH-1, lactate continues to accumulate, leading to cytoplasmic acidosis and a loss in the capacity to maintain ATP levels (Roberts *et al.*, 1984b). These data have been primarily collected for root tissues of a limited number of crop plants and this mechanism may or may not operate similarly in other plant tissues. The production of lactate was considered to be inadequate, in some cases, to account for the initial cytoplasmic acidification (Drew, 1997). Roberts *et al.* (1984a) also speculated that H^+ transport from the cytoplasm to the vacuole via the action an H^+-ATPase on the vacuolar membrane may be disabled by O_2 stress. The reduced activity of this enzyme may result in an inability of the tissue to maintain the H^+ gradient that exists between the cytoplasm and the much more acidic vacuole. In any case, acidosis, while strongly linked to tissue damage, may not be directly responsible for cell death. Cell death often lags acidosis by several hours and less severe levels of acidosis (i.e. tissues having higher cytosolic pH) are still associated with cell death when the duration of acidosis is extended (Roberts *et al.*, 1984a).

Apart from the general metabolic upset that may result from the induction of fermentation by low levels of O_2, significant effects on the sensory quality of some tissues can result. The accumulation of ethanol can lead to the development of off-flavors in many tissues. In apple, pear, plum and berry fruit, off-flavors were detected in approximately two to five days of anoxia at room temperature and 7 to 20 days at $0°C$ (Richardson and Kosittrakun, 1995). The off-flavor was qualitatively related to the content of fermentation products in the tissues. Under conditions of hypoxia when O_2 levels were sufficiently low to induce fermentation, but at a lower rate than anoxia, off-flavor development was slower. In some cases, the off-flavors could be partially reversed by holding the produce in air for several days. The reversal was hypothesized to be dependent on the metabolism of the fermentation products. Ethanol, in particular, can sometimes be rapidly metabolized by incorporation into ethyl esters (Berger and Drawert, 1984), which tend to evaporate from the fruit tissues because they are less soluble in aqueous solutions than ethanol. However, the production of ethyl esters can alter the aroma of the product (Mattheis *et al.*, 1991).

Carbon dioxide also has the potential to induce fermentation, a process originally termed 'CO$_2$-zymasis' (Thomas, 1925). Even in the presence of saturating amounts of O$_2$, CO$_2$ can induce significant rates of fermentation. There tends to be relatively more acetaldehyde produced in CO$_2$-induced fermentation relative to fermentation resulting from O$_2$ deprivation as long as O$_2$ is present. Thomas (1925) demonstrated for apple fruit that under an atmosphere comprised solely of CO$_2$, the rate of fermentation and acetaldehyde production was similar to that of pure nitrogen. However, if as little as 5 kPa O$_2$ were present, the rate of acetaldehyde increased dramatically when the remainder of the atmosphere was CO$_2$. The accumulation of acetaldehyde in these studies was associated with tissue injury. Carbon dioxide can also influence the fermentation threshold, defined as the concentration of O$_2$ below which fermentation results. The fermentation threshold of blueberry fruit increases as CO$_2$ partial pressure is increased in the range of 0 to 20 kPa (Beaudry, 1993) but the same result was not obtained for asparagus spears (Silva, 1998). Unlike O$_2$, CO$_2$ probably does not affect the terminal oxidase of the respiratory chain, but rather may influence the activity of the TCA cycle.

6.2.1.2   Tricarboxylic acid cycle

Provided there is sufficient O$_2$ to permit the further oxidation of pyruvate, it will enter the mitochondrion and undergo oxidation in the TCA or citric acid cycle (figure 6.2). The TCA cycle, first described by Krebs and Johnson (1937), serves to extract the remaining 75% of the energy of the original glucose molecule contained in two molecules of pyruvate and provides the cell with carbon skeletons that feed into a number of important metabolic pathways. As in the glycolytic pathway, much of the carbon that enters the respiratory pathway is used in the formation of other cellular constituents, so each glucose molecule is not necessarily completely oxidized. Important among the classes of molecules generated from these carbon skeletons include proteins (from oxaloacetate and α-ketoglutarate), porphyrins from (α-ketoglutarate), terpenes, fatty acids and lipids (from acetyl CoA) (Taiz and Zeiger, 1998).

In the TCA cycle, pyruvate can be completely oxidized to CO$_2$. During this process, four molecules of NADH are formed, one molecule of flavin adenine nucleotide (FADH$_2$) and one molecule of ATP. NADH formation is catalyzed by four dehydrogenases, three of which also catalyze the release of CO$_2$. Each NADH molecule formed has the potential to generate three ATP under normal circumstances and each molecule of FADH$_2$ can be used to form 2 ATP. The TCA cycle, therefore, can potentially form about 30 molecules of ATP per molecule of glucose entering the respiratory pathway.

The step connecting glycolysis with the TCA cycle initiates the catabolism of pyruvate and involves the production of NADH and the release of CO$_2$ in

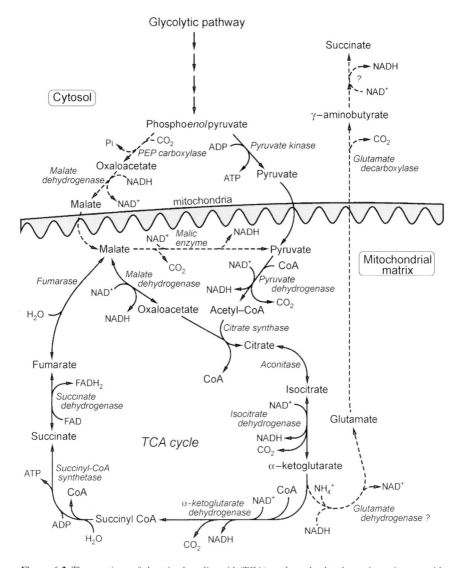

Figure 6.2 The reactions of the tricarboxylic acid (TCA) cycle and related reactions. Arrows with single heads represent essentially irreversible reactions; those with two heads represent freely reversible reactions. Dashed lines indicate pathways probably used under conditions of plant stress. Question marks indicate reactions that have not been fully characterized or verified.

the process of forming one molecule of acetyl CoA. This step is catalyzed by pyruvate dehydrogenase (PDH); PDH is actually an enzyme complex employing three enzymatic steps and a number of cofactors including thiamine pyrophosphate, FAD and lipoate. This step can be inhibited by the products of the reaction, citrate and acetyl CoA. Several steps in the remainder of the cycle are also under regulatory influence of intermediates and products of the cycle. ATP is inhibitory to aconitase, α-ketoglutarate dehydrogenase and malate dehydrogenase. NADH inhibits isocitrate dehydrogenase and malate dehydrogenase, oxaloacetate inhibits succinate dehydrogenase, malonic acid competitively inhibits succinate dehydrogenase, and AMP and ADP stimulate α-ketoglutarate dehydrogenase. The inhibitory influence of citrate on ATP-PFK and PFK should be recalled as it may provide a feedback mechanism for balancing the carbon flux through the glycolytic with that of the TCA cycle.

An enzyme of some significance in the mitochondrion is NAD^+ malic enzyme, which catalyzes the conversion of malate to pyruvate with the production of NADH and the release of CO_2. The presence of this enzyme allows the complete oxidation of organic acids in the absence of pyruvate from glycolysis, which normally acts as substrate to the TCA cycle. Thus, malate formed by cytosolic malate dehydrogenase can enter the mitochondrion (via a dicarboxylate transporter) and be converted to either oxaloacetate (OAA) by malate dehydrogenase or to pyruvate by malic enzyme, thereby supporting all the reactions of the TCA cycle. In fact, the presence of malic enzyme allows any component of the TCA cycle to act as a substrate in respiration, although only a few do. For example, succinate, which can be derived externally to the mitochondria from the breakdown of lipid materials in the glyoxysomes, can enter the TCA cycle and feed respiration.

The influence of reduced O_2 partial pressure on the TCA cycle largely takes place via the effect of substrate limitation of the terminal oxidases of the electron transport pathway. In a study on banana fruit, McGlasson and Wills (1972) found that four days of exposure to low levels of O_2 were sufficient to inhibit respiration by approximately 50%, caused a modest depletion in succinate, and a slight accumulation of pyruvate, oxaloacetate and α-ketoglutarate. The data were consistent with oxygen limiting the TCA cycle at the conversions of pyruvate (or OAA) to citrate and α-ketoglutarate to succinate. Nevertheless, it is primarily the lack of O_2 to act as electron acceptor that slows carbon flux through this portion of the respiratory chain, rather than the direct influence of O_2 on any one component of the pathway. Generally, low O_2 tends to reduce the rate of depletion of TCA cycle acids malate and citrate, which form the bulk of the acids associated with the tart flavor of many fruits. Low O_2 apparently does not influence the expression of genes for TCA cycle enzymes (Sachs *et al.*, 1996; Andrews *et al.*, 1994b; Chang *et al.*, 2000).

Carbon dioxide has the potential to alter respiration, but the direction and magnitude of the effect is species-dependent. Kubo *et al.* (1990) noted that CO_2 reduced the respiration rate (O_2 uptake) of ethylene-producing crops, caused an increase in the respiration of crops that produced ethylene in response to CO_2 and had no effect on the respiration rate of plant tissues whose ethylene production rate was not influenced by CO_2. They concluded that the influence of CO_2 on O_2 uptake was probably mediated by its action on ethylene. Carbon dioxide is also able to suppress respiration in some crops long after the plant material has been removed from the atmosphere (Herner, 1987; Li and Kader, 1989).

Many tissues are damaged by excessively high levels of CO_2 (Herner, 1987; Watkins, 2000). Carbon dioxide is known to influence the operation of the TCA cycle at the level of the cycle's enzymes rather than the electron transport pathway as does O_2. *In vivo* studies have demonstrated that CO_2 inhibits succinate dehydrogenase, which catalyzes the conversion of succinate to fumarate (Hulme, 1956; Ranson *et al.*, 1960; Frenkel and Patterson, 1973). Among the horticulturally relevant crops, for which elevated CO_2 has been shown to lead to an accumulation of succinate, are: apples (Hulme, 1956), strawberries (Fernández-Trujillo *et al.*, 1999), pear (Frenkel and Patterson, 1973), peaches and apricots (Wankier *et al.*, 1970) and lettuce (Ke *et al.*, 1993b). Carbon dioxide-induced succinate was not detected in banana, however, perhaps because the CO_2 exposure level was only 5% (McGlasson and Wills, 1972). Bananas typically accumulate 2 to 8 kPa CO_2 in the internal gas spaces during the course of ripening (Beaudry *et al.*, 1989) and may not be especially sensitive to the levels of CO_2 applied. Hulme (1956) speculated that CO_2 damage to apple tissues might be mediated by succinate and, indeed, small quantities of succinate have been found to damage apple fruit (Neal and Hulme, 1958). In contrast, Fernández-Trujillo *et al.* (1999) found that those cultivars of strawberry that experienced an accumulation of succinate in response to high levels of CO_2 appeared to experience less fermentative metabolism, which can be interpreted to mean they were more tolerant of the applied CO_2.

It has been speculated that the mechanism for the inhibition of the TCA cycle by CO_2 is a result of a CO_2-induced drop in the pH of the mitochondrial matrix, which would have the effect of reducing the activity of several of the TCA cycle dehydrogenases (Ke *et al.*, 1993a). Ke *et al.* (1993b) detected a drop in glutamate and an increase in γ-aminobutyrate concentrations in lettuce exposed to CO_2, which they interpreted as evidence of a pH shift in the mitochondria inhibiting succinate dehydrogenase, thereby favoring the diversion of α-ketoglutarate to glutamate (possibly via the mitochondrial enzyme glutamate dehydrogenase) and, by the action of glutamate decarboxylase, to γ-aminobutyrate. The pH optimum of glutamate decarboxylase, a cytosolic enzyme (Wu *et al.*, 1988), is approximately 6.0 (Beevers, 1951; Streeter and Thompson, 1972b; Ke *et al.*, 1993a), well below the typical cytosolic pH of 7.0 to 7.5. The same pH-based mechanism was earlier suggested for the adaptation of tissues to O_2 stress by Streeter and Thompson (1972a,b). Gamma-aminobutyrate has been shown to accumulate in response to limiting O_2 levels in a wide range of tissues including radish (Streeter and Thompson, 1972a,b), apple (Hulme and Arthington, 1950) and beetroot (Westall, 1950). The diversion of α-ketoglutarate to γ-aminobutyrate may play a role in conserving nitrogen and may help ameliorate the drop in pH in the mitochondria by removing a carboxylic acid group from α-ketoglutarate (Drew, 1997). It would also have the effect of removing reductant from the mitochondria and providing NADH for fermentation.

Carbon dioxide may also inhibit other components of the TCA cycle. A drop in mitochondrial pH sufficient to inhibit succinate dehydrogenase would also probably inhibit PDH, which has a similar pH-dependence (Ke *et al.*, 1995) thereby achieving a more global inhibition of the TCA cycle. In mitochondria from apple, Shipway and Bramlage (1973) demonstrated a concentration-dependent decline in the capacity for consumption of all TCA cycle intermediates except malate. This finding is consistent with enhanced malate consumption (Wankier *et al.*, 1970; Ke *et al.*, 1993a) and the accelerated decline sometimes detected in titratable acids (Siriphanich and Kader, 1986) induced by stress levels of CO_2.

In theory, CO_2 could reduce the pH of the cell sap through its dissociation of carbonic acid to bicarbonate and hydrogen ion. Lebermann *et al.* (1968), however, reported an increase in cellular pH in broccoli, while Lakshminarayana and Subramanyan (1970) found the opposite in mango. Reasons for the discrepancies may be related to differing responses to CO_2 or to the time lag between the period of exposure and the measurement of pH. For lettuce, the pH of the cytosol was found to drop by 0.4 pH units and that of the vacuole by 0.1 pH units while the tissue was under the influence of the CO_2 (Siriphanich and Kader, 1986). However, when the tissue was removed to air, the pH actually increased to a level greater than those maintained in air, possibly in response to a loss in acids during exposure. A decrease in the pH of the cytosol was also noted for avocado treated with high levels of CO_2 (Hess and Kader, 1993; Ke *et al.*, 1995; Lange and Kader, 1997). The magnitude of the response (0.4 to 0.6 pH units) was dependent on the concentration of CO_2.

These data suggest that stress levels of CO_2 induce the CO_2-zymasis effect noted by Thomas (1925) by interfering with TCA cycle operation, perhaps through acidification of the mitochondria as suggested by Ke *et al.* (1993a). The drop in pH would be expected to inhibit PDH and succinate dehydrogenase, as previously noted, thereby inhibiting ATP formation. In asparagus, ATP levels were reduced by a CO_2 partial pressure of 20 kPa, even in the presence of near-ambient O_2 levels (Silva, 1998). If a decline in the cytosolic pH is accompanied by a loss in energy capture and the induction of fermentation, then the reduced pH should inhibit the formation of lactate and carbon flux should proceed through pyruvate decarboxylase via the ethanolic fermentation pathway. The induction of fermentation by stress levels of CO_2 is indeed accompanied by relatively little lactate accumulation in a number of tissues (Ke *et al.*, 1993a). If O_2 levels are sufficient to avoid induction of ADH, then acetaldehyde formation should be promoted relative to that of ethanol, much as Thomas (1925) originally noted.

The normal function of the TCA cycle is required for maintaining the energy status of plant cells. This entails the successful transfer of electrons from the TCA cycle intermediates to the terminal electron acceptor, O_2. The implication is that if stress levels of O_2 and CO_2 interfere sufficiently, energy starvation could result. Central to the recovery of the energy contained in the sugar molecules is the synthesis of ATP and the function of the mitochondrial electron transport chain.

6.2.1.3 Mitochondrial electron transport chain

Oxidation of carbon in glycolysis and the TCA cycle is coupled to the reduction of the pyridine nucleotide NAD^+ to NADH as previously described. Under normal circumstances, NADH is subsequently reoxidized by the mitochondrial electron transport chain (ETC) in reactions coupled to the reduction of O_2 to H_2O and the concomitant production of ATP (Douce, 1985).

As has been mentioned, stress levels of O_2 and CO_2 are associated with shifts in the pools of reductant (NADH and NADPH) and ATP. In peas, brief anaerobic periods resulted in a build-up of NADH and a depletion of NAD (Barker *et al.*, 1967). In avocado, reduced O_2 caused a drop in NAD^+ levels of 60% or more, resulting in a ten-fold increase in the NADH:NAD ratio (Ke *et al.*, 1995). Carbon dioxide had almost no effect on NADH levels, however, as long as the O_2 level was in the aerobic range. In asparagus, 20 kPa CO_2 caused a 25% drop in ATP (Silva, 1998). A similar CO_2-induced depression in ATP was noted for avocado (Hess and Kader, 1993). However, severely low O_2 caused as much as a 90% decline in steady-state ATP levels in asparagus (Silva, 1998). A similar depletion (75 to 90%) of ATP was noted for peas held for short periods in anaerobic conditions (Barker *et al.*, 1967). In each case, the adenylate pool shifted toward the accumulation of ADP and AMP. Thus, in absence of adequate O_2, reoxidation of NADH to NAD^+ is stifled and the dehydrogenase reactions of the TCA cycle slow down, ATP synthesis is prevented, carbon flux through the TCA cycle is inhibited and pyruvate is diverted from the TCA cycle to fermentation reactions.

The mitochondrial ETC is a complicated arrangement of ca. 40 redox centers, 50 polypeptides and significant amounts of phospholipids in the inner membrane of the mitochondria. One single chain has a molecular weight of 1.52×10^6 Da. The respiratory components of the electron transport pathways are arranged into four discrete multiprotein units, generally referred to as complex I, II, III and IV. Complex I catalyzes the transfer of electrons from internal NADH to ubiquinone, forming ubiquinol. Complex II transfers electrons from succinate to ubiquinone, again forming ubiquinol. Complex III, also known as b–c_1 complex, transfers electrons from ubiquinol to cytochrome c. Complex IV, generally known as cytochrome c oxidase (CytOx), catalyzes the transfer of electrons from cytochrome c to O_2. Except for CytOx, all the components of these protein complexes are hydrophobic and are soluble in the 'fluid' lipid bilayer medium of the mitochondrial inner membrane (Mitchell, 1966). Ubiquinone in the lipid bilayer of the mitochondrial membrane behaves as a homogenous pool connecting the respiratory chain enzymes in a substrate like fashion. Traditionally, various components of mitochondrial ETC can be dissected with the use of specific inhibitors. The inhibitors include: rotenone and amytal for complex I; malonate and thenoyltrifluoroacetone for complex II; antimycin for complex III; and cyanide, carbon monoxide and azide for complex IV.

In addition to these four well-studied protein complexes that are common in the animal and plant kingdoms, the plant respiratory chain has additional components which bypass some sites of energy conservation normally associated with respiration (Douce and Neuburger, 1989; Vedel *et al.*, 1999). The non-phosphorylative pathways in plants include several NADPH dehydrogenases mediating rotenone-insensitive oxidation of matrix or cytosolic NADPH and the cyanide-resistant alternate oxidase (AltOx) (Meeuse, 1975). The rotenone-insensitive dehydrogenases are poorly characterized but functionally they bypass complex I, the first site of energy conservation in electron transport (Soole and Menz, 1995). AltOx catalyzes the oxidation of ubiquinol and the reduction of O_2 to H_2O, bypassing the last two sites of energy conservation normally associated with the cytochrome c pathway (Vanlerberghe and McIntosh, 1997). The AltOx pathway results in relatively low conservation of energy compared with the CytOx pathway and, if engaged, has important implications for the regulation of respiration in plants and harvestable plant products. However, unlike CytOx, which is almost always present and functioning, AltOx protein is not ubiquitous and is not necessarily engaged when present. AltOx activity is commonly modulated at the molecular level or by allosteric effectors such as the promoter pyruvate and other organic acids (Vanlerberghe and McIntosh, 1997).

Evidence for the presence of AltOx and its active participation in respiration during development (e.g. fruit ripening, anthesis) or during stress adaptation (e.g. wounding, chilling, osmotic stress and drought) in horticultural commodities is increasing. The capacity for oxidation of malate via the alternate pathway by isolated avocado mitochondria increases during fruit ripening (Moreau and Romani, 1982). AltOx protein accumulates during ripening of mango (Cruz-Hernádez and Gómez-Lim, 1995) and apple (Dupque and Arrabaça, 1999). Hiser *et al.* (1996) altered the capacity of alternative respiration by overexpressing the alternative oxidase protein in transgenic potato, suggesting that its capacity is primarily dependent on the amount of AltOx protein (Vanlerberghe and McIntosh, 1997). Dupque and Arrabaça (1999) have also shown that climacteric rise in respiration of apple fruit during cold storage was associated with an enhanced capacity of the AltOx pathway, but not of the CytOx pathway. The increase in fruit temperature during ripening for mango and banana correlated with the increase in alternate respiration of their respective fruit slices (Kumar *et al.*, 1990; Kumar and Sinha, 1992). However, these data on the activity or engagement of the alternative pathway have been obtained using respiratory inhibitors, a method that has been seriously questioned (Day *et al.*, 1996).

The reduced capacity for ATP synthesis resulting from the engagement of the AltOx pathway seems to run counter to the requirements for energy conservation and so its function has been difficult to define in many tissues. Only in flowers of the Araceae has a role for the AltOx been clearly defined. In this tissue, the AltOx is fully engaged and is used to generate a temperature increase of the spadix, resulting in the volatilization of odiferous compounds needed to attract pollinating insects (Laties, 1982). A role for the AltOx pathway during postharvest life of fresh produce has yet to emerge. Several hypotheses have been put forward.

In situations where the cytochrome pathway is restricted, alternative pathway might have a function when increased demand for ATP exists. Such a function has been suggested for potato tuber callus and soybean cells when ATP production coupled to the cytochrome pathway is not sufficient (van der Plas and Wagner, 1980; de Klerk-Kiebert *et al.*, 1982). The alternative respiration pathway of electron transport could provide a mechanism whereby respiratory carbon flux is maintained under conditions in which the availability of ADP and/or P_i, both critical for regulating carbon flux (Dry *et al.*, 1987), might be restrictive.

It has also been proposed that the alternative pathway plays a protective role against oxidative damage by avoiding over-reduction of the mitochondrial ETC and consequent production of reactive oxygen species (Millar *et al.*, 1993; Purvis and Shewfelt, 1993; Wagner and Krab, 1995). Active oxygen species production in mitochondria at the flavoprotein region of the internal NADH dehydrogenase and at ubiquinone have been reported (Rich and Bonner, 1978). It has been proposed that as glycolytic substrate flux becomes too high for the cytochrome chain, reduction levels of the respiratory complexes increase and organic acids will accumulate. These organic acids are suggested to activate the alternate oxidase, thereby decreasing the redox levels and the potential for free radical formation (Wagner and Krab, 1995).

The AltOx pathway may have an anabolic role as well. Under conditions of relatively low ATP requirements such as fruit ripening, the operation of the alternate pathway may allow more rapid production of TCA cycle intermediates, especially acetyl CoA, which has the potential to serve as a precursor to ripening-linked processes in fruit such as wax, pigment, terpene and ester formation.

While the precise role for alternate respiration in most tissues is yet to be established, evidence suggests that it can compete with CytOx for electrons. While molecular techniques offer a great deal of flexibility in regulating these pathways, the relative contributions to the total electron flow *in vivo* have not been accurately quantified so far for horticultural products. Taking the lead from research on organisms such as cyanobacteria (Radmer and Kok, 1976; Mir *et al.*, 1995), mass spectrometric methods have been developed that simultaneously measure the discrimination of ^{16}O and ^{18}O between competing reactions to determine the flux of O_2 through each. Guy *et al.* (1989) have shown that alternate oxidase discriminated against heavy labeled O_2 substantially more than did cytochrome oxidase. Recently, Beverly *et al.* (1999) have developed an isotopic ^{18}O and ^{16}O discrimination method for measuring *in vivo* activities of cytochrome and alternate oxidases. Such a noninvasive method should prove helpful in quantify the relative contributions of two O_2 reduction pathways to total mitochondrial electron flux and to identify factors that influence this flux.

Cytochrome c oxidase has a very high affinity for O_2, having a K_m of 0.15 kPa O_2 or less in the atmosphere external to the cell, while AltOx has much lower affinity for O_2, as reflected by its K_m of 1 to 3 kPa (Mapson and Burton, 1962; Solomos, 1977). Thus, levels of O_2 that slow electron flux through AltOx would not necessarily reduce flux through the cytochrome c pathway. Thus reduced O_2 levels could preferentially reduce the influence of AltOx on respiratory activity as well as reactions that are dependent on metabolic events or conditions derived from the activity of this pathway.

It is worth mentioning that the control of O_2 uptake may not be strictly controlled at the level of CytOx and AltOx. There are a large number of oxidases present in plant tissues, many of which have the potential to contribute measurably to O_2 uptake. More relevant to the present discussion, perhaps, is a proposed O_2 sensor that regulates carbon flux (Mapson and Burton, 1962; Solomos, 1997). While proof for this system is lacking, such a system would help explain the rather broad nature of O_2-dependent curves obtained for glycolytic processes in some tissues (Silva, 1998; Solomos, 1997). In some instances, skin and flesh resistances have been suggested to cause broad O_2-dependent respiratory curves (Tucker and Laties, 1985), whereas these effects have been discounted in other analyses (Knee, 1991).

The survival and self-maintenance of harvested plant tissues is heavily reliant on the coordinated biochemical activity of the three segments of the respiratory chain: glycolysis, the TCA cycle and the mitochondrial ETC. Restriction in the availability of O_2 and the superabundance of CO_2 have been demonstrated to affect each of these components, either directly or indirectly, and provide a fundamental basis for their influence on some of the components of fruit quality. From a practical perspective, however, observations at the whole tissue level provide additional insight as to the responses of harvested plant products to atmosphere modification.

6.2.2 *Whole tissue respiratory responses to O_2 and CO_2*

As has been noted, the potential for the application of modified atmospheres to preserve the quality of harvested plant products is dependent upon beneficial as well as adverse responses, both of which may be elicited simultaneously. Thus, it is the response of the tissue as a whole that must be considered when contemplating the application of modified atmospheres. The nature of the physical route of gas entry and escape from the tissues is important in determining the actual level of O_2 and CO_2 in a plant organ when exposed to storage atmospheres. In addition, it is important to know the potential for the applied atmosphere to reduce metabolic activity via respiratory restriction without inducing fermentation.

6.2.2.1 Gas exchange

The critical importance of barriers to gas exchange in the process of respiration has been recognized for more than a century (Devaux, 1891). Oxygen must first diffuse from the external atmosphere through the surface layers of the harvested plant organ to the interstices of the plant tissue. Following this, the O_2 must cross the barriers imposed by the cell wall and plasmalemma, dissolve into the cytosol, diffuse to the mitochondria and diffuse through the mitochondrial membranes to the matrix of the mitochondria, and there to act as substrate for the mitochondrial ETC. Carbon dioxide released from the TCA cycle in the mitochondrial matrix must journey along the same path, but in the opposite direction, to exit the tissues.

In that the terminal oxidases of the mitochondrial ETC operate in solution, some comment on the relationship between the intercellular atmosphere and the concentration of O_2 and CO_2 in the cell solution is in order. The affinity of CytOx for dissolved O_2 in potato tubers is quite high and the K_m has been estimated to range from 3 µM to as low as 0.07 µM O_2 (Burton, 1974). An internal O_2 partial pressure of as little as 1 kPa would be in equilibrium with between 12 to 15 µM O_2, depending on the temperature of the system (table 6.1), and this is at least fourfold higher than the K_m. The enzyme would be expected to be operating at near saturating rates under such circumstances. Burton (1974) also measured the concentration of O_2 in the cell sap and found it to be approximately 94% of the theoretical equilibrium with the O_2 in the intercellular space, suggesting that cellular structures posed very little barrier and that diffusion into and through the cell solution was essentially unrestricted.

The primary barrier to gas exchange typically resides in the skin of the fruit or vegetable, although in some tissues the flesh can impose a significant restriction (Burton, 1974; Burg and Burg, 1965; Banks and Kays, 1988). Early investigations by Devaux (1891) and subsequent studies (Burg and Burg, 1965; Burton, 1982) indicated that the primary route of gas exchange was through the openings on the surface of fruits and vegetables. The cuticle is also permeable to O_2 and CO_2 and may allow transmission of a significant percentage of these gases (Devaux, 1891; Marcellin, 1974; Cameron and Yang, 1982), with CO_2 moving through the cuticle more readily than O_2 (Devaux, 1891; Marcellin, 1974; Yearsley, 1996). Relative to the skin, the internal (tissue) resistances to gas exchange tend to be much less, contributing to O_2 and CO_2 gradients to only a minor degree, although there are instances in which flesh resistance is significant (e.g. melon, avocado and ripe banana).

Table 6.1 Equilibrium concentrations of free, dissolved CO_2 and O_2 in the solution of the cell sap for different gas partial pressures under the assumption that the solubility of CO_2 and O_2 in the sap is about the same as the solubility in a 0.4 M sucrose solution (Isenberg, 1979)

	Partial pressure (kPa) in gas phase (O_2 or CO_2)				
	1	5	10	15	20
Temperature (°C)	Equilibrium concentration in sap (M $\times$ 10^{-4})				
CO_2					
0	6.9	34.3	68.6	103.0	137.5
5	5.7	28.5	57.0	85.5	114.0
10	4.8	23.8	47.7	71.5	95.5
15	4.1	20.5	41.0	61.5	82.0
20	3.5	17.6	35.1	52.6	70.2
O_2					
0	1.9	9.6	19.1	28.7	38.2
5	1.7	8.3	16.6	25.0	33.3
10	1.5	7.4	14.8	22.3	29.7
15	1.4	6.7	13.5	20.2	27.0
20	1.2	6.0	12.0	18.1	24.1

As a consequence of skin and flesh barriers to O_2 and CO_2 diffusion, O_2 levels are lower and CO_2 levels higher in the intercellular spaces than in the external atmosphere. In apple fruit, typical gradients are 1 to 3 kPa for both gases at 0°C (Yearsley, 1996). Temperature has a marked influence on the magnitude of this gradient. For some apple cultivars, fruit having an gradient of 1 kPa O_2 at 0°C had a gradient of as much as 8 or 9 kPa at 24°C (Yearsley, 1996). The increase in the gradient is caused by the large increase in the rate of respiration with increasing temperature. The diffusive resistance of the skin and flesh do not change appreciably within the physiological range of temperatures (Nobel, 1983); thus the increase in respiration with temperature drives the formation of a proportionally larger gradient in a manner consistent with Fick's law (Burg and Burg, 1965). In addition to the effect of temperature, the respiratory gases themselves can impact the gas gradient. As O_2 levels are reduced in the external environment to the extent that the internal levels of O_2 become limiting to the terminal oxidases of the mitochondrial electron transport pathway, the reduction in O_2 demand and CO_2 production will lead to a drop in the gradients for O_2 and CO_2, respectively.

Variability in the resistance to gas exchange may present problems in the application of modified atmospheres. Product-to-product variability can lead to significant variability in the internal levels of O_2 and CO_2, with the influence of variability increasing as resistance (and the resultant gradient) increases (Banks, 1984). This phenomenon has been responsible in large measure for limiting the success of coating materials used to reduce internal O_2 levels and thereby reduce the rate of ripening in climacteric crops. In particular, variability in skin and tissue resistance is of serious concern when O_2 levels approach the fermentation threshold.

The nature of the O_2 affinity of the two oxidases suggests that tissues that depend strictly on CytOx for O_2 uptake would have a respiratory dependence on O_2 that would saturate at very low levels of O_2, while tissues that possessed a large dependence on AltOx respiration would be expected to have a much broader curve, saturating at quite high O_2 levels. These curves can be fitted with mathematical expressions. A standard means of mathematically describing the dependence of O_2 uptake on its substrate (O_2) is the Michaelis–Menten model (Cameron *et al.*, 1994; Hertog *et al.*, 1998), which is primarily applied to specific enzymatic reactions. The model is expressed as follows:

$$r_{O_2} = \frac{V_{max}\, p_{i,O_2}}{K_m\, p_{i,O_2}} \qquad (6.2)$$

where r_{O_2} is the O_2 uptake (mol kg^{-1} s^{-1}), V_{max} (mol kg^{-1} s^{-1}) is the maximal rate of O_2 uptake, p_{i,O_2} is the partial pressure (Pa) of O_2 in the fruit and K_m is the O_2 partial pressure at 50% of V_{max}. For whole fruit, the skin and/or flesh can pose significant resistance to gas flux. The term 'apparent K_m' or $K_{1/2}$ is used to refer the external O_2 level at 50% of V_{max} and is preferred to K_m, which has purely a biochemical interpretation and rightly refers to the concentration of O_2 dissolved in the solution of the mitochondrial matrix. The $K_{1/2}$ for various plant parts often ranges between 0.25 to 5 kPa O_2 (Cameron *et al.*, 1994; Hertog *et al.*, 1998; Joles, 1993). Because of the effect of temperature on respiration and the gradient of gases across the skin and flesh, the $K_{1/2}$ increases as temperature increases. A high $K_{1/2}$ results in a rather broad curve describing the dependence of O_2 uptake on O_2 partial pressure in comparison to that of single cells and other tissues with little diffusive resistance (figure 6.3).

6.2.2.2 Respiratory patterns

The $K_{1/2}$ has been used in models of respiratory responses of plant material to O_2 (Cameron *et al.*, 1994, 1995; Hertog *et al.*, 1998) and has been used to evaluate the potential performance of plant tissues (Beaudry, 2000). Beaudry (2000) reasoned that a significant reduction in respiratory (i.e. metabolic) activity will add value via its impact on product storability sufficient to offset the expenses incurred in the application of the atmosphere. The point at which investment in the technology is balanced by the added value is difficult to identify, but Beaudry (2000) suggested the 50% reduction in respiration coincident with the $K_{1/2}$ would suffice as an estimate of commercially valuable response that could be compared across commodities. Complementing this beneficial O_2 target is the O_2 threshold for the generally undesirable response of fermentation, which can be used to define the lower O_2 limit (Cameron *et al.*, 1994; Gran and Beaudry, 1993; Yearsley, 1996). The fermentation threshold does not always coincide

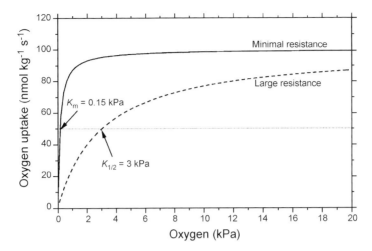

Figure 6.3 Hypothetical respiratory responses to O_2 for cytochrome *c* oxidase, which exhibits a K_m of approximately 0.15 kPa representative of single cells or tissues with minimal skin and flesh resistance and a tissue with significant skin resistance to gas exchange that yields an apparent K_m ($K_{1/2}$) of 3 kPa O_2.

with the lower O_2 limit in commercial practice, as in the case of lettuce and mixed salads when the benefit of reduced browning outweighs the loss in flavor and other aspects of quality (Cameron *et al.*, 1995; Peiser *et al.*, 1997; Smyth *et al.*, 1998).

Beaudry (2000) suggested that the difference between the $K_{1/2}$ for O_2 uptake and the fermentation threshold can be used as a criterion for deciding whether low O_2 will provide a beneficial response with respect to respiratory reduction. If, for instance, the fermentation threshold is much lower than $K_{1/2}$, then a greater than 50% reduction in respiration and attendant metabolic activities can be achieved without the threat of fermentation. The O_2 level between the fermentation threshold and the $K_{1/2}$ was termed the 'safe working atmosphere'. If, on the other hand, the fermentation threshold is equal to or greater than the $K_{1/2}$, then little or no advantage due to reduced metabolic activity was expected to be achieved by reducing O_2 since the tissues would be compromised by fermentative activity. In the latter case, there would be no safe working atmosphere.

An evaluation of the respiratory behavior of strawberry revealed a $K_{1/2}$ of approximately 1 kPa O_2 at 20°C, but fermentation, as judged by an increase in the respiratory quotient, occurs below ca. 1.2 kPa O_2 (figure 6.4). Thus, no safe working atmosphere exists for strawberry, implying that a reduction in respiration by O_2 limitation would be of no practical benefit. Consistent with the lack of a safe working atmosphere, low O_2 storage is not recommended to extend shelf-life (table 6.2). Interestingly, no safe working atmosphere was found for O_2–response curves of cut lettuce published by Smyth *et al.*, (1998),

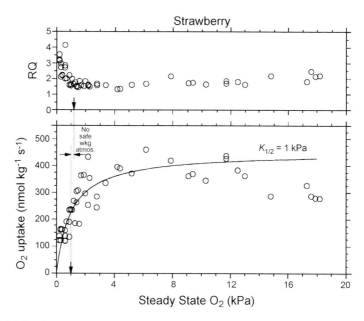

Figure 6.4 Respiratory response of strawberry to O_2 when held at 22°C and the associated respiratory quotient (RQ). The vertical arrow in the upper graph depicts the lower O_2 limit based on the increase in fermentative activity at lower O_2 partial pressure. The vertical arrow in the lower graph indicates the $K_{1/2}$ of the fitted line. The fact that the lower O_2 limit exceeds the $K_{1/2}$ is taken to indicate that there is no safe working atmosphere.

Table 6.2 Oxygen limits below which injury can occur for selected horticultural crops held at typical storage temperatures (from Beaudry, 2000; Gorny, 1997; Kader, 1997a,b; Kupferman, 1997; Richardson and Kupferman, 1997; Saltveit, 1997) (those commodities in bold are considered to have very good to excellent potential to respond to low O_2.)

O_2 (kPa)	Commodities
0.5 or less	Chopped greenleaf, redleaf, Romaine and iceberg lettuce, spinach, sliced pear, **broccoli**, mushroom
1	Broccoli florets, chopped butterhead lettuce, sliced apple, Brussels sprouts, cantaloupe, cucumber, crisphead lettuce, onion bulbs, apricot, avocado, **banana**, cherimoya, atemoya, sweet cherry, cranberry, grape, **kiwifruit**, litchi (lychee), nectarine, peach, plum, rambutan, sweetsop
1.5	**Most apples, most pears**
2	Shredded and cut carrots, artichoke, **cabbage**, cauliflower, celery, bell and chilli pepper, sweetcorn, tomato, blackberry, durian, fig, mango, olive, papaya, pineapple, pomegranate, raspberry, strawberry
2.5	Shredded cabbage, blueberry
3	Cubed or sliced cantaloupe, **low permeability apples and pears**, grapefruit, persimmon
4	Sliced mushrooms
5	Green snap beans, lemon, lime, orange
10	Asparagus
14	Orange sections

despite the fact that low O_2 is an important tool in quality preservation. The benefit of low O_2 in that tissue, however, is in retarding the browning response referred to as pinking, rather than slowing metabolism.

'Empire' apple fruit had no safe working atmosphere at the preclimacteric stage when maintained in that condition by 1-methylcyclopropene (1-MCP) (figure 6.5). Ripening fruit, however, had a safe working atmosphere nearly 6.5 kPa in breadth between 2.5 and 9 kPa O_2. A similar response to O_2 was also found for 'Jonathan' apple fruit (data not shown). The rather broad safe working atmosphere of ripening apple fruit suggests that significant respiratory suppression can be achieved by reduced O_2 levels. This is consistent with the successful application of low O_2, long known to enhance the storability of apple fruit two- to threefold relative to air storage (Kidd and West, 1927, 1945; Fidler, 1965). The lack of a safe working atmosphere in preclimacteric apple fruit suggests that reduced O_2 at this stage operates primarily by reducing ethylene synthesis and perception and, indeed, 1-MCP treatment at this stage will prevent fruit ripening even at room temperature as long as the 1-MCP is present (Mir et al., 2001). The increase in the $K_{1/2}$ and V_{max} may reflect the induction and

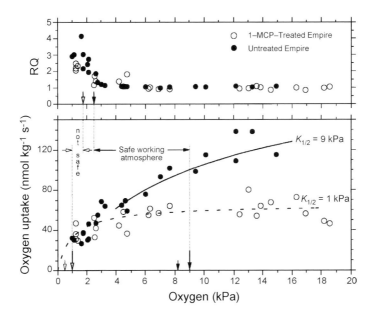

Figure 6.5 Respiratory response of unripe (1-methylcyclopropene-treated) and ripening (untreated) apple to O_2 when held at 22°C and the associated respiratory quotient (RQ). The vertical arrow in the upper graph depicts the lower O_2 limit based on the increase in fermentative activity at lower O_2 partial pressure. The vertical arrow in the lower graph indicates the $K_{1/2}$ of the fitted line. The fact that the lower O_2 limit exceeds the $K_{1/2}$ for unripe fruit is taken to indicate that there is no safe working atmosphere for fruit of this developmental stage. The fact that the $K_{1/2}$ exceeds the lower O_2 limit for ripening fruit is taken to indicate that there is a substantial safe working atmosphere for fruit of this developmental stage.

extensive use of the alternative oxidase as ripening commenced, an observation supported by previous research on apple and other climacteric crops (Cruz-Hernández and Gómez-Lim, 1995), but not yet confirmed.

The safe working atmosphere was proposed as a tool for developing a mechanistic interpretation of respiratory responses rather than fruit storability. Respiratory responses need to be viewed in terms of the impact of O_2 and CO_2 on other physiological phenomena such as ethylene synthesis and action, aroma, fungal activity and so on.

6.3   Ethylene biology

The effect of O_2 on reducing respiration demonstrated by early workers in the field was recognized as not having a mechanistic foundation and subsequent work by Burg and Burg (1967) suggested that, in addition to altering respiration, atmosphere composition markedly affected ethylene action. More on this topic will be found in Chapter 8 of this book, however, it is important to juxtapose the preceding discussion regarding the impact of O_2 and CO_2 on respiratory metabolism with their profound effects on ethylene biology.

Ethylene biosynthesis is dependent on the presence of O_2 (Abeles *et al.*, 1992; Makhlouf *et al.*, 1989b). The $K_{1/2}$ of 1-aminocyclopropane carboxylic acid oxidase (ACO), the enzyme responsible for the last step in the enzymatic production of ethylene from 1-aminocyclopropane carboxylic acid (ACC) has been variously reported as falling within the range of 1.4 to 10 kPa (Abeles *et al.*, 1992). The requirement for O_2 is dependent on the concentration of the other substrate; as ACC levels increase, the K_m of the enzyme for O_2 declines.

Much data suggest that O_2 is required for plants to respond to ethylene and elevated CO_2 inhibits ethylene action, but some ethylene responses for a few plant materials are relatively insensitive to O_2 and CO_2 (Abeles *et al.*, 1992; Burg and Burg, 1967; Yip *et al.*, 1988). However, in most, if not all climacteric fruit crops, low O_2 and high CO_2 reduce the rate of ripening by a number of measures including texture, aroma and pigmentation changes.

In climacteric fruit crops, prior to the onset of the climacteric, the primary influence of low O_2 may be to suppress ripening through its ability to restrict ethylene action, as opposed to general metabolic suppression via respiratory inhibition. Evidence comes from instances in which O_2 levels that would normally not affect respiration would still reduce the rate of ripening. For example, in 'Empire' apple, 3 kPa O_2 at 22°C has little effect on respiration (see figure 6.3) yet ripening can be significantly retarded by partial pressures of O_2 as high as 6 kPa O_2 (Sfakiotakis and Dilley, 1973). Certainly, suppression of ethylene action prior to the onset of the ethylene climacteric can very effective at preventing ripening. Application of 1-MCP, a specific inhibitor of ethylene binding (Sisler and Blankenship, 1996) completely prevents ripening of climacteric fruit without atmosphere modification when applied in the preclimacteric or early climacteric stages (Fan *et al.*, 1999; Golding *et al.*, 1998; Mir *et al.*, 2001).

6.4 Secondary metabolic pathways

Low O_2 has important effects on metabolism other than those on respiration and ethylene action that can have significant impacts on quality of plant products at both the distributor and consumer level. Low O_2 reduces the rate of degreening due to chlorophyll loss and inhibits browning reactions catalyzed by polyphenol oxidase (PPO). Chlorophyll loss, a desirable trait for many climacteric fruits, results in a quality loss for many vegetable products. Chlorophyll degradation in green vegetables can be inhibited by low O_2 (Saltveit, 1997), a response that is probably partly mediated by O_2 on ethylene-mediated promotion of senescence and perhaps by the direct action of O_2 limiting the reaction of pheophorbide *a* oxygenase (Matile *et al.*, 1999). Degreening can be inhibited by low O_2 (Makhlouf *et al.*, 1989a), which reduces ethylene synthesis (Makhlouf *et al.*, 1989b), enhanced by added ethylene or the ethylene analog propylene (Tian *et al.*, 1994) and inhibited by the specific inhibitor of ethylene action, 1-MCP.

Low O_2 is used in modified atmosphere packages of lettuce to reduce browning of the cut surfaces and has been evaluated for its potential to reduce browning in fruit slices (Lakakul, 1994). Low O_2 slows or prevents browning, apparently via its action on PPO (Mayer and Harel, 1979; Vámos-Vigyázó, 1981). The K_m for O_2 has been reported to be between 6 to 10 kPa (Mapson and Burton, 1962; Mayer and Harel, 1979). Smyth *et al.* (1998) demonstrated that O_2 levels below 2 kPa and above the fermentation threshold of approximately 0.5 kPa reduced the rate of browning in lettuce.

The production of volatile esters—which contribute to characteristic aromas of a number of fruit, including apple, banana, pear, peaches, strawberries and other fruits—is dependent on O_2 and is, in some cases, altered by CO_2 (Beaudry, 1999; Lidster *et al.*, 1983; Golias, 1984; Shamaila *et al.*, 1992; Fellman *et al.*, 1993; Mattheis and Fellman, 2000; Song *et al.*, 1997). This topic is discussed in detail in Chapter 4. Aroma compounds that confer characteristic odors are generally suppressed by low O_2, in part by the action of O_2 on ethylene action in climacteric fruits, but is also likely to be via action of O_2 on oxidative processes, including respiration, required for substrate production. Many volatiles that do not contribute to aroma are also suppressed by low O_2. In apple and pear, for instance, low O_2 reduces the accumulation of α-farnesene, a semi-volatile sesquiterpene that induces superficial scald (Chen *et al.*, 1993; Huelin and Murray, 1966). In general, most products recover from moderate low O_2 suppression of aroma volatiles and eventually develop characteristic flavors.

Table 6.3 Carbon dioxide partial pressures above which injury will occur for selected horticultural crops (modified from Herner, 1987; Kader, 1997a; Saltveit, 1997)

CO_2 (kPa)	Commodity
2	Lettuce (crisphead), pear
3	Artichoke, tomato
5	Apple (most cultivars), apricot, cauliflower, cucumber, grape, nashi, olive, orange, peach (clingstone), potato, pepper (bell)
7	Banana, bean (green snap), kiwifruit
8	Papaya
10	Asparagus, Brussels sprouts, cabbage, celery, grapefruit, lemon, lime, mango, nectarine, peach (freestone), persimmon, pineapple, sweetcorn
15	Avocado, broccoli, litchi (lychee), plum, pomegranate, sweetsop
20	Cantaloupe (muskmelon), durian, mushroom, rambutan
25	Blackberry, blueberry, fig, raspberry, strawberry
30	Cherimoya

6.5 Conclusions

Oxygen and CO_2 remain as powerful tools to assist in the preservation of many fruits. These vital gases have significant potential to modify respiratory activity. However, this altered respiratory activity probably has relatively little direct impact on the storability of most fruits and vegetables. Where a significant positive impact of low O_2 and elevated CO_2 does exist, it is often coupled with atmospheric influence on ethylene action. It is interesting to note that climacteric crops form the bulk of those tissues for which low O_2 is recommended and commercially utilized (table 6.2).

The difference between the responsiveness of climacteric and non-climacteric crops to atmospheric modification may center around two interrelated phenomena: 1. the capacity of low O_2 and high CO_2 to reduce the influence of ethylene; and 2. the inhibition of respiration by low O_2 relative to its potential (climacteric) maximum. Low O_2 markedly slows the onset of fruit ripening in climacteric tissues, probably by suppressing ethylene action directly and by minimizing the accumulation of ethylene in the fruit tissues. Once climacteric fruits have begun to ripen and respiration becomes elevated, low O_2 can also reduce respiratory and associated metabolic activity. There is no evidence to suggest, however, that low O_2 can reduce respiration of climacteric fruit kept in a preclimacteric state, as by continuous treatment with 1-MCP, and thereby improve storability. It may be that respiratory suppression by low O_2 and benefits accrued from the resulting reduced metabolic activity are conditional, depending on the ethylene-induced climacteric rise in respiration.

In many cases, the useful effects of low O_2 and elevated CO_2 (tables 6.2 and 6.3) are offset by their negative effects on quality attributes, such as fermentation and aroma loss. The effects of O_2 and CO_2 on respiration, ethylene biosynthesis, ethylene action and the biochemistry of quality changes in fruits are still not completely understood. The interrelationships between these pathways leave further room for exploration with the object of improving fruit quality.

References

Abeles, F.B., Morgan, P.W. and Saltveit, M.E. Jr (1992) *Ethylene in Plant Biology*, 2nd edn. Academic Press, San Diego.

Andrews, D.Z., Cobb, B.G., Johnson, J.R. and Drew, M.C. (1993) Hypoxic and anoxic induction of alcohol dehydrogenase in roots and shoots of seedlings of *Zea mays*: *Adh* transcripts and enzyme activity. *Plant Physiology*, **101**, 407-414.

Andrews, D.L., Drew, M.C., Johnson, J.R. and Cobb, B.G. (1994a) The response of maize seedlings of different ages to hypoxic and anoxic stress. *Plant Physiology*, **105**, 53-60.

Andrews, D.L., MacAlpine, D.M., Johnson, J.R., Kelley, P.M., Cobb, B.G. and Drew, M.C. (1994b) Differential induction of mRNAs for the glycolytic and ethanolic fermentative pathways by hypoxia and anoxia in maize seedlings. *Plant Physiology*, **106**, 1575-1582.

ap Rees, T., Green, J.H. and Wilson, P.M. (1985) Pyrophosphate:fructose 6-phosphate 1-phospho-transferase and glycolysis in nonphotosynthetic tissues of higher plants. *Biochemical Journal*, **227**, 299-304.

Avigad, G. and Dey, P.M. (1997) Carbohydrate metabolism: storage carbohydrates, in *Plant Biochemistry* (eds P.M. Dey and J.B. Harborne), Academic Press, San Diego.

Banks, N.H. (1984) Studies of the banana fruit surface in relation to the effects of Tal-Prolong coating on gaseous exchange. *Scientia Horticulturae*, **24**, 279-286.

Banks, N.H. and Kays, S.J. (1988) Measuring internal gases and lenticel resistance to gas diffusion in potato tubers. *Journal of the American Society for Horticultural Science*, **113**, 577-580.

Barker, J., Khan, M.A.A. and Solomos, T. (1967) Studies in the respiratory and carbohydrate metabolism of plant tissues, XXI, The mechanism of the Pasteur effect in peas. *New Phytologist*, **66**, 577-596.

Beaudry, R.M. (1993) Effect of carbon dioxide partial pressure on blueberry fruit respiration and respiratory quotient. *Postharvest Biology and Technology*, **3**, 249-258.

Beaudry, R.M. (1999) Effect of O_2 and CO_2 partial pressure on selected phenomena affecting fruit and vegetable quality. *Postharvest Biology and Technology*, **15**, 293-303.

Beaudry, R.M. (2000) Responses of horticultural commodities to low oxygen: limits to the expanded use of MAP. *HortTechnology*, **10**, 491-500.

Beaudry, R.M., Severson, R.F., Black, C.C. and Kays, S.J. (1989) Banana ripening: implications of changes in glycolytic intermediate concentrations, glycolytic and gluconeogenic carbon flux, and fructose 2, 6-bisphosphate concentration. *Plant Physiology*, **91**, 1436-1444.

Beevers, H. (1951) An L-glutamic acid decarboxylase from barley. *Biochemical Journal*, **48**, 132-137.

Berard, J. (1821) Memoire sur la maturation des fruits. *Annales de Chimie et de Physique*, **16**, 152-183, 225-251.

Berger, R.G. and Drawert, F. (1984) Changes in the composition of volatiles by the post-harvest application of alcohols to Red Delicious apples. *Journal of the Science of Food and Agriculture*, **35**, 1318-1325.

Beverley, K.H., Owen, K.A., Day, D.A., Millar, A.H., Menz, R.I. and Farquhar, G.D. (1999) Calculation of the isotope discrimination factor for studying plant respiration. *Australian Journal of Plant Physiology*, **26**, 773-780.

Brown, W. (1922) On the germination and growth of fungi at various temperatures and in various concentrations of oxygen and carbon dioxide. *Annals of Botany*, **36**, 257-283.

Brownleader, M.D., Harborne, J.B. and Dey, P.M. (1997) Carbohydrate metabolism: primary metabolism of monosaccharides, in *Plant Biochemistry* (eds P.M. Dey and J.B. Harborne), Academic Press, San Diego.

Burg, S.P. and Burg, E.A. (1965) Gas exchange in fruits. *Physiologia Plantarum*, **18**, 870-884.

Burg, S.P. and Burg, E.A. (1967) Molecular requirements for the biological activity of ethylene. *Plant Physiology*, **42**, 114-152.

Burton, W.G. (1974) Some biophysical principles underlying the controlled atmosphere storage of plant material. *Annals of Applied Biology*, **78**, 149-168.

Burton, W.G. (1982) *Post-harvest Physiology of Food Crops*, Longman, New York.

Cameron, A.C. and Yang, S.F. (1982) A simple method for determination of resistance to gas diffusion in plant organs. *Plant Physiology*, **70**, 21-23.

Cameron, A.C., Beaudry, R.M., Banks, N.H. and Yelanich, M.V.M. (1994) Modified-atmosphere packaging of blueberry fruit: modeling respiration and package oxygen partial pressures as a function of temperature. *Journal of the American Society for Horticultural Science*, **119**, 534-539.

Cameron, A.C., Talasila, P.C. and Joles, D.J. (1995) Predicting the film permeability needs for modified-atmosphere packaging of lightly processed fruits and vegetables. *HortScience*, **30**, 25-34.

Ceponis, M.J. and Cappellini, R.A. (1985) Reducing decay in fresh blueberries with controlled atmospheres. *HortScience*, **20**, 228-229.

Chang, W.W.P., Huang, L., Shen, M., Webster, C., Burlingame, A.L. and Roberts, J.K.M. (2000) Patterns of protein synthesis and tolerance of anoxia in root tips of maize seedlings acclimated to a low-oxygen environment, and identification of proteins by mass spectrometry. *Plant Physiology*, **122**, 295-317.

Chen, P.M., Varga, R.J. and Xiao, Y.Q. (1993) Inhibition of α-farnesene biosynthesis and its oxidation in the peel tissue of 'd'Anjou' pears by low-O_2/elevated-CO_2 atmospheres. *Postharvest Biology and Technology*, **3**, 215-223.

Cruz-Hernández, A. and Gómez-Lim, M.A. (1995) Alternative oxidase from mango (*Mangifera indica* L.) is differentially regulated during fruit ripening. *Planta*, **197**, 569-576.

Davies, D.D., Grego, S. and Kenworthy, P. (1974) The control of the production of lactate and ethanol by higher plants. *Planta*, **118**, 297-310.

Day, D.A., Krab, K., Lambers, H., Moore, A.L., Siedow, J.N., Wagner, A.K., and Wiskich, J.T. (1996) The cyanide-resistant oxidase: to inhibit or not to inhibit, that is the question. *Plant Physiology*, **110**, 1-2.

de Klerk-Kiebert, Y.M., Kneppers, T.J.A. and van der Plas, L.H.W. (1982) Influence of chloramphenicol on growth and respiration of soybean (*Glycine max* L.) suspension cultures. *Physiologia Plantarum*, **55**, 98-102.

Devaux, H. (1891) Etude experimental sur l'aeration des tissus massifs. *Annales des Sciences Naturalles Botanique, Series 7*, **14**, 279-395.

Douce, R. (1985) *Mitochondria in Higher Plants. Structure, Function and Biogenesis*. Academic Press, Orlando, Florida.

Douce, R. and Neuburger, M. (1989) The uniqueness of plant mitochondria. *Annual Review of Plant Physiology and Plant Molecular Biology*, **40**, 371-314.

Drew, M. (1997) Oxygen deficiency and root metabolism: injury and acclimation under hypoxia and anoxia. *Annual Review of Plant Physiology and Plant Molecular Biology*, **48**, 223-250.

Dry, I.B., Bryce, J.H. and Wiskich, J.T. (1987) Regulation of mitochondrial respiration, in *The Biochemistry of Plants: A Comprehensive Treatise, Vol. 11: Biochemistry of Metabolism* (ed. D.D. Davies), Academic Press, New York, pp. 213-252.

Dupque, P. and Arrabaça, J.D. (1999) Respiratory metabolism during cold storage of apple fruit. II. Alternate oxidase is induced at the climacteric. *Physiologia Plantarum*, **107**, 24-31.

Edwards, S., Bich-Ty, N., Do, B. and Roberts, J.K.M. (1998) Contribution of malic enzyme, pyruvate kinase, phosphoenolpyruvate carboxylase, and the Krebs cycle to respiration and biosynthesis and to intracellular pH regulation during hypoxia in maize root tips observed by nuclear magnetic resonance imaging and gas chromatography-mass spectrometry. *Plant Physiology*, **116**, 1073-1081.

Fan, X., Blankenship, S.M. and Mattheis, J.P. (1999) 1-Methylcyclopropene inhibits apple ripening. *Journal of the American Society for Horticultural Science*, **124**, 690-695.

Fellman, J.K., Mattinson, D.S., Bostick, B.C., Mattheis, J.P. and Patterson, M.E. (1993) Ester biosynthesis in 'Rome' apples subjected to low-oxygen atmospheres. *Postharvest Biology and Technology*, **3**, 201-214.

Fernández-Trujillo, J.P., Nock, N.F. and Watkins, C.B. (1999) Fermentative metabolism and organic acid concentrations in fruit of selected strawberry cultivars with different tolerances to carbon dioxide. *Journal of the American Society for Horticultural Science*, **124**, 696-701.

Fidler, J.C. (1965) Controlled atmosphere storage of apples, in *Proceedings of the Institute for Refrigeration*, National College for Heating, Ventilation, Refrigeration and Fan Engineering, London, pp. 1-7.

Frenkel, C. and Patterson, M.E. (1973) Effect of carbon dioxide on activity of succinic dehydrogenase in 'Bartlett' pears during cold storage. *HortScience*, **8**, 395-396.

Givan, C.V. (1968) Short-term changes in hexose phosphates and ATP in intact cells of *Acer pseudoplatanus* L. subjected to anoxia. *Plant Physiology*, **43**, 948-952.

Givan, C.V. (1999) Evolving concepts in plant glycolysis: two centuries of progress. *Biological Review*, **74**, 277-309.

Golding, J.B., Shearer, D., Wyllie, S.G. and McGlasson, W.B. (1998) Application of 1-MCP and propylene to identify ethylene-dependent ripening processes in mature banana fruit. *Postharvest Biology and Technology*, **14**, 87-98.

Golias, J. (1984) Biogenesis of volatile flavor compounds in apples in a low oxygen atmosphere. *Acta Universidad de Agricultura Faculty Agronomica (BRNO)*, **32**, 95-100.

Gorny, J.R. (1997) A summary of CA and MA requirements and recommendations for fresh-cut (minimally-processed) fruits and vegetables, in *CA '97 Proceedings, Vol. 5: Fresh-cut Fruits and Vegetables and MAP*. (ed. J. Gorny), Postharvest Horticulture Series No. 19, University of California, Davis, pp. 30-66.

Gran, C.D. and Beaudry, R.M. (1993) Determination of the low oxygen limit for several commercial apple cultivars by respiratory quotient breakpoint. *Postharvest Biology and Technology*, **3**, 259-267.

Guy, R.D., Berry, J.A., Fogel, M.L. and Hoering, T.C. (1989) Differential fractionation of oxygen isotopes by cyanide-resistant and cyanide-sensitive respiration in plants. *Planta*, **177**, 483-491.

Hattori, T. and Nakamura, K. (1988) Genes coding for the major tuberous root protein of sweet potato: identification of putative regulators sequence in the $5'$ upstream region. *Plant Molecular Biology*, **11**, 417-426.

Hatzfeld, W.D. and Stitt, M. (1991) Regulation of glycolysis in heterotrophic suspension cultures of *Chenopodium rubrum* in response to proton fluxes at the plasmalemma. *Physiologia Plantarum*, **81**, 103-110.

Herner, R.C. (1987) High CO_2 effects on plant organs, in *Postharvest Physiology of Vegetables* (ed. J. Weichman), Marcel Dekker, New York, pp. 239-253.

Hertog, M.L.A.T.M., Peppelenbos, H.W., Evelo, R.G. and Tijskens, L.M.M. (1998) A dynamic and generic model of gas exchange of respiring produce: the effects of oxygen, carbon dioxide and temperature. *Postharvest Biology and Technology*, **14**, 335-349.

Hess, B., Ke, D. and Kader, A.A. (1993) Changes in intracellular pH, ATP, and glycolytic enzymes in 'Hass' avocado in response to low O_2 and high CO_2 stresses, in *CA '93, Proceedings from the Sixth International Controlled Atmosphere Conference, NRAES-71, Vol. 1* (ed. M.E. Saltveit), NRAES, Ithaca, NY, pp. 1-10.

Hiser, C., Kapranov, P. and McIntosh, L. (1996) Genetic modification of respiratory capacity in potato. *Plant Physiology*, **110**, 277-286.

Hole, D.J., Cobb, B.G., Hole, P.S. and Drew, M.C. (1992) Enhancement of anaerobic respiration in root tips of *Zea mays* following low-oxygen (hypoxic) acclimation. *Plant Physiology*, **99**, 213-218.

Huelin, F.E. and Murray, K.E. (1966) α-Farnesene in the natural coating of apples. *Nature*, **210**, 1260-1261.

Hulme, A.C. (1956) Carbon dioxide injury and presence of succinic acid in apples. *Nature*, **178**, 218-219.

Hulme, A.C. and Arthington, W. (1950) Amino acids of the apple fruit. *Nature*, **165**, 716-717.

Isenberg, F.M.R. (1979) Controlled atmosphere storage of vegetables. *Horticultural Reviews*, **1**, 337-394.

Johnson, J.R., Cobb, B.G. and Drew, M.C. (1994) Hypoxic induction of anoxia tolerance in roots of Adh1 null *Zea mays* L. *Plant Physiology*, **105**, 61-67.

Joles, D.W. (1993) *Modified-atmosphere packaging of raspberry and strawberry fruit: characterizing the respiratory response to reduced O_2, elevated CO_2 and changes in temperature*. MS thesis, Michigan State University, East Lansing, MI 48824, USA.

Kader, A.A. (1997a) A summary of CA requirements and recommendations for fruits other than apples and pears, in *CA '97 Proceedings, Vol. 3: Fruits Other than Apples and Pears* (ed. A. Kader), Postharvest Horticulture Series No. 17, University of California, Davis, pp. 1-36.

Kader, A.A. (1997b) Biological bases of O_2 and CO_2 effects on postharvest life of horticultural perishables, in *CA '97 Proceedings, Vol. 4: Vegetables and Ornamentals* (ed. M.E. Saltveit), Postharvest Horticulture Series No. 18, University of California, Davis, pp. 160-163.

Kato-Noguchi, H. and Watada, A.E. (1996a) Low-oxygen atmosphere increases fructose 2, 6-bis-phosphate in fresh-cut carrots. *Journal of the American Society for Horticultural Science*, **121**, 307-309.

Kato-Noguchi, H. and Watada, A.E. (1996b) Regulation of glycolytic metabolism in fresh-cut carrots under low oxygen atmosphere. *Journal of the American Society for Horticultural Science*, **121**, 123-126.

Kays, S.J. (1997) *Postharvest Physiology of Perishable Plant Products*, Van Nostrand Reinhold, New York.

Ke, D., Mateos, M. and Kader, A.A. (1993a) Regulation of fermentative metabolism in fruits and vegetables by controlled atmospheres, in *CA '93, Proceedings from the Sixth International Controlled Atmosphere Conference, NRAES-71, Vol. 1* (ed. M.E. Saltveit), NRAES, Ithaca, New York, pp. 63-77.

Ke, D., Mateos, M., Siriphanich, J., Li, C. and Kader, A.A. (1993b) Carbon dioxide action on metabolism of organic and amino acids in crisphead lettuce. *Postharvest Biology and Technology*, **3**, 235-247.

Ke, D., Yahia, E., Hess, B., Zhou, L. and Kader, A.A. (1995) Regulation of fermentative metabolism in avocado fruit under oxygen and carbon dioxide stresses. *Journal of the American Society for Horticultural Science*, **120**, 481-490.

Kerbel, E.L., Kader, A.A. and Romani, R.J. (1988) Effects of elevated CO_2 concentrations on glycolysis in intact 'Bartlett' pear fruit. *Plant Physiology*, **86**, 1205-1209.

Kidd, F. and West, C. (1927) A relation between the concentration of oxygen and carbon dioxide in the atmosphere, rate of respiration, and length of storage of apples. *Food Investigation Board Report of London for 1925*, pp. 41-42.

Kidd, F. and West, C. (1945) Respiratory activity and duration of life of apples gathered at different stages of development and subsequently maintained at constant temperature. *Plant Physiology*, **20**, 467-504.

King, G.A., Davis, K.M., Stewart, R.J. and Borst, W.M. (1995) Similarities in gene expression during the postharvest-induced senescence of spears and natural foliar senescence of asparagus. *Plant Physiology*, **108**, 125-128.

Knee, M. (1991) Fruit metabolism and practical problems of fruit storage under hypoxia and anoxia, in *Plant Life Under Oxygen Deprivation* (eds M.B. Jackson, D.D. Davies and H. Lambers), SPB Academic Publishing, The Hague, pp. 229-243.

Koch, K.E., Nolte, K.D., Duke, E.R., McCarty, D.R. and Avigne, W.T. (1992) Sugar levels modulate differential expression of maize sucrose synthase genes. *Plant Cell*, **4**, 59-69.

Krebs, H.A. and Johnson, W.A. (1937) The role of citric acid in intermediate metabolism in animal tissues. *Enzymologia*, **4**, 148-156.

Kubo, Y., Inaba, A. and Nakamura, R. (1990) Respiration and C_2H_4 production in various harvested crops held in CO_2-enriched atmospheres. *Journal of the American Society for Horticultural Science*, **115**, 975-978.

Kumar, S. and Sinha, S.K. (1992) Alternative respiration and heat production in ripening banana fruits (*Musa paradisiaca* var. Mysore Kadali). *Journal of Experimental Botany*, **43**, 1639-1642.

Kumar, S., Patil, B.C. and Sinha, S.K. (1990) Cyanide resistant respiration is involved in temperature rise in ripening mangoes. *Biochemical and Biophysical Research Communications*, **168**, 818-822.

Kupferman, E. (1997) Controlled atmosphere storage of apples, in *CA '97 Proceedings, Vol. 2: Apples and Pears* (ed. E.J. Mitcham), Postharvest Horticulture Series No. 16, University of California, Davis, pp. 1-30.

Lakakul, R. (1994) *Modified-atmosphere packaging on apple slices: modeling respiration and package oxygen partial pressure as a function of temperature and film characteristics.* MS thesis, Michigan State University, East Lansing, MI, USA.

Lakshminarayana, S. and Subramanyan, H. (1970) CO_2 injury and fermentative decarboxylation in mango fruit at low temperature storage. *Journal of Food Science*, **7**, 148-152.

Lange, D. and Kader, A. (1997) Elevated carbon dioxide exposure alters intracellular pH and energy charge in avocado fruit tissue. *Journal of the American Society of Horticultural Science*, **122**, 253-257.

Laties, G.G. (1982) The cyanide-resistant, alternative path in higher plant respiration. *Annual Review of Plant Physiology*, **33**, 519-555.

Lebermann, K.W., Nelson, A.I. and Steinberg, M.P. (1968) Postharvest changes of broccoli stored in modified atmospheres. *Food Technology*, **22**, 490-493.

Li, C. and Kader, A.A. (1989) Residual effects of controlled atmospheres on postharvest physiology and quality of strawberries. *Journal of the American Society for Horticultural Science*, **114**, 629-634.

Lidster, P.D., Lightfoot, H.J. and McRae, K.B. (1983) Production and regeneration of principal volatiles in apples stored in modified atmospheres and in air. *Journal of Food Science*, **48**, 400-402, 410.

Loescher, W.H. and Everard, J.D. (1996) Metabolism of carbohydrates in sinks and source: sugar alcohols, in *Photoassimilate Distribution in Plants and Crops: Source-Sink Relationships* (eds E. Zamski and A.A. Schaffer), Marcel Dekker, New York, pp. 185-207.

Makhlouf, J., Willemot, C., Arul, J., Castaigne, F. and Emond, J.-P. (1989a) Long-term storage of broccoli under controlled atmosphere. *HortScience*, **24**, 637-639.

Makhlouf, J., Willemot, C., Arul, J., Castaigne, F. and Emond, J.-P. (1989b) Regulation of ethylene biosynthesis in broccoli flower buds in controlled atmospheres. *Journal of the American Society for Horticultural Science*, **114**, 955-958.

Mapson, L.W. and Burton, W.G. (1962) The terminal oxidases of potato tuber. *Biochemistry Journal*, **82**, 19-25.

Marcellin, P. (1974) Conditions physiques de la circulation des gaz respiratories à travers la masses des fruits et maturation, in *Colloques Internationaux C.N.R.S. No. 238: Facteurs et Régulation de la Maturation Des Fruits*, Centre National de la Recherche Scientifique, Paris, pp. 241-251.

Matile, -P., Hortensteiner, -S. and Thomas, -H. (1999) Chlorophyll degradation. *Annual Review of Plant Physiology and Plant Molecular Biology*, **50**, 67-95.

Mattheis, J.P. and Fellman, J.K. (2000) Impacts of modified atmosphere packaging and controlled atmospheres on aroma, flavor, and quality of horticultural commodities. *HortTechnology*, **10**, 507-510.

Mattheis, J.P., Buchanan, D.A. and Fellman, J.K. (1991) Change in apple fruit volatiles after storage in atmospheres inducing anaerobic metabolism. *Journal of Agricultural and Food Chemistry*, **39**, 1602-1605.

Mayer, A.M. and Harel, E. (1979) Polyphenol oxidases in plants. *Phytochemistry*, **18**, 193-215.

McCarty, D.R., Shaw, D.R. and Hannah, L.C. (1986) The cloning, genetic mapping, and expression of the constitutive sucrose synthase locus of maize. *Proceedings of the National Academy of Sciences of the USA*, **83**, 9099-9103.

McGlasson, W.B. and Wills, R.B.H. (1972) Effects of oxygen and carbon dioxide on respiration, storage life and organic acids of green bananas. *Australian Journal of Biological Sciences*, **25**, 35-42.

Meeuse, B.J.D. (1975) Thermogenic respiration in Aroids. *Annual Review of Plant Physiology*, **26**, 117-126.

Millar, A.H., Wiskich, J.T., Whelan, J. and Day, D.A. (1993) Organic acid activation of the alternative oxidase of plant mitochondria. *FEBS Letters*, **329**, 259-262.

Mir, N.A., Salon, C. and Canvin, D.T. (1995) Inorganic carbon-stimulated photoreduction is suppressed by N assimilation in air grown cells of *Synechococcus* UTEX 625. *Plant Physiology*, **109**, 1295-1300.

Mir, N.A., Curell, E., Khan, N., Whitaker, M. and Beaudry, R.M. (2001) Harvest maturity, storage temperature, and 1-MCP application frequency alter firmness retention and chlorophyll fluorescence of 'Redchief Delicious' apple fruit. *Journal of the American Society for Horticultural Science*, **125**, 618-624.

Mitcham, E., Zhou, S. and Kader, A. (1997) Potential of CA for postharvest insect control in fresh horticultural perishables, in *CA '97 Proceedings, Vol. 1: CA Technologies and Disinfestation Studies* (eds J. Thompson and E. Mitcham), Postharvest Horticulture Series No. 15, University of California, Davis, pp. 78-90.

Mitchell, P. (1966) Chemiosmotic coupling in oxidative and photosynthetic phosphorylation. *Biological Review*, **41**, 445-502.

Moreau, F. and Romani, R. (1982) Malate oxidation and cyanide-insensitive respiration in avocado mitochondria during climacteric cycle. *Plant Physiology*, **70**, 1385-1390.

Neal, G.E. and Hulme, A.C. (1958) The organic acid metabolism of Bramley's seedling apple peel. *Journal of Experimental Botany*, **9**, 142-157.

Nobel, P. (1983) *Biophysical Plant Physiology and Ecology*, W.H. Freeman and Company, New York.

Peiser, G., López-Gálvez, G. and Cantwell, M. (1997) Changes in off-odor volatiles of salad products during storage, in *CA '97 Proceedings, Vol. 5: Fresh-cut Fruits and Vegetables and MAP* (ed. J. Gorny), Postharvest Horticulture Series No. 19, University of California, Davis, pp. 23-28.

Plaxton, W. (1996) The organization and regulation of plant glycolysis. *Annual Review of Plant Physiology and Plant Molecular Biology*, **47**, 185-214.

Podestá, F.E. and Plaxton, W.C. (1994) Regulation of cytosolic carbon metabolism in germinating *Ricinus communis* cotyledons. II. Properties of phosphoenol pyruvate carboxylase and cytosolic pyruvate kinase associated with the regulation of glycolysis and nitrogen assimilation. *Planta*, **194**, 381-387.

Purvis, A.C. and Shewfelt, R.L. (1993) Does the alternate pathway ameliorate chilling injury in sensitive plant tissues? *Physiologia Plantarum*, **88**, 712-718.

Radmer, R.J. and Kok, B. (1976) Photoreduction of O_2 primes and replaces CO_2 assimilation. *Plant Physiology*, **58**, 336-340.

Ranson, S.L., Walker, D.A. and Clarke, I.D. (1960) Effects of CO_2 on mitochondrial enzymes from *Ricinus*. *Biochemical Journal*, **76**, 216-220.

Rich, P.R. and Bonner, W.D. (1978) The sites of superoxide anion generation in higher plant mitochondria. *Archives of Biochemistry and Biophysics*, **188**, 206-213.

Richardson, D.G. and Kosittrakun, M. (1995) Off-flavor development of apples, pears, berries, and plums under anaerobiosis and partial reversal in air, in *Fruit Flavors: Biogenesis, Characterization and Authentication* (eds R.L. Rouseff and M.L. Leahy), American Chemical Society, Washington DC, pp. 211-275.

Richardson, D.G. and Kupferman, E. (1997) Controlled atmosphere storage of pears, in *CA '97 Proceedings, Vol. 2: Apples and Pears* (ed. E.J. Mitcham), Postharvest Horticulture Series No. 16, University of California, Davis, pp. 31-35.

Roberts, J.K.M., Callis, J., Jardetzky, O., Walbot, V. and Freeling, M. (1984a) Cytoplasmic acidosis as a determinant of flooding intolerance in plants. *Proceedings of the National Academy of Sciences of the USA*, **81**, 6029-6033.

Roberts, J.K.M., Callis Wemmer, D., Walbot, V. and Jardetzky, O. (1984b) Mechanism of cytoplasmic pH regulation in hypoxic maize root tips and its role in survival under hypoxia. *Proceedings of the National Academy of Sciences of the USA*, **81**, 3379-3383.

Sachs, M.M., Freeling, M. and Okimoto, R. (1980) The anaerobic proteins of maize. *Cell*, **20**, 761-767.

Sachs, M.M., Subbaiah, C.C. and Saab, I.N. (1996) Anaerobic gene expression and flooding tolerance in maize. *Journal of Experimental Botany*, **47**, 1-15.

Saltveit, M.E. (1997) A summary of CA and MA recommendations for harvested vegetables, in *CA '97 Proceedings, Vol. 4: Vegetables and Ornamentals* (ed. M.E. Saltveit), Postharvest Horticulture Series No. 18, University of California, Davis, pp. 98-117.

Schwartz, D. (1969) An example of gene fixation resulting from selective advantage in suboptimal conditions. *American Naturalist*, **103**, 479-481.

Sfakiotakis, E.M. and Dilley, D.R. (1973) Induction of autocatalytic ethylene production in apple fruits by propylene in relation to maturity and oxygen. *Journal of the American Society for Horticultural Science*, **98**, 504-508.

Shamaila, M., Powire, W.D. and Skura, B.J. (1992) Analysis of compounds from strawberry fruit stored under modified atmosphere packaging (MAP). *Journal of Food Science*, **5**, 1173-1176.

Shipway, M.R. and Bramlage, W.J. (1973) Effects of carbon dioxide on activity of apple mitochondria. *Plant Physiology*, **51**, 1095-1098.

Silva, S. (1998) *Regulation of glycolytic metabolism in asparagus spears (Asparagus officinalis L.)*, PhD thesis, Department of Horticulture, Michigan State University, East Lansing, MI, USA.

Siriphanich, J. and Kader, A.A. (1986) Changes in cytoplasmic and vacuolar pH in harvested lettuce tissue as influenced by CO_2. *Journal of the American Society for Horticultural Science*, **111**, 73-77.

Sisler, E. and Blankenship, S. (1996) US Patent No. 5, 518, 988, *Method of counteracting an ethylene response in plants*, May 21.

Smeekens, S. (2000) Sugar-induced signal transduction in plants. *Annual Review of Plant Physiology and Plant Molecular Biology*, **51**, 49-81.

Smyth, A.B., Song, J. and Cameron, A.C. (1998) Modified-atmosphere packaged cut iceberg lettuce: effect of temperature and O_2 partial pressure on respiration and quality. *Journal of Agricultural and Food Chemistry*, **46**, 4556-4562.

Solomos, T. (1977) Cyanide-resistant respiration in higher plants. *Annual Review of Plant Physiology*, **28**, 279-297.

Solomos, T. (1997) Effects of hypoxia on the senescence of horticultural crops, in *CA '97 Proceedings, Vol. 4: Vegetables and Ornamentals* (ed. M.E. Saltveit), Postharvest Horticulture Series No. 18, University of California, Davis, pp. 138-148.

Song, J., Deng, W., Fan, L., Verschoor, J. and Beaudry, R.M. (1997) Aroma volatile and quality changes in modified atmosphere packaging, in *CA '97 Proceedings, Vol. 5: Fresh-cut Fruits and Vegetables and MAP* (ed. J. Gorny), Postharvest Horticulture Series No. 19, University of California, Davis, pp. 89-95.

Soole, K.L. and Menz, R.I. (1995) Functional molecular aspects of the NADH dehydrogenases of plant mitochondria. *Journal of Bioenergetics and Biomembranes*, **27**, 397-406.

Streeter, J.G. and Thompson, J.F. (1972a) Anaerobic accumulation of γ-aminobutyric acid and alanine in radish leaves (*Raphanus sativus* L.). *Plant Physiology*, **49**, 572-578.

Streeter, J.G. and Thompson, J.F. (1972b) *In vivo* and *in vitro* studies on γ-aminobutyric acid metabolism with the radish plant (*Raphanus sativus* L.). *Plant Physiology*, **49**, 579-584.

Taiz, L. and Zeiger, E. (1998) *Plant Physiology*, 2nd Ed. Sinauer Associates, Inc., Sunderland Massachusetts.

Thomas, M. (1925) The controlling influence of carbon dioxide, V, A quantitative study of the production of ethyl alcohol and acetaldehyde by cells of the higher plants in relation to concentration of oxygen and carbon dioxide. *Biochemical Journal*, **19**, 923-947.

Tian, M.S., Downs, C.G., Lill, R.E. and King, G.A. (1994) A role for ethylene in the yellowing of broccoli after harvest. *Journal of the American Society for Horticultural Science*, **119**, 276-281.

Tsukaya, H., Ohshima, T., Naito, S., Chino, M. and Komeda, Y. (1991) Sugar-dependent expression of the CHS-A gene for chalcone synthase from *Petunia* in transgenic *Arabidopsis*. *Plant Physiology*, **97**, 1414-1421.

Tucker, M. and Laties, G. (1985) The dual role of oxygen in avocado respiration: kinetic analysis and computer modeling of diffusion-affected respiratory oxygen isotherm. *Plant Cell and Environment*, **9**, 117-127.

Vámos-Vigyázó, L. (1981) Polyphenol oxidase and peroxidase in fruits and vegetables. *CRC Critical Reviews of Food Science and Nutrition*, **15**, 49-127.

van der Plas, L.H.W. and Wagner, M.J. (1980) Influence of ethanol on alternative oxidase in mitochondria from callus-forming potato tuber discs. *Physiologia Plantarum*, **49**, 121-126.

Vanlerberghe, G.C. and McIntosh, L. (1997) Alternate oxidase: from gene to function. *Annual Review of Plant Physiology and Plant Molecular Biology*, **48**, 703-734.

Vedel, F., Lalanne, E., Sabar, M., Chetrit, P. and De Paepe, R. (1999) The mitochondrial respiratory chain and ATP synthase complexes: composition, structure and mutational studies. *Plant Physiology and Biochemistry*, **37**, 629-643.

Wagner, A.M. and Krab, K. (1995) The alternative respiration pathway in plants: role and regulation. *Physiologia Plantarum*, **95**, 318-325.

Wankier, B.N., Salunkhe, D.K. and Campbell, W.F. (1970) Effects of controlled atmosphere storage on biochemical changes in apricot and peach fruit. *Journal of the American Society for Horticultural Science*, **95**, 604-609.

Watkins, C.B. (2000) Responses of horticultural commodities to high carbon dioxide as related to modified atmosphere packaging. *HortTechnology*, **10**, 501-506.

Wenzler, H., Migneri, G., Fisher, L. and Park, W. (1989) Sucrose-regulated expression of a chimeric potato tuber gene in leaves of transgenic tobacco plants. *Plant Molecular Biology*, **13**, 347-354.

Westall, R.G. (1950) Isolation of gamma-aminobutyric acid from beetroot (*Beta vulgaris*). *Nature*, **165**, 717-718.

Winter, H. and Huber, S.C. (2000) Regulation of sucrose metabolism in higher plants: localization and regulation of activity of key enzymes. *Critical Reviews in Plant Science*, **19**, 31-67.

Wu, J., Denner, L.A., Lin, C.T. and Hwang, B. (1988) Glutamate decarboxylase, in *Glutamine and Glutamate in Mammals, Vol. 1* (ed. E. Kvamme), CRC Press, Boca Raton.

Yearsley, C.W. (1996) *Internal lower oxygen limits of apple fruit*, PhD dissertation, Massey University, Palmerston North, New Zealand.

Yip, W.-K., Jiao, X.-Z. and Yang, S.F. (1988) Dependence of *in vivo* ethylene production rate on 1-aminocyclopropane-1-carboxylic acid and oxygen concentrations. *Plant Physiology*, **88**, 553-558.

Zeng, Y., Wu, Y., Avigne, W.T. and Koch, K.E. (1998) Differential regulation of sugar-sensitive sucrose synthases by hypoxia and anoxia indicate complementary transcriptional and posttranscriptional responses. *Plant Physiology*, **116**, 1573-1583.

7   Mechanical injury

Michael Knee and A. Raymond Miller

Now it is Autumn and the falling fruit
and the long journey towards oblivion.
The apples falling like great drops of dew
to bruise themselves an exit from themselves

D.H. Lawrence

7.1   Introduction

The ancestral species of the fleshy fruits that we enjoy today probably evolved as vehicles for dispersal of seeds by reptiles or mammals (van der Pijl, 1982). Many drupe and berry fruits that may once have been dispersed by reptiles are now more commonly eaten by birds. Because birds cannot peel or chew, the edible tissues of bird-dispersed fruits tend to be readily accessible and easily disrupted. Mammals are often more adept at peeling away hard outer tissues and can masticate tough tissue. So it seems likely that ancestral melons, peaches, apples, prunes, bananas, mangoes and pineapples, for example, co-evolved with mammals as dispersal agents. Citrus fruits would seem to belong in this group but apparently they are mostly taken by birds (van der Pijl, 1982).

The physical properties of the ancestral fruits were related to the dispersal of their seeds by non-human animals. Commercial production of fruits requires different physical attributes so that the soft tissues survive their dispersal through the marketing chain in a state suitable for human consumption (the seeds will be discarded). We have all experienced the conflict between our eating preferences and marketing requirements. We may have enjoyed a tree-ripe peach or apple and made the sad comparison with the commodities trading under the same names in the grocery store. At times, agricultural scientists have been accused of developing fruit varieties and postharvest handling technology more for the convenience of the shipper than for the benefit of consumers (Hightower, 1976).

It can be instructive to see the results of attempts to introduce to the market fruits that are not far removed from their wild progenitors. Pawpaw (*Asimina triloba*) is a native of the deciduous forests of the eastern USA whose fruit has a melting texture when ripe. It occasionally turns up in specialty grocery stores in the hard green condition; even in this state its skin is often discolored by the minor abrasions that it experiences during distribution. This chapter will address how fruit anatomy and physiology interact with mechanical stresses in the postharvest environment to influence fruit quality.

7.2 Fruit anatomy

Many insights into the physical properties of fruits can be obtained from nearly 200 years of research on their cellular anatomy, summarized by Roth (1977). The edible parts of fleshy fruits contain masses of mainly parenchyma cells. These typically exceed 100 µm in at least one dimension, whereas their cell walls are about 1 µm thick and usually unlignified. Because of these features they are among the most fragile of plant cells. The rigidity of fruit tissues is conferred by the turgor pressure generated by the osmotic potential of the organic acids and sugars in the vacuole that occupies most of the cell volume. The parenchyma cells in many fruits are more or less isodiametric. They are elongated along the radial axis of the fruit in citrus, banana and raspberry. Radial arrangements of the cells themselves may be obvious as in kiwifruit, peach and strawberry or only apparent on microscopic examination as in the apple. A further feature contributing to the weakness of fruit parenchyma tissue is that the cells are often loosely packed leaving extensive intercellular spaces between them. As a consequence of cell shape and arrangement fruit tissues may be mechanically anisotropic, responding differently to mechanical stresses imposed from different directions (Khan and Vincent, 1990).

The fragility of the parenchyma cells can be offset by other features of fruit structure. Fruit epidermis generally consists of small, tightly packed cells with a thick outer cuticle. The surface of many fruits is smooth and waxy and this could alleviate effects of frictional contact between fruits and other surfaces. By contrast, surface irregularities such as the achenes on strawberries or the pubescence on peaches increase frictional resistance and are likely to aggravate some kinds of damage. Beneath the epidermis the hypodermis may contain up to six layers of collenchyma, contributing to the strength of the 'skin'. The strength of the skin in fruits like apple gives some protection against puncture but less against impact where the inelastic outer cell layers merely transmit the force to the underlying parenchyma cells. Some fruits have a much thicker rind, conferring a higher degree of protection to all kinds of mechanical stress to the underlying tissues than is possible with only a few cell layers. The loose cell packing in the albedo layer of citrus rind seems to be an almost ideal impact-absorbing feature. It is difficult to imagine global distribution of citrus fruits without that outer protection.

Another feature that enhances mechanical stability is the presence of structural elements such as vascular strands and fibers or sclereids. These are more prevalent in drupes (peach, plum and mango) and pomes (apple and pear) than in berries (grape, tomato and banana).

7.3 Fruit cells

Vincent (1990) summarizes earlier work on the strength of plant cells. Whereas cellulose is very strong, breaking at 25 GPa, the strongest plant cells break at around 0.9 GPa. These are fibers with a high content of highly ordered cellulose microfibrils. As already noted fruit parenchyma cells are some of the weakest plant cells. Their cell walls contain a low proportion of cellulose microfibrils randomly arranged in a gel matrix of pectic polysaccharides. There are few reports of the strength of this kind of cell. Carpita (1985) found that cells in a carrot suspension culture burst with an internal pressure of 3 MPa, corresponding to a stress of 450 MPa on the wall itself. Harker and Hallet (1992, 1994) reported that apple and kiwifruit cells burst at an internal pressure of 0.5 to 1 MPa, whereas nectarines burst at 0.25 to 0.5 MPa (Harker and Sutherland, 1993). According to Carpita's (1985) equation relating bursting pressure and breaking stress on the cell wall, a pressure of 1 MPa corresponds to a stress of 50 MPa for a spherical cell with a radius of 100 µm and a cell wall thickness of 1 µm.

Fruit tissues are injured under much lower stresses (of the order of 0.1 MPa) than the apparent strength of their cell walls. Vincent (1990) concludes that plant tissues under compression can fail because of 'buckling of cell walls rather than tensile failure of cellulose'. This change of shape and volume would require extrusion of water through the cell wall and this would be dependant on the permeability of the plasma membrane and cell wall to water. Steudle and Wieneke (1985) report a value of 10^{-13} m Pa^{-1} s^{-1} for the hydraulic conductivity of apple cells. According to Wu and Pitts (1999) this would allow for extrusion under semi-static loading with a strain rate of $0.001 \ s^{-1}$. However, rapid loading such as occurs during impact would not allow time for extrusion and cell breakage would be more likely.

Cells may be broken because the stress is concentrated on particular regions of the cell wall. Under both compression and shear, stress will be experienced by the contact areas between cells. Under shear the cell would break if the adhesion between cells were stronger than the cell walls. Adhesion depends on the properties of the middle lamella, which is weaker than the cell wall because it contains pectin only and no cellulose. However the area over which forces are distributed is larger for the middle lamella than for the cell wall. Tensile testing of immature fruit tissues indicated breaking stresses of 0.34 MPa for apple, 0.9 MPa for kiwifruit and 0.25 MPa for nectarine (Harker and Hallett, 1992, 1994; Harker and Sutherland, 1993). The strength of the middle lamella was higher than this for two reasons. Because of the presence of intercellular spaces, cell-to-cell contact areas comprise only a proportion of the cross-sectional area of tissue. Since failure in all cases was by cell breakage, the breaking stress of the middle lamella was not reached in the tests.

Cells separate in ripe fruits at strains of 0.08 MPa for apple and 0.01 MPa for the other fruits (Harker and Hallett, 1992, 1994; Harker and Sutherland,

1993). Cells in ripe fruit under mechanical stress may have more opportunity to slide against one another in order to relax the stress without breaking cells. Interestingly, Harker and Hallett (1992) and Steudle and Wieneke (1985) using very different methods concluded that the walls of apple cells become stronger during ripening.

Up to this point we have considered fruit tissues as 'St Venant bodies' whose components would separate from each other whenever a certain strain was reached. However, other factors modify the stress–strain relationship for fruit tissues. Cell walls are elastic and cells can absorb energy within the limits of their elastic extensibility. Much of the elasticity of a cell is taken up by the turgor pressure of the cell contents. To complicate matters further the modulus of elasticity is itself pressure-dependent (Steudle and Wieneke, 1985). High turgor pressure makes cells brittle so that impact energy must be absorbed by means other than elastic deformation. Conversely, if turgor pressure is lowered the cells can be deformed elastically and return to their original shape after stress is relieved. When the limit of elasticity is reached, plastic deformation becomes the dominant effect. This can occur at several levels: microfibrils may slip in the matrix of the cell wall; cells may slip against one another along the line of the middle lamella; and finally cells may be broken and the tissue disrupted. Because fruit tissues display viscoelastic properties, the extent to which permanent deformation (injury) occurs depends on the rate at which force is applied. Slow loading allows plastic deformation to occur when the elastic limit is reached so that more deformation or strain occurs before failure than under fast loading.

Fruits differ from most other plant organs that are used as foods in that they often undergo profound developmental changes in their physical properties. With the exception of the citrus group, ripening usually involves loss of cellular cohesion and often weakening of the primary cell wall through the action of various polysaccharide degrading enzymes (see Chapter 3). Many immature fruits contain starch reserves that are hydrolyzed to sugars during ripening. This increases the osmotic potential of the cell contents, which could lead to an increase in turgor pressure if the external water potential were high. Because of their weak cell walls, fruit cells are unusually prone to rupture in solutions of low osmotic strength (Simon, 1977). However, in the intact fruit it seems likely that there is some equilibration of inter- and intracellular solutes so that turgor forces do not disrupt the cells as wall pressure declines with ripening. As a consequence of cell wall and turgor changes the internal tissues become less elastic and more plastic with ripening. Although the tissue becomes weaker with ripening it is not inevitable that more injury will occur with a given mechanical stress. The mode of failure under mechanical stress may change so that cells separate and less cell breakage occurs in the ripe fruit than in the unripe fruit (Harker and Hallett, 1992, 1994; Harker and Sutherland, 1993).

Table 7.1 Levels of fruit structure and corresponding components affecting mechanical strength of fruits

Level of structure	Component	Critical feature(s)
Molecular	Solutes	Concentration
	Lipids	Composition
	Structural polysaccharides	Composition, chain-length
	Structural proteins	Composition
Molecular assemblage	Cell membranes	Fluidity
	Middle lamella	Bonding
	Cell wall	Matrix and microfibrils
Cellular	Cell	Shape
	Cell wall	Thickness
	Cell to cell	Contact areas
Tissue	Epidermis	Cracks and pores
	Collenchyma	Number of layers
	Parenchyma	Density
	Vascular bundles	Thickness and number
	Sclerenchyma	Arrangement
	Air space	Proportion
Whole fruit		Tissue arrangement, shape, size, surface texture

In mechanical terms, fruits are complex biological composites whose metabolism may cause changes in physical properties over time. Because of their composite character their physical properties are determined at several levels of structure from the component molecules to the whole organ (table 7.1). While it is possible to describe, at least in qualitative terms, how features at every level influence response to mechanical stress, we do not know how to use information about each feature to predict the extent of damage resulting from a specific stress acting on a particular fruit. Mechanical damage is an emergent property of the whole fruit and empirical approaches have dominated research on the avoidance of injury in postharvest handling. We are beginning to be able to describe how physical properties of tissue influence their behavior under particular kinds of stress. Also it is now possible to observe changes in the mechanical properties of plant tissue resulting from the manipulation of the activity of single genes (Langley *et al.*, 1994).

7.4   Causes of injury

Fruits can be injured in several ways while still on the plant. During growth they may come into contact with other fruits or other parts of the plant such as branches, causing abrasion, puncture and bruising. Herbivorous animals, slugs, insects, birds and mammals, can puncture the skin and consume a proportion of the tissue. Damage is prevented by controlling the animal in question. Weather is another important cause of damage: wind can aggravate damage caused by contact with other plant parts; hail causes impact bruises that can be locally devastating for fruit production. Apart from predation, preharvest physical damage is infrequent or sporadic and not easily controlled. Affected fruits will be culled in the packing house.

The focus of this chapter is on damage caused by human handling. This begins at harvest, which still occurs by hand for most fruits. Of course, harvesting itself is an injury. Usually the aim is to separate the fruit from the plant by a break or cut at the base of the pedicel. If there is an abscission zone it is usually desirable that the break should occur at this point. The ethylene releasing compound, 2-chloroethanephosphonic acid (ethephon) is widely used on fruits to promote abscission and facilitate the harvesting process. Some fruits, particularly the citrus group, do not have an abscission zone in the pedicel. Citrus fruits tend to abscise from the calyx but this is undesirable since the wound is a site of entry for pathogens. The same danger exists when pedicels are pulled out of other fruits when they are picked or during handling after harvest. If the fruit does not detach easily from the plant it can be injured by the force of the hand. This is particularly true when two or three fingertips are used to pull, and so growers plead for the use of a whole hand grip and a bending movement to concentrate the force on the pedicel.

After harvest the fruit begins its long, bumpy ride from the field or orchard to the consumer. It may pass through several containers and modes of transport along the way. Every time there is a change of container there is the danger of impact with other fruits, containers and equipment used to sort and pack the fruit. Punctures, cuts, bruises and abrasions can result, depending on how contact is made and on the nature of the surface with which contact is made. For delicate fruits such as strawberries it is highly desirable to avoid changing containers by harvesting directly into the pack that will be offered to the consumer. The same dangers of injury arise from rapid changes of momentum, such as bumps during transport. Transport introduces a different kind of stress through vibration, a series of small but possibly cumulative loads applied to the fruit that cause abrasion or bruising. During transport or storage fruits are also subject to compression from other fruits above them. Whereas the largest and lightest container may be economical and convenient for handling, it must be strong enough to bear the load when stacked and the upper limit on size is set by the strength (resistance to compression) of the fruit that is carried.

7.5   Losses caused by mechanical injury

The most comprehensive data on incidence of mechanical damage (and other causes of loss) of fruits comes from surveys of some of the produce passing through New York terminal markets between 1972 and 1984 (table 7.2). There is no reason to think that similar crops would arrive in better condition in

Table 7.2 Incidence of mechanical injuries on lots of fruit passing through New York terminal markets 1972 to 1984 (data summarized from: Cappellini *et al.*, 1986, 1987, 1988a,b,c; Ceponis *et al.*, 1986a,b, 1987, 1988; Wells *et al.*, 1994.) (only data for injuries that affected more than 1% of fruit lots are shown)

Fruit	Number of lots	Injury	Percentage with injury		
			Lots	All fruit	Affected lots
Apple	4453	Bruise	75.1	5.54	7.38
		Cut/puncture	3.3	0.10	3.10
Apricot	204	Bruise	44.6	3.79	8.49
		Cut/puncture	6.9	0.23	3.36
Avocado	889	Bruise	6.4	0.40	6.23
Cherry	2455	Bruise	35.0	2.37	6.76
		Cut/puncture	1.3	0.06	4.41
Grape	8100	Crush	33.7	1.25	3.73
		Shatter	47.3	2.26	4.77
		Wet/sticky	42.3	2.33	5.51
Grapefruit	4910	Bruise	4.7	0.23	4.82
Mango	717	Bruise	8.5	0.84	9.84
Nectarine	2576	Bruise	72.7	5.31	7.31
		Cut/puncture	3.2	0.12	3.87
Orange	9104	Bruise	4.8	0.29	6.03
Papaya	209	Bruise	14.8	1.59	10.73
Peach	2610	Bruise	86.9	9.19	10.58
		Cut/puncture	1.0	0.03	3.00
Pear	4409	Bruise	17.5	0.85	4.86
		Cut/puncture	1.9	0.06	3.46
Pineapple	677	Bruise	14.2	1.36	9.57
Plum	3079	Bruise	25.7	1.38	5.38
		Cut/puncture	16.3	1.03	6.30
Strawberry	1777	Bruise	69.2	7.98	11.53
Tomato	9059	Bruise	14.0	0.66	4.73
Watermelon	894	Bruise	35.3	3.42	9.66
		Abrasion	3.2	0.31	9.48

other markets or that the conditions will have changed substantially in the intervening years. Bruising was the most common injury recorded on all fruit types. Punctures or cuts were recorded on about half the types of fruit and the incidence was generally lower than for bruising. Some injuries appeared on only one kind of fruit, such as abrasion on water melons and 'rolling' on pears.

For apple, nectarine, peach and strawberry more than half of the fruit samples inspected showed some bruising, although the overall incidence did not exceed 10% for any type of fruit. Bruising seems to be almost unavoidable in postharvest handling of peaches but 13% of samples showed no evidence of bruising. For other fruits the incidence of bruising in affected batches was up to ten times higher than the average incidence for each type of fruit. The affected batches were either unusually susceptible to bruising or suffered some unusual abuse during handling. Cuts and punctures were common only on plums and to a lesser extent, apricots. Again there was a large difference between incidence on affected batches and average incidence for each fruit type. Grapes were the only fruit distributed in bunches in these studies and thus the only fruit to exhibit 'shatter' which is the separation of fruits from the pedicels.

In the New York market studies mechanical injury was never the most common defect except on peaches. Usually one or more kinds of fungal disease exceeded mechanical injury. The fungi were often gray mold (*Botrytis*) or blue mold (*Penicillium*). Like most postharvest pathogens, these organisms cannot infect healthy tissue. Typically they enter through dead or wounded tissue before parasitizing the rest of the fruit (see Chapter 9). It is likely that minor mechanical injuries were not counted by the market inspectors and were only seen through their consequences in fungal infections. So mechanical injury could be the most important cause of defects and disease on most if not all kinds of fruit. If mechanical injury could be avoided (and it is a big 'if') there would be less need for fungicides to prevent disease and there would be much less loss of fruit.

Further studies estimated the loss of selected fruits at retail and consumer levels. Generally, less than 1% of grapefruit was culled because of mechanical injury at the retail level (Ceponis and Cappellini, 1985b). A low percentage (1 to 3%) of cherries were discarded because of mechanical injury at retail and consumer levels (Ceponis and Butterfield, 1981). Around 1% of nectarines were estimated to be lost at wholesale and consumer levels because of mechanical injury, but at the retail level the wastage was 4% (Ceponis and Cappellini, 1985a). The authors attributed this wastage to the excessive handling of nectarines in self-service displays. Since these are cull figures they reflect the proportion of fruit that was unsaleable or inedible rather than the totals affected by mechanical damage.

7.6   Fruit impact injuries

7.6.1   General features of impact injury

Impacts are the most common source of mechanical injury to fruits in the market. There has been extensive research to determine how injury occurs and how it can be avoided through improvements in handling from the point of harvest onwards. During an impact energy is delivered rapidly to the fruit surface and elastic strain may be more important than plastic deformation up to the point of failure. Thus it has been argued that fruits can be treated as essentially elastic bodies for the purpose of predicting impact injuries (Horsfield *et al.*, 1972). If the impact is purely elastic the absorbed energy is returned as the fruit rebounds from the surface and no injury occurs. The energy level that leads to bruising varies according to the impact surface. Round surfaces will concentrate more of the energy in a limited area than flat surfaces, causing damage at lower energy levels.

Soft impact surfaces, such as other fruits or padding will themselves deform during impact allowing the fruit itself to absorb and return energy over a larger surface than during contact with a firm surface. 'Instrumented spheres' have been designed to monitor impacts during harvesting and postharvest handling of fruits. Because other factors modify the relationship between impact energy and injury these instruments do not directly predict how much injury will occur (Pang *et al.*, 1994).

7.6.2 Energy and impact injury

The impact energy that is not returned through elastic recovery does other work. In apples it seems that it causes cellular damage because several authors have found that bruise volume is directly related to the energy absorbed during impact (Holt and Schoorl, 1977; Topping and Luton, 1986; Brusewitz and Barsch, 1989; Pang *et al.*, 1992). Whereas in apples energy absorbed accounts for over 90% of variation in bruise volume, in peaches less than 20% of variation could be accounted for in this way (Brusewitz *et al.*, 1991; Maness *et al.*, 1992). The low correlation for peaches probably reflects the low probability of bruising at impact energies that would cause bruising on most apples. For example, Brusewitz and Barstch (1989) reported that more than 50% of McIntosh apples were bruised at impact energies of 0.08 J but Maness *et al.* (1992) found that only 2.3% of peaches were bruised at an average impact energy of 0.07 J. The peach study included four cultivars at three stages of maturity from harvest maturity to firm ripe and even at 0.25 J only 27% of fruits were bruised. Data presented by Vergano *et al.* (1991) for five peach cultivars at normal harvest maturity suggest that 50% were bruised at an impact energy of 1 J (approximately equivalent to 100 g dropping from a height of 1 m). Hung and Prussia (1989) included peaches that had been stored in their study and found that 55% were bruised at an impact energy of 0.65 J.

When the data presented by Maness *et al.* (1992) were recalculated, excluding samples for which no bruising was recorded, the correlation (r^2) between bruise volume and energy absorbed was 0.50. This was higher than the correlation for bruise volume and total energy ($r^2 = 0.39$) and could be even higher if the calculation were performed only for the fruit within samples that showed damage. So it seems that the relationship between energy absorbed and bruise volume first proposed by Holt and Schoorl (1977) for apples does apply to peaches. Hung and Prussia (1989) define bruise susceptibility as the slope of the regression of bruise volume on energy absorbed. They show values of 500 to 1000 mm^3 J^{-1} for peaches around harvest time. This contrasts with figures of 5000 to 12,000 mm^3 J^{-1} for apples reported by Holt and Schoorl (1977), Brusewitz and Bartsch (1989) and Pang *et al.* (1992). Bruise susceptibility increased to 5000 mm^3 J^{-1} or more for peaches after four week's storage (Hung and Prussia, 1989).

7.6.3   Cellular structure and impact injury

Horsfield *et al.* (1972) proposed that bruising in peaches occurred through shear some distance below the skin of the fruit. Khan and Vincent (1990) argue that apple tissue is unlikely to fail in this way under compression because radial airspaces in the fruit parenchyma allow stress to be relieved by lateral movement of cells, so 'simple' compressive failure occurs. Apples contain 15–20% airspace whereas peaches are likely to resemble nectarines in having less than 10% airspace (Hatfield and Knee, 1988; Harker and Hallett, 1992; Harker and Sutherland, 1993). Failure through shear is more likely in peaches because lateral movement of cells is restricted by their lower proportion of airspace and the radial arrangement of vascular strands. Compression failure may be a sudden process in which energy is absorbed by a mass of breaking cells simultaneously. Shear can be a more gradual process where cells absorb energy by breaking over time. There may be more opportunity for 'crack stopping' and distribution of damage without obvious bruising in this process (Vincent, 1990). It may be relevant that apples tend to show sudden failure under compression, whereas peaches fail more gradually as stress increases (Fridley and Adrian, 1966).

7.6.4   Factors affecting impact damage

The energy exchanged at impact for a falling body increases in proportion to its mass. Although larger fruits may be desirable for profitability, they are more vulnerable to impact damage when they are dropped. Experiments on the effects of impact on fruits can be done in two ways: a fixed mass can be dropped onto a fruit or the fruit itself can be dropped onto a surface. The first method has the advantage that the impact force will be constant throughout the experiment whereas the second method may be more representative of what happens to fruits after harvest. In research on peaches that were dropped, fruit mass had a significant effect on bruise volume (but not on bruise incidence) (Brusewitz *et al.*, 1991; Maness *et al.*, 1992).

Fruit mass is a confounding factor in attempts to compare bruise susceptibility of fruit cultivars by dropping them (Maness *et al.*, 1992). Seasonal and cultural variations are further potential confounding factors and it is desirable that fruits from different sites and growing seasons should be included when varietal comparisons are made. These requirements are seldom met. Schoorl and Holt (1977) used constant impact energy to test three apple varieties in two seasons and found no consistent difference in bruise volume. Postharvest researchers commonly use a penetrometer with an 8 or 11 mm cylindrical plunger to assess the maturity and texture of fruits. There seems to be no relationship between resistance to the penetrometer plunger and susceptibility to bruising of different apple varieties (Topping and Luton, 1986; Klein, 1987). Klein (1987) reports low correlations between bruise dimensions and fruit weight or density for apples of 20 varieties that were dropped from 10 and 40 cm. However, this analysis appears to be based on data for individual apples, so a large part of the variance can be attributed to variability in measurements. Recalculation from the average values reported by Klein (1987) shows the expected strong correlations with fruit weight for bruise diameter ($r = 0.661$) and bruise volume ($r = 0.611$). Inclusion of fruit density in the models improves the correlation with bruise diameter ($r = 0.749$) and volume ($r = 0.661$). Both models suggest that density is inversely correlated with bruise dimensions. This is consistent with the argument developed above that the air spaces in apples make them prone to failure under some kinds of mechanical stress.

Fruit to fruit impacts may be the most common source of damage in postharvest handling, but most research has focused on contact between fruits and hard surfaces. Apparently, classical elasticity theory for colliding spheres gives an adequate description of apple-to-apple impact (Pang *et al.*, 1992). In this situation the total bruise volume for the two apples is proportional to the energy absorbed on impact but because of variation between apples one may fail before the other so that it absorbs most of the energy. The other apple may rebound elastically without failure. Pang *et al.* (1992) commented that bruise area is commercially more important than bruise volume because it will decide whether an apple is down-graded. They suggest that this should be related to the area of contact in the collision between apples. Their data were consistent with predictions of elasticity theory that impact energy is proportional to contact area to the power of 2.5.

Studman *et al.* (1997) observed that the green side of an apple is more likely to be bruised and will suffer more damage in an impact than the red side. This effect was seen in apple-to-apple impacts, illustrating the sensitivity to differences in failure properties noted by Pang *et al.* (1992) for this kind of impact. It is also interesting to note that Smith (1938) found that there was more intercellular space between the parenchyma cells on the green side of an apple than on the red. Peaches are more susceptible to damage on the fruit shoulder or on the suture than elsewhere, but there seems to be no difference between the red and the green sides (Brusewitz *et al.*, 1991; Schulte *et al.*, 1994).

7.6.5 Impact injury and harvest maturity

Because of the changes in physical properties of many fruits during development, changes in bruise susceptibility are to be expected. For commercial purposes there is interest in the changes in bruise susceptibility around harvest time as one factor in the decision of when to harvest. Bruise susceptibility after storage and ripening is relevant to marketing operations. Brusewitz *et al.* (1991) separated peach fruits soon after harvest into three ripeness categories and found an increase in bruise volume with ripeness. Maness *et al.* (1992)

observed a similar increase for several varieties of peaches but the percentage of impacts leading to bruises did not increase with ripeness. Hung and Prussia (1989) observed that mature peaches had lower bruise incidence than immature peaches at harvest. There was no difference in bruise volume resulting from impacts on ripe and unripe fruits. Vergano *et al.* (1991) found no difference in bruise incidence among maturity classes of peaches soon after harvest. Diener *et al.* (1979) reported a 30% decrease in bruise volume for Golden Delicious apples over a 27-day harvest period. Klein (1987) observed a 10% increase in bruise volume expressed as a proportion of fruit mass when apples were harvested over a period of approximately 30 days. Thus there is little evidence in support of the general belief that allowing fruit to mature on the tree before harvest increases the risk of bruising.

7.6.6   Ripening and impact injury

Fridley and Adrian (1966) reported that the 'impact yield energy' of apricots, peaches and pears decreased at least fourfold during ripening whereas it remained more constant for apples. Hung and Prussia (1989) observed that the ratio of bruise volume to impact energy increased approximately fivefold for peaches in storage. Holt and Schoorl (1977) found that bruise volume at constant impact energy increased for Jonathan and Delicious apples in storage whereas it remained more constant for Granny Smith apples. Klein (1987) reported that bruise volumes, averaged across two cultivars of apples, decreased during storage. The relationship between turgor pressure and the brittleness of fruit tissue was noted above and several authors have found that fruits suffer less bruising on impact if their turgor is first reduced by promoting weight loss. This was reported by Horsfield *et al.* (1972) and Brusewitz *et al.* (1992) for peaches, Garcia *et al.* (1995) for apples and Banks and Joseph (1991) for bananas. The turgor effect is not large enough to offset the increase in bruise susceptibility for a fruit such as the peach that softens considerably during storage and ripening. Apples soften less and a small increase in bruise susceptibility could be offset by the effect of change in turgor. This could explain the inconsistencies in the literature with respect to changes in bruise susceptibility during storage of apples.

7.6.7   Temperature and impact injury

Temperature might be expected to influence the physical properties of fruit tissue, but reports of the effects on bruising are inconsistent. According to Saltveit (1984) two varieties of apple showed progressively higher bruise volumes from 0 to 30°C. Klein (1987) agrees with Schoorl and Holt (1977) that there is no effect of temperatures from 0 to 25°C on bruising of apples. However the data presented by Schoorl and Holt (1977) show that bruise volumes were smaller at 30°C than at lower temperatures. Van Lancker (1979) demonstrated that the elastic modulus of Golden Delicious apples decreased between 3 and 33°C. Because the deformation and contact areas for impact with a metal sphere increased over the same range, he concluded that damage would be inversely related to temperature. Lidster and Tung (1980) found that higher proportions of cherries were injured by dropping at 0°C than at higher temperatures. Horsfield *et al.* (1972) reported that bruise volumes were increased when peaches were cooled from an unspecified temperature to 7°C.

7.7 Compression damage

Because of their viscoelastic character, fruit tissues can be injured by lower forces when applied over a long time than when subjected to rapid loading and unloading, such as occurs on impact. However more deformation (strain) may be required to cause failure under slow than under rapid loading conditions (Chen and Sun, 1984). The rate dependence of failure stress was confirmed for whole peaches and apricots by Fridley and Adrian (1966), cylinders of apple tissue by Chen and Sun (1984), cylinders of pear tissue by Baritelle and Hyde (2000) and whole cherries by Lidster and Tung (1979). In short-term impacts fruits behave like elastic bodies, returning to their original shape after temporary deformation. Thus elasticity theory can provide an adequate description of the physics of fruit impacts (Horsfield *et al.*, 1972). Over longer periods of stress plastic changes in fruit structure occur so that tissues do not return to their original dimensions. Some of the applied energy can be used in causing extrusion of water from cells, movement of cells within the tissue, creep of microfibrils in the cell wall matrix and permanent deformations of cell shape. These changes would reduce the energy available for actual breakage of cells and bruising but the probability of cell failure at a given stress is likely to increase with time. Once cells start to fail the energy stored in the strained tissue could be used to extend the failure to other cells. So the increase in probability of failure with time could result in failure at lower energies than cause failure in impacts.

Pitt (1992) summarizes data on failure stress for various fruits. Although values around 40 MPa were recorded for some fruits, typical values exceeded 100 MPa. Presumably, the low values represented ripe fruits and the higher values were for fruits at harvest maturity. Fruits are most likely to experience compression damage when handled in bulk. The bins used to transfer apples from the orchard to storage or packing facilities are typically 0.6 m deep. Timm *et al.* (1998) put force sensors in the base of such bins before loading with apples and recorded forces typically from 15 to 50 N during transport. Such forces would need to be concentrated on an area of 1 mm^2 to exceed the failure strains summarized by Pitt (1992). Failure under compression seems possible when fruit stalks are pressed against fruit surfaces but unlikely for other contacts between fruits under the static loads recorded by Timm *et al.* (1998). These loads can be exceeded when boxes of fruit are stacked if the boxes themselves do not support the weight of the fruit so that fruits in the bottom box have to support all those above.

Deformation without failure is a likely consequence of long-term compression of fruits in packages. Although citrus fruits are resistant to bruising they are prone to deformation under compression. Sarig and Nahir (1973) observed that Valencia oranges were permanently deformed by a 20 N load, whereas 50 N loads caused deformation in grapefruit (Rivero *et al.*, 1979). Kawada and Kitagawa (1984) found that grapefruit were more susceptible to deformation when compressed laterally than longitudinally. Apples are also more resistant to compression in the longitudinal than in the lateral direction (Khan and Vincent, 1991). However, the presence of the stalk makes it advantageous to pack apples on their sides in tray packs.

7.8   Vibration

Movement of fruit in containers during transport is probably a more serious cause of damage than simple compression. McLaughlin and Pitt (1984) showed that the probability of bruising of apple tissue increased with repeated cyclic loading at stress levels that caused little or no damage in a single cycle. This is analogous to the fatigue experienced by metals under cyclic loading. Fruits are most likely to suffer vibration injury during transport and fruits at the top of containers tend to experience the maximum acceleration (O'Brien *et al.*, 1978). The frequency and amplitude of vibration will depend on the size and construction of the container and the suspension system of the vehicle. Vibrational stress will be highest when the frequency coincides with the natural resonant frequency of the fruit. Because peaches are more elastic than apricots and they tend to resonate with the frequencies experienced during truck transport they suffered more vibrational injury (O'Brien *et al.*, 1963). Plastic bins caused less damage than wooden and vehicles with air suspension caused less than those with metal spring suspension for Golden Delicious apples (Timm *et al.*, 1996). Peaches are damaged when accelerations exceed 0.67 g (O'Brien *et al.*, 1978). The effects of vibration are more complex than impacts between fruits. Frictional forces are another source of injury. Vergano *et al.* (1992) found that plastic film or paper between peaches decreased bruising when the fruit were subjected to low frequency (6 Hz) vibrations.

7.9   Prevention of damage

Identification of the sources of injury described above allows one to begin developing strategies to reduce wounding in a commercial setting. Obvious preventative measures that could be implemented include: 1. minimizing injuries by mechanical harvesters; 2. reducing the number of times fruit are dropped or transferred; 3. reducing the height of drops; 4. reducing the number of abrupt directional changes; 5. removing sharp edges from containers or conveyors; 6. maintaining constant velocities between conveyor belts; 7. minimizing compression as fruit is funneled into narrow spaces; 8. padding sharp edges in the cleaning and sorting lines; 9. padding the bottom of receiving containers and inner surfaces of transport; and 10. packing fruit between layers of foam padding in shipping cartons. However, as stressed by Shewfelt (1986), the best means to minimize losses due to mechanical injury is through motivated employees who are careful and conscientious in their handling of the produce.

7.10   Detection of injury

Understanding types of injury and the response mechanisms to injury enables us to develop methods that identify and segregate wounded fruit on an individual or lot basis. Although the methods developed will probably be specific to a particular fruit, they must be general enough to detect the majority or all of the affected fruit and not be subject to variation caused by environmental factors or changes with growth and development. In some cases, simple visual inspection may suffice. For other produce, a number of different biochemical and physical methods have been investigated. Destructive or nondestructive methods may be useful when sampling representative fruit to gain an overall understanding of quality in a large shipment. However, non-destructive methods are a necessity if one wishes to separate wounded from undamaged fruit on an individual basis, which is probably the case since most shipments could contain a relatively high percentage of undamaged individuals among those that are wounded. Other factors that need to be considered when developing an evaluation method are the time required to conduct each evaluation and whether the method is amenable to mechanization and computerization. Instantaneous or near instantaneous measurements would be preferred in high volume packing houses and processing facilities. Inspection rates of four to ten produce items per second per lane have been proposed (Abbott *et al.*, 1998). In this same setting, rapid measurements coupled with predetermined specifications to detect wounded fruit would allow computer-aided decision-making and mechanical separation of fruits that do not meet the specifications. Analogous systems are already in operation to grade many kinds of produce for color, size and shape. A challenge to overcome for detection of much wounded produce is the internal, chemical, or textural nature of the undesirable character.

   Both physiological/biochemical and physical methods have been proposed to detect injury. Physiological/biochemical methods include assay for wound-inducible enzymes, wound metabolite analysis, volatile analysis and histochemical methods (for review see Miller, 1992, 2001). Physiological and biochemical methods are often destructive, time-consuming and not easily adapted to mechanization and computerization. Several physical methods to evaluate the quality of produce, including mechanical injury-related disorders, have also been investigated. The advantages of these methods are their adaptation to mechanization and computerization and their non-destructive nature. Research has been conducted to assess the applicability of firmness/textural analysis, acoustic impulse, resonance frequency, ultrasonics, X-ray, gamma ray, magnetic resonance, dielectric properties, fluorescence, refreshed delayed light emission (RDLE), light reflectance and light transmission (for review see Abbott et al., 1998; Dull, 1986).

7.11 Metabolism in injured tissue

7.11.1 Phenolics

The vacuoles of fruit cells typically contain a complex mixture of phenolic compounds. Many of these are potential substrates of catechol oxidase (phenolase, polyphenoloxidase) which is located in the plastids of intact cells. Cell breakage is inevitably accompanied by the disruption of the tonoplast so that substrates and enzyme are brought into contact. In the presence of molecular oxygen, catecholase attacks O-diphenol groups forming quinones and water. The quinones tend to polymerize naturally to dark colored compounds. The pH optimum of catechol oxidase is typically pH 5 to 6 and consequently the reaction may be retarded by the organic acids that occur in the vacuoles of most fruits (Mayer and Harel, 1981). The quinone groups can be reduced to hydroxyls by ascorbate that is also present in many fruits. Because of the complexity of the substrates and the reaction conditions it is not surprising that many attempts to relate browning potential of different fruits to enzyme activity or phenolic compounds have been inconclusive (e.g. Coseting and Lee, 1987; Klein, 1987). Amiot et al. (1992) analyzed the changes in phenolic composition in relation to color changes in bruised tissue of 11 apple cultivars. They found that overall browning was related to degradation of hydroxycinnamic acids and flavan-3-ols. Chlorogenic acid is usually the predominant hydroxycinnamic acid derivative in apples and is an excellent substrate for catechol oxidase. Although its oxidation products are soluble and only weakly colored, it seems to participate in co-oxidation reactions, involving the flavan-3-ols, that produce more strongly colored, insoluble products.

Bruises in apples darken within minutes but the intensity of the brown color develops for 10 to 20 h, after which it decreases (Samin and Banks, 1993). The dry weight of bruised tissue decreases by about 95% within 10 h, presumably because solutes are absorbed into surrounding healthy tissue. The fading of the bruised area is associated with a more gradual loss of water by evaporation or absorption into the tissue (Samin and Banks, 1993).

Skin darkening is a problem of peaches and nectarines associated with abrasion damage that is restricted to outer cell layers of the fruit (Chrisosto, 1993). This seems to be caused by access of metal ions and, possibly, alkaline solutions to anthocyanins in the disrupted cells.

7.11.2   Respiration

Stimulation of respiration by wounding of plant tissues has been observed in many organs and plant species. In many fruits the oxidation of phenolic compounds by catechol oxidase causes a transient increase in oxygen uptake by the damaged tissue, but there are more lasting effects on the cellular respiration of adjacent, undamaged tissue. Bruising caused increased respiration in apples (Robitaille and Janick, 1973), tomatoes (MacLeod *et al.*, 1976), cranberries (Massey *et al.*, 1982) and mangosteens (Ketsa and Koolpluksee, 1993). Massey *et al.* (1982) suggested that the rate of carbon dioxide production could be used as an indicator of mechanical abuse for cranberries. Mao *et al.* (1995) demonstrated that vibrational stress caused respiration to increase in figs although there was no visible sign of damage.

7.11.3   Ethylene synthesis

The literature on wound ethylene production is at least as extensive as that on wound respiration. Early research raised the possibility that ethylene could be produced by degradation of lipids after cell disruption (Galliard, 1978). Following the demonstration that ACC is the normal precursor for ethylene synthesis in plant tissues, it has become clear that wounding activates both the synthesis of ACC and its conversion to ethylene (Adams and Yang, 1979). Thus there are wound-inducible isoforms of ACC synthase and ACO (Lincoln *et al.*, 1993; Barry *et al.*, 1996). Wound ethylene must be produced in cells adjacent to the area of damage because these enzymes are inactivated by cell breakage.

Climacteric fruits produce large amounts of ethylene during ripening and wounding at this time may not increase—or may even decrease ethylene production (Robitaille and Janick, 1973; Ketsa and Koolpuksee, 1993). Stimulation of ethylene synthesis is more likely in preclimacteric or early climacteric fruits and such responses to bruising have been reported in apples, tomatoes and kiwifruit (Lougheed and Franklin, 1974; MacLeod *et al.*, 1976, Mencarelli *et al.*, 1996). Vibrational stress increased ethylene production by figs, even though there was no obvious damage (Mao *et al.*, 1995). De Vries *et al.* (1995) used laser-based instrumentation to monitor ethylene production from local wounds on the surface of tomato fruits. They observed increases within minutes after wounding followed by a return to basal levels in 2 to 3 h and later by a climacteric rise in ethylene production.

Ripening can be accelerated by ethylene in all climacteric and some non-climacteric fruits. Because of the induction of ethylene production, mechanical injury would be expected to promote ripening in these ethylene-responsive fruits. Wound ethylene appears to promote ripening in fruit slices that have been used as model systems for the study of this phase of plant development (McGlasson and Pratt, 1964). Although wounding has been known to promote ripening of figs since antiquity, there have been few direct demonstrations of its effects on whole fruit ripening (Zeroni *et al.*, 1972; MacLeod *et al.*, 1976; Mencarelli *et al.*, 1996).

7.11.4   Other metabolic changes

Oxidation of phenolic compounds, accelerated ethylene synthesis and respiration are common responses of all plant tissues to wounding. They are associated with metabolism that limits the damage to the plant and inhibits microbial invasion of the wound site. Histochemical tests revealed accumulations of callose, suberin, tannins and pectic substances some days after the skins of mature pears were punctured (Spotts *et al.*, 1998). These accumulations were associated with an increased resistance to fungal pathogens and occurred in tissues adjacent to the wound site. In whole plants there may be metabolic changes throughout the plant after local wounding and there has been considerable research on signal transduction and changes in gene expression following various kinds of mechanical perturbation in vegetative plant tissue (Bowles, 1993).

7.11.5   Summary time line of changes following injury

Based on results from numerous published reports, Bostock and Stermer (1989) developed a general time line of physiological, morphological and metabolic changes that occur after wounding. This time line was subsequently modified by Miller (2001) to apply to vegetables that were mechanically injured, but can be applied easily to injured fruits. To summarize, injury is perceived at the cell membrane, resulting in a disruption of homeostasis and the generation and transduction of injury signals. These signals induce changes in the cytoplasm, organelles, membranes and cell wall, which ultimately lead to cell division, wound healing and the restoration of homeostasis. Perception and signaling probably occur within seconds of injury, whereas the induction and response by cellular components may take minutes to hours. Restoration of homeostasis may take up to several days.

Clearly, many cellular processes are affected in response to injury. Due to genetic and external factors, particular fruits will probably respond differently to mechanical injury. However, as suggested above, one can envision a general pattern. Also, in addition to specific changes in phenolic metabolism, respiration and ethylene synthesis outlined in other sections of this chapter, one can observe the generation of reactive oxygen species, membrane depolarization, changes in ion balance, increased cell wall hydrolysis, loss of compartmentalization and formation/disappearance of watersoaked lesions (summarized by Bostock and Stermer, 1989).

7.12 Conclusions

Unlike other parts of the plant, fleshy fruits are terminal structures that evolved to be palatable to herbivorous animals. Their defensive responses to mechanical injury may be attenuated and ineffectual, relative to those in vegetative organs. There would be selective advantage in discouraging consumption before seed maturity. High levels of phenolic compounds and catechol oxidase are present in immature fruit, but these decline with maturation and ripening (Mayer and Harel, 1981; Ozawa *et al.*, 1987). Conversely, ripening generates a series of sensory signals that encourage consumption. Human consumption of fruits and their distribution through the global market have set new challenges in the life of fruits. It seems unlikely that breeding and biotechnology can generate fruits that resist gross physical abuse without compromising the qualities that make them attractive to consumers. The prevention of bruises, puncture wounds and compression damage requires engineering rather than biological research. Smaller and more superficial injuries may have no impact on consumer quality but they can open the way to pathogen attack. Manipulation of the signaling and response pathways to these could make a major contribution to postharvest technology.

Acknowledgement

Supported in part by state and federal funds appropriate to The Ohio State University, Ohio Agricultural Research and Development Center. Article number: HCS-01-09.

References

Abbott, J.A., Lu, R., Upchurch, B.L. and Stroshine, R.L. (1998) Technologies for nondestructive quality evaluation of fruits and vegetables. *Horticultural Reviews*, **20**, 1-120.

Adams, D.O. and Yang, S.F. (1979) Ethylene biosynthesis: identification of 1-aminocyclopropane-1-carboxylic acid as an intermediate in the conversion of methionine to ethylene. *Proceedings of the National Academy of Sciences of the USA*, **76**, 170-174.

Amiot, M.J., Tacchini, M., Aubert, S. and Nicolas, J. (1992) Phenolic composition and browning susceptibility of various apple cultivars at maturity. *Journal of Food Science*, **57**, 958-962.

Banks, N.H. and Joseph, M.J. (1991) Factors affecting resistance of banana fruit to compression and impact bruising. *Journal of the Science of Food and Agriculture*, **56**, 315-323.

Baritelle, A. and Hyde, G.M. (2000) Strain rate and size effects on pear tissue failure. *Transactions of the ASAE* (American Society of Agricultural Engineers), **43**, 95-98.

Barry, C.S., Blume, B., Bouzayen, M., Cooper, W., Hamilton, A.J. and Grierson, D. (1996) Differential expression of the 1-aminocyclopropane-1-carboxylic acid oxidase gene family of tomato. *Plant Journal*, **9**, 525-535.

Bostock, R.M. and Stermer, B.A. (1989) Perspectives on wound healing in resistance to pathogens. *Annual Review of Phytopathology*, **27**, 343-371.

Bowles, D.J. (1993) Local and systemic signals in the wound response. *Seminars in Cell Biology*, **4**, 103-111.

Brusewitz, G.H. and Barstch, J.A. (1989) Impact parameters related to post harvest bruising of apples. *Transactions of the ASAE* (American Society of Agricultural Engineers), **32**, 953-957.

Brusewitz, G.H., McCollum, T.G. and Zhang, X. (1991) Impact bruise resistance of peaches. *Transactions of the ASAE* (American Society of Agricultural Engineers), **34**, 962-965.

Brusewitz, G.H., Zhang, X. and Smith, M.W. (1992) Picking time and postharvest cooling effects on peach weight loss, impact parameters, and bruising. *Applied Engineering in Agriculture*, **8**, 84-90.

Cappellini, R.A., Ceponis, M.J. and Lightner, G.W. (1986) Disorders in table grape shipments to the New York Market, 1972–1984. *Plant Disease*, **70**, 1075-1079.

Cappellini, R.A., Ceponis, M.J. and Lightner, G.W. (1987) Disorders in apple and pear shipments to the New York Market, 1972–1984. *Plant Disease*, **71**, 852-855.

Cappellini, R.A., Ceponis, M.J. and Lightner, G.W. (1988a) Disorders in cucumber, squash and watermelon shipments to the New York Market, 1972–1985. *Plant Disease*, **72**, 81-85.

Cappellini, R.A., Ceponis, M.J. and Lightner, G.W. (1988b) Disorders in avocado, mango and pineapple shipments to the New York Market, 1972–1985. *Plant Disease*, **72**, 270-273.

Cappellini, R.A., Ceponis, M.J. and Lightner, G.W. (1988c) Disorders in apricot and papaya shipments to the New York Market, 1972–1985. *Plant Disease*, **72**, 366-368.

Carpita, N.C. (1985) Tensile strength of cell walls of living cells. *Plant Physiology*, **79**, 485-488.

Ceponis, M.J. and Butterfield, J.E. (1981) Cull losses in western sweet cherries at retail and consumer levels in Metropolitan New York. *HortScience*, **16**, 324-326.

Ceponis, M.J. and Cappellini, R.A. (1985a) Wholesale, retail and consumer level losses of nectarines in metropolitan New York. *HortScience*, **20**, 90-91.

Ceponis, M.J. and Cappellini, R.A. (1985b) Wholesale, retail and consumer level losses in grapefruit marketed in metropolitan New York. *HortScience*, **20**, 93-95.

Ceponis, M.J., Cappellini, R.A. and Lightner, G.W. (1986a) Disorders in citrus shipments to the New York Market, 1972–1984. *Plant Disease*, **70**, 1162-1165.

Ceponis, M.J., Cappellini, R.A. and Lightner, G.W. (1986b) Disorders in tomato shipments to the New York Market, 1972–1984. *Plant Disease*, **70**, 261-265.

Ceponis, M.J., Cappellini, R.A., Wells, J.M. and Lightner, G.W. (1987) Disorders in plum, peach, and nectarine shipments to the New York Market, 1972–1985. *Plant Disease*, **71**, 947-952.

Ceponis, M.J., Cappellini, R.A. and Lightner, G.W. (1988) Disorders in sweet cherry and strawberry shipments to the New York Market, 1972–1984. *Plant Disease*, **72**, 472-475.

Chen, P. and Sun, Z. (1984) Critical strain failure criterion: pros and cons. *Transactions of the ASAE* (American Society of Agricultural Engineers), **27**, 278-281.

Coseting, M.Y. and Lee, C.Y. (1987) Changes in polyphenoloxidase and polyphenol concentrations in relation to the degree of browning. *Journal of Food Science*, **52**, 985-989.

Crisosto, C.H., Johnson, R.S. and Luza, J. (1993) Incidence of physical damage on peach and nectarine skin discoloration development: anatomical studies. *Journal of the American Society for Horticultural Science*, **118**, 786-800.

de Vries, H.S.M., Harren, F.J.M. and Reuss, J. (1995) In situ real-time monitoring of wound-induced ethylene in cherry tomatoes by two infrared laser-driven systems. *Postharvest Biology and Technology*, **6**, 275-285.

Diener, R.G., Elliot, K.C., Nesselroad, P.E., Ingle, M., Adams, R.E. and Blizzard, S.H. (1979) Bruise energy of peaches and apples. *Transactions of the ASAE* (American Society of Agricultural Engineers), **22**, 287-290.

Dull, G.G. (1986) Nondestructive evaluation of quality of stored fruits and vegetables. *Food Technology*, **40**, 106-110.

Fridley, R.B. and Adrian, P.A. (1966) Mechanical properties of peaches, pears, apricots and apples. *Transactions of the ASAE* (American Society of Agricultural Engineers), **9**, 135-142.

Galliard, T. (1978) Lipolytic and lipoxygenase enzymes in plants and their action in wounded tissues, in *Biochemistry of Wounded Plant Tissues* (ed. G. Kahl), Walter de Gruyter, Berlin, pp. 155-202.

Garcia, J.L., Ruiz-Altisent, M. and Barreiro, P. (1995) Factors influencing mechanical properties and bruise susceptibility of apples and pears. *Journal of Agricultural Engineering Research*, **61**, 11-18.

Harker, F.R. and Hallett, I.C. (1992) Physiological changes associated with development of mealiness of apple fruit during storage. *HortScience*, **27**, 1291-1294.

Harker, F.R. and Hallett, I.C. (1994) Physiological and mechanical properties of kiwifruit tissue associated with textural change during cool storage. *Journal of the American Society for Horticultural Science*, **119**, 987-993.

Harker, F.R. and Sutherland, P.W. (1993) Physiological changes associated with fruit ripening and the development of mealy texture during storage of nectarines. *Postharvest Biology and Technology*, **2**, 269-277.

Hatfield, S. and Knee, M. (1988) Effects of water loss on apples in storage. *International Journal of Food Science and Technology*, **23**, 575-583.

Hightower, J. (1976) Hard tomatoes, hard times: the failure of the Land Grant college complex, in *Radical Agriculture* (ed. R. Merrill), New York University Press, New York, pp. 87-110.

Holt, D. and Schoorl, J.E. (1977) Bruising and energy dissipation in apples. *Journal of Texture Studies*, **7**, 421-432.

Horsfield, B.C., Fridley, R.B. and Claypool, L.L. (1972) Application of theory of elasticity to the design of fruit harvesting and handling equipment for minimum bruising. *Transactions of the ASAE* (American Society of Agricultural Engineers), **15**, 746-750.

Hung, Y.-C. and Prussia, S.E. (1989) Effect of maturity and storage time on the bruise susceptibily of peaches (cv. Red Globe). *Transactions of the ASAE* (American Society of Agricultural Engineers), **32**, 1377-1382.

Kawada, K. and Kitagawa, H. (1984) Deformation of 'Marsh' grapefruit as affected by fruit orientation at packing. *Proceedings of the Florida State Horticultural Society*, **97**, 138-140.

Ketsa, S. and Koolpluksee, M. (1993) Some physical and biochemical characteristics of damaged pericarp of mangosteen fruit after impact. *Postharvest Biology and Technology*, **2**, 209-215.

Khan, A.A. and Vincent, J.F.V. (1990) Anisotropy of apple parenchyma. *Journal of the Science of Food and Agriculture*, **52**, 455-466.

Khan, A.A. and Vincent, J.F.V. (1991) Bruising and splitting of apple fruit under uni-axial compression and the role of the skin in preventing damage. *Journal of Texture Studies*, **22**, 251-263.

Klein, J.D. (1987) Relationship of harvest date, storage conditions, and fruit characteristics to bruise susceptibility of apple. *Journal of the American Society for Horticultural Science*, **112**, 113-118.

Langley, K.R., Martin, A., Stenning, R., Murray, A.J., Hobson, G.E., Schuch, W.W. and Bird, C.R. (1994) Mechanical and optical assessment of the ripening of tomato fruit with reduced polygalacturonase activity. *Journal of the Science of Food and Agriculture*, **66**, 547-554.

Lidster, P.D. and Tung, M.A. (1979) Identification of deformation parameters and fruit response to mechanical damage in sweet cherry. *Journal of the American Society for Horticultural Science*, **104**, 808-811.

Lidster, P.D. and Tung, M.A. (1980) Effects of fruit temperatures at time of impact damage and subsequent storage temperature and duration on the development of surface disorders in sweet cherries. *Canadian Journal of Plant Science*, **60**, 555-559.

Lincoln, J.E., Campbell, A.D., Oetiker, J., Rottman, W.H., Oeller, P.W., Shen, N.F. and Theologis, A. (1993) LE-ACS4, a fruit ripening and wound induced 1-aminocyclopropane-1-carboxylate synthase gene of tomato (*Lycopersicon esculentum*). *Journal of Biological Chemistry*, **268**, 19422-19430.

Lougheed, E.C. and Franklin, E.W. (1974) Ethylene production increased by bruising of apples. *HortScience*, **9**, 192-193.

MacLeod, R.F., Kader, A.A. and Morris, L.L. (1976) Stimulation of ethylene and CO_2 production of mature-green tomatoes by impact bruising. *HortScience*, **11**, 604-606.

Maness, N.O., Brusewitz, G.H. and McCollum, T.G. (1992) Impact bruise resistance comparison among peach cultivars. *HortScience*, **27**, 1008-1011.

Mao, L., Ying, T., Xi, Y. and Zhen, Y. (1995) Respiration rate, ethylene production, and cellular leakage of fig fruit following vibrational stress. *HortScience*, **30**, 145.

Massey, L.M. Jr, Chase, B.R. and Starr, M.S. (1982) Effect of rough handling on CO_2 evolution from 'Howes' cranberries. *HortScience*, **17**, 57-58.

Mayer, A.M. and Harel, E. (1981) Polyphenol oxidases in fruits–changes during ripening, in *Recent Advances in the Biochemistry of Fruit and Vegetables* (eds J. Friend and M.J.C. Rhodes), Academic Press, London, pp. 161-180.

McGlasson, W.B. and Pratt, H.K. (1964) Effects of wounding on respiration and ethylene production by cantaloupe fruit tissue. *Plant Physiology*, **39**, 128-132.

McLaughlin, N.B. and Pitt, R.E. (1984) Failure characteristics of apple tissue under cyclic loading. *Transactions of the ASAE* (American Society of Agricultural Engineers), **27**, 311-320.

Mencarelli, F., Massantini, R. and Botondi, R. (1996) Influence of impact surface and temperature on the ripening response of kiwifruit. *Postharvest Biology and Technology*, **8**, 165-177.

Miller, A.R. (1992) Physiology, biochemistry and detection of bruising (mechanical stress) in fruits and vegetables. *Postharvest News and Information*, **3**, 53N-58N.

Miller, A.R. (2001) Harvest and handling injury: physiology, biochemistry, and detection, in *Postharvest Physiology and Pathology of Vegetables* (eds J. Brecht and J. Barz), Marcel-Dekker, New York (in press).

O'Brien, M., Claypool, L.L., Leonard, S.J., York, G.K. and MacGillivray, J.H. (1963) Cause of fruit bruising on transport trucks. *Hillgardia*, **35**, 113-124.

O'Brien, M., Fridley, R.B. and Claypool, L.L. (1978) Food losses in harvest and handling systems for fruits and vegetables. *Transactions of the ASAE* (American Society of Agricultural Engineers), **21**, 386-390.

Ozawa, T., Lilley, T.H. and Haslam, E. (1987) Polyphenol interactions: astringency and the loss of astringency in ripening fruit. *Phytochemistry*, **26**, 2937-2942.

Pang, W., Studman, C.J. and Ward, G.T. (1992) Bruising damage in apple-to-apple impact. *Journal of Agricultural Engineering Research*, **52**, 229-240.

Pang, D.W., Studman, C.J. and Banks, N.H. (1994) Apple bruising thresholds for an integrated sphere. *Transactions of the ASAE* (American Society of Agricultural Engineers), **37**, 893-897.

Pitt, R.E. (1992) Viscoelastic properties of fruits and vegetables, in *Viscoelastic Properties of Foods* (eds M.A. Rao and J.F. Steffe), Elsevier Applied Science, London, pp. 49-76.

Rivero, L.G., Grierson, W. and Soule, J. (1979) Resistance of 'Marsh' grapefruit to deformation as affected by picking and handling methods. *Journal of the American Society for Horticultural Science*, **104**, 551-554.

Robitaille, H.A. and Janick, J. (1973) Ethylene production and bruise injury in apple. *Journal of the American Society for Horticultural Science*, **98**, 411-413.

Roth, I. (1977) *Fruits of Angiosperms*, Gebruder Borntraeger, Berlin.

Saltveit, M.E. Jr (1984) Effects of temperature on firmness and bruising of 'Starkrimson Delicious' and 'Golden Delicious' apples. *HortScience*, **19**, 550-551.

Samin, W. and Banks, N.H. (1993) Color changes in apple bruises over time. *Acta Horticulturae*, **343**, 304-306.

Sarig, Y. and Nahir, D. (1973) Deformation characteristics of 'Valencia' oranges as an indicator of firmness. *HortScience*, **8**, 391-392.

Schoorl, D. and Holt, J.E. (1977) The effects of storage time and temperature on the bruising of Jonathan Delicious and Granny Smith apples. *Journal of Texture Studies*, **8**, 409-416.

Schulte, N.L., Timm, E.J. and Brown, G.K. (1994) 'Redhaven' peach impact damage thresholds. *HortScience*, **29**, 1052-1055.

Shewfelt, R.L. (1986) Postharvest treatment for extending the shelf-life of fruits and vegetables. *Food Technology*, **40**, 70-80, 89.

Simon, E.W. (1977) Leakage from fruit cells in water. *Journal of Experimental Botany*, **28**, 1147-1152.

Smith, W.H. (1938) Anatomy of the apple fruit. *Report of the Food Investigation Board, Great Britain, Department of Scientific and Industrial Research 1937*, pp. 139-142.

Spotts, R.A., Sanderson, P.G., Lennox, C.L., Sugar, D. and Cervantes, L.A. (1998) Wounding, wound healing and staining of mature pear fruit. *Postharvest Biology and Technology*, **13**, 27-36.

Steudle, E. and Wieneke, J. (1985) Changes in water relations and elastic properties of apple fruit cell during growth and development. *Journal of the American Society for Horticultural Science*, **110**, 824-829.

Studman, C.J., Brown, G.K., Timm, E.J., Schulte, N.L. and Vreede, M.J. (1997) Bruising on blush and non-blush sides in apple-to-apple impacts. *Transactions of the ASAE* (American Society of Agricultural Engineers), **40**, 1655-1163.

Timm, E.J., Brown, G.K. and Armstrong, P.R. (1996) Apple damage in bulk bins during semi-trailer transport. *Applied Engineering in Agriculture*, **12**, 369-377.

Timm, E.J., Bollen, A.F., Dela Rue, B.T. and Woodhead, I.M. (1998) Apple damage and compressive forces in bulk bins during orchard transport. *Applied Engineering in Agriculture*, **14**, 165-172.

Topping, A.J. and Luton, M.T. (1986) Cultivar differences in the bruising of English apples. *Journal of Horticultural Science*, **61**, 9-13.

van der Pijl, L. (1982) *Principles of Dispersal in Higher Plants*, Springer-Verlag, Berlin.

Van Lancker, J. (1979) Bruising of unpeeled apples and potatoes in relation with temperature and elasticity. *Lebensmittel Wissenschaft und Technologie*, **12**, 157-161.

Vergano, P.J., Testin, R.F. and Newall, W.C. Jr (1991) Peach bruising: susceptibility to impact, vibration and compression abuse. *Transactions of the ASAE* (American Society of Agricultural Engineers), **34**, 2110-2116.

Vergano, P.J., Testin, R.F., Choudhari, A.C. and Newall, W.C. Jr (1992) Peach vibration bruising: the effect of paper and plastic films between peaches. *Journal of Food Quality*, **15**, 183-197.

Vincent, J.F.V. (1990) Fracture properties of plants. *Advances in Botanical Research*, **17**, 235-287.

Wells, J.M., Butterfield, J.E. and Ceponis, M.J. (1994) Diseases, physiological disorders, and injuries of plums marketed in metropolitan New York. *Plant Disease*, **78**, 642-644.

Wu, N. and Pitts, M.J. (1999) Development and validation of a finite element model of an apple fruit cell. *Postharvest Biology and Technology*, **16**, 1-8.

Zeroni, M., Ben-Yehoshua, S. and Galil, J. (1972) Relationship between ethylene and the growth of *Ficus sycomorus*. *Plant Physiology*, **50**, 378-381.

8 Ethylene synthesis, mode of action, consequences and control

Christopher B. Watkins

8.1 Introduction

Ethylene is one of several plant growth regulators that affect plant growth, development and senescence, but, from a postharvest perspective, its role as a principal regulator of fruit ripening is the most important. Ethylene is a simple gaseous hydrocarbon that can diffuse into and out of plant tissues, from both endogenous and exogenous (non-biological and biological) sources, and it can profoundly affect quality factors of horticultural products such as color, texture and flavor. These effects can be beneficial or deleterious depending on the product and its uses. Strategies for use or avoidance of ethylene during ripening, harvest, storage, transport and handling operations have been widely developed for commercial operations. Also, the importance of ethylene in affecting ripening and/or senescence has been reflected in extensive research efforts on ethylene biosynthesis and action. Consequently, a large amount of literature exists, including reviews on ethylene biosynthesis and perception, and its interaction with fruit ripening and quality (Abeles *et al.*, 1992; Biale and Young, 1981; Brady, 1987; Fluhr and Mattoo, 1996; Jiang and Fu, 2000; Lelievre *et al.*, 1997a; Mattoo and Suttle, 1991; McGlasson, 1985; Pech *et al.*, 1994; Saltveit, 1999; Yang and Hoffman, 1984).

This chapter outlines current understanding of ethylene effects on fruit ripening, effects on quality, ethylene biosynthesis, ethylene perception and current methods for manipulating ethylene responses. While the focus here is on fruit, much of the information is also relevant to vegetables, flowers and other senescing tissues.

8.2 Fruit ripening and interactions with ethylene

The term 'fruit' represents a diverse collection of morphological, biochemical and physiological types, and the actual morphological part of fleshy fruits that is consumed varies widely (Kays, 1997). Fruit ripening, by definition, involves the transition of a physiologically mature plant organ from an inedible state to one possessing required aesthetic and/or food quality characteristics. Ripening can be separated into two distinct types based on fruit wall structure: the fleshy fruits such as the tomato, apple and strawberry; and the sclerenchymatous or dry fruits such as rice, wheat and peas. Most research on ethylene and ripening has been concerned with fleshy fruits because of their greater perishability.

8.2.1 Non-climacteric and climacteric fruit

The concept of non-climacteric and climacteric fruit ripening is fundamental to any discussion of ethylene in postharvest systems. The term 'climacteric' was coined by Kidd and West (1925) to describe the rise in respiration rate that accompanied the maturation phase in apple. Subsequently, climacteric and non-climacteric categories were developed on the basis of the presence or absence of a respiratory rise during ripening (Biale and Young, 1981). Autocatalytic ethylene production is invariably associated with increased respiration in climacteric fruit, although it can precede, coincide or follow the respiratory rise depending on the fruit under investigation (Biale and Young, 1981). The metabolic importance of the respiratory rise that accompanies autocatalytic ethylene production is uncertain (Brady, 1987; Solomos, 1988) but, from an applied perspective, keeping respiration rates, and therefore carbohydrate utilization, at minimal levels is desirable for maintaining fruit quality.

Typically the presence or absence of a climacteric is now based on evidence of autocatalytic ethylene production rather than respiration. Diagnostically, climacteric fruit can be separated from non-climacteric fruit by responses of respiration and/or ethylene production to exogenous ethylene or its analogs such as propylene. In climacteric fruit, ethylene will advance the timing of the climacteric, autocatalytic production will continue after removal of ethylene, and in contrast to non-climacteric fruit the magnitude of the respiratory rise is independent of the concentration of applied ethylene.

A list of climacteric and nonclimacteric fruit, updated from Biale and Young (1981) and Kays (1997), is shown in table 8.1. Allocation of fruit into these categories has sometimes been controversial since experimental conditions during storage that stimulate respiration and ethylene production, such as chilling or decay, can result in incorrect classifications. Species, such as the water melon, that were initially classified as climacteric (Mizano and Pratt, 1973) are now classified as non-climacteric (Elkashif *et al.*, 1989). Cultivars of the same species may appear in both categories, e.g. Asian pear (Downs *et al.*, 1991), pepper (Villavicencio *et al.*, 1999) and the tomato in which the ripening inhibitor (*rin*) and non-ripening (*nor*) ripening mutants are classified as non-climacteric (Tigchelaar *et al.*, 1978). For fruit such as the blackberry and raspberry, there is still debate as to classification (Burdon and Sexton, 1990a,b, 1993; Perkins-Veazie and Nonnecke, 1992; Perkins-Veazie *et al.*, 2000; Walsh *et al.*, 1983), perhaps because of difficulties in separating ethylene involvement during abscission and ripening processes, as well as on- and off-vine differences.

Other difficulties in categorization include the effects of temperature, where, for example, kiwifruit exhibits climacteric behavior when ripened at 20°C but

Table 8.1 Climacteric and non-climacteric classification of fruit

Climacteric

Common name	Scientific name	Reference
Apple	*Malus pumila* Mill.	Biale, 1960
Apricot	*Prunus armeniaca* L.	Biale, 1960
Asian pear	*Pyrus serotina* Rehder	Downs *et al.*, 1991
Atemoya	*Annona squamosa* × *cherimola*	Brown *et al.*, 1988
Avocado	*Persea americana* Mill.	Biale, 1960
Banana	*Musa* L.	Biale, 1960
Biriba	*Rollinia deliciosa* Safford	Biale and Barcus, 1970
Bitter melon	*Momordica charantia* L.	Kays and Hayes, 1978
Blackberry	*Rubus* L.	Walsh *et al.*, 1983
Blueberry, lowbush	*Vaccinium angustifolium* Ait.	Ismael and Kender, 1969
Blueberry, highbush	*Vaccinium corymbosum* L.	Ismael and Kender, 1969
Blueberry, rabbiteye	*Vaccinium ashei* Reade	Lipe, 1978
Breadfruit	*Arcocarpus altilis* (Parkins) Fosb.	Biale and Barcus, 1970
Cantaloupe	*Cucumis melo* L. Cantalupensis group	Lyons *et al.*, 1962
Cherimoya	*Annona cherimola* Mill.	Biale, 1960
Corossol sauvage	*Rollinia orthopetala* A. DC.	Biale, 1976
Date palm	*Phoenix dactylifera* L.	Abbas and Ibrahim, 1996
Durian	*Durio zibethinus* Murray	Ketsa and Daengkanit, 1998
Feijoa	*Feijoa selloviana* O. Berg.	Biale, 1960
Fig	*Ficus carica* L.	Marei and Crane, 1971
Goldenberry	*Physalis peruviana* L.	Trinchero *et al.*, 1999
Guava	*Psidium guajava* L.	Akamine and Goo, 1979
Guava, 'Purple Strawberry'	*Psidium littorale* var. *longipes* (O. Berg.) Fosb.	Akamine and Goo, 1979
Guava, 'Strawberry'	*Psidium littorale* Raddi.	Akamine and Goo, 1979
Guava, 'Yellow Strawberry'	*Psidium littorale* var. *littorale* Fosb	Akamine and Goo, 1979
Honeydew melon	*Cucumis melo* L. Inodorus group	Pratt and Goeschl, 1968
Jujube	*Ziziphus sativa* Mill	Kader *et al.*, 1982
Kiwifruit, Chinese gooseberry	*Actinidia deliciosa* (A. Chev) C.F. Liang et A.R. Ferguson var. *deliciosa*	Pratt and Reid, 1974
Mammee apple	*Mammea americana* L.	Akamine and Goo, 1978
Mango, common	*Mangfera indica* L.	Biale, 1960
Mango, African	*Irvingia gabonensis* Baillon & Irvingiaceae	Aina and Oladunjoye, 1993
Oilseed rape	*Brassica napus* L.	Meakin and Roberts, 1990
Papaw	*Asimina triloba* (L.) Dunal.	Biale, 1960
Papaya	*Carica papaya* L.	Biale, 1960
Passion fruit	*Passiflora edulis* Sims	Biale, 1960
Peach	*Prunus persica* (L.) Batsch	Biale, 1960
Pear, European	*Pyrus communis* L.	Biale, 1960
Pear, Chinese	*Pyrus bretschneideri* R.	Tian *et al.*, 1987
Persimmon	*Diospyros kaki* L.	Reid, 1975
Plum	*Prunus americana* Marsh.	Biale, 1960
Raspberry	*Rubus idaeus* L.	Burdon and Sexton, 1990a

Table 8.1 (continued)

Climacteric

Common name	Scientific name	Reference
Sapote	*Casimiroa edulis* Llave	Biale, 1960
Saskatoon	*Amelanchier alnifolia* Nutt.	Rogiers *et al.*, 1998
Soursop	*Annona muricata* L.	Biale and Barcus, 1970
Sweetsop, sugar apple	*Annona squamosa* L.	Brown *et al.*, 1988
Tomato	*Lycopersicon esculentum* Mill.	Biale, 1960

Non-climacteric

Asian pear	*Pyrus serotina* Rehder	Downs *et al.*, 1991
Blackberry	*Rubus* L.	Lipe, 1978; Perkins-Veazie *et al.*, 2000
Cacao	*Theobroma cacao* L.	Biale and Barcus, 1970
Cactus pear	*Opuntia amyclaea* Tenore	Lakshminarayana and Estrella, 1978
Carambola	*Averrhoa carambola* L.	Lam and Wan, 1983
Cashew	*Anacardium occidentale* L.	Biale and Barcus, 1970
Cherry, sour	*Prunus cerasus* L.	Blanpied, 1972
Cherry, sweet	*Prunus avium* L.	Biale, 1960
Cucumber	*Cucumis sativus* L.	Biale, 1960
Grape	*Vitus vinifera* L.	Biale, 1960
Grapefruit	*Citrus paradisi* Macf.	Biale, 1960
Java plum	*Syzygium cumini* (L.) Skeels	Akamine and Goo, 1979
Lemon	*Citrus jambhiri* Lush.	Biale, 1960
Litchi (lychee)	*Litchi chinensis* Sonn.	Akamine and Goo, 1979
Loquat	*Eriobotrya japonica* Lindl.	Zheng *et al.*, 1993
Mountain apple	*Syzygium malaccense* (L.) Merrill & Perry	Akamine and Goo, 1979
Olive	*Olea europaea* L.	Maxie *et al.*, 1960
Orange	*Citrus sinensis* (L.) Osb.	Biale, 1960
Pepper	*Capsicum annuum* L.	Saltveit, 1977
Pineapple	*Ananas comosus* (L.) Merr.	Biale, 1960
Pitaya, yellow	*Selenicereus megalanthus* Scum. Ex Vaupel	Nerd and Mizrahi, 1999
Pomegranate	*Punica granatum* L.	Ben-Arie *et al.*, 1984
Rambutan	*Nephelium lappaceum* L.	O'Hare, 1995
Raspberry	*Rubus idaeus* L.	Perkins-Veazie and Nonnecke, 1992
Rose apple	*Syzygium jambos* (L.) Alston	Akamine and Goo, 1979
Satsuma manadarin	*Citrus reticulata* Blanco	Reid, 1975
Star apple	*Chysophyllum cainito* L.	Pratt and Mendoza, 1980
Strawberry	*Fragaria xananassa* Duch.	Biale, 1960
Strawberry, wild	*Fragaria vesca* L.	Nam *et al.*, 1999
Surinan cherry	*Eugenia uniflora* L.	Akamine and Goo, 1979
Tomato (*nor-*, *rin-*, *cnr-*)	*Lycopersicon esculentum* Mill.	Thompson *et al.*, 1999; Tigchelaar *et al.*, 1978
Tree tomato, tamarillo	*Cyphomandra betacea* (Cav.) Sendtu	Pratt and Reid, 1976
Watermelon	*Citrullus lanatus* (Thunb.) Mansf.	Elkashif *et al.*, 1989

not at 10°C (Antunes *et al.*, 2000). Within a species, considerable differences in 'climacteric physiology' among cultivars or selections can exist. In apple, early season cultivars usually have high ethylene production rates and ripen quickly, while late season ones have low ethylene production rates and ripen slowly (Chu, 1988; Sunako *et al.*, 1999; Watkins *et al.*, 1989a). Melon (Pratt *et al.*, 1977), papaya (Zhang and Paull, 1990) and peach cultivars (Brovelli *et al.*, 1999) also vary in ethylene production rates and slow ripening nectarine genotypes have been identified (Brecht and Kader, 1984). Abdi *et al.* (1997) have described 'suppressed climacteric' plums, which do not produce sufficient ethylene to coordinate ripening but show characteristic responses to propylene.

Fruits also vary greatly in the duration of the ethylene climacteric—bananas and avocados, for example, exhibit a sharp rise and fall of ethylene compared with other fruit (Hoffman and Yang, 1980; Liu *et al.*, 1999). The timing of ethylene production relative to fruit ripening is variable; it is associated with the initiation of ripening in tomato (Grierson and Tucker, 1983) but occurs towards the end of the softening process in kiwifruit (Wang *et al.*, 2000). The presence of a respiratory climacteric in melon before harvest has been questioned by Shellie and Saltveit (1993). Hadfield *et al.* (1995), however, did not detect differences in respiratory behavior of melons pre- and postharvest, and cultivar, or factors such as fruit respiration, may influence the ability to detect a preharvest respiratory climacteric (Andrews, 1995; Knee, 1995). Ethylene production may be the more reliable criterion for separating climacteric from non-climacteric fruit.

Collectively, the literature suggests that climacteric and non-climacteric categories represent an oversimplification. Nevertheless, the commercial importance of the categories lies in the impact of ethylene responses on handling, packaging and storage conditions employed to maintain quality. The climacteric is typically associated with a number of closely coordinated catabolic and anabolic events resulting in rapid perishability. In contrast, non-climacteric fruit ripen more slowly with separation of ripening events, although notable exceptions exist, where, for example, non-climacteric strawberry ripens rapidly in a coordinated fashion, apparently in the absence of control by ethylene (Perkins-Veazie, 1995). The scientific importance for climacteric fruits resides in the promise that many aspects of fruit ripening may be manipulated by controlling ethylene production or perception.

8.2.2   *Ethylene and fruit quality*

The rates of ethylene production of ripe climacteric fruits have a wide range (Knee *et al.*, 1985; Thompson *et al.*, 2000), while those of non-climacteric fruits, by definition, remain low in the absence of physiological or pathological injury. However, the quality of fruit of both categories can be profoundly affected by ethylene. In addition to listings of ethylene responsiveness of many commodities (Thompson *et al.*, 2000), more detailed information on responses of individual fruits to the gas are available (Abeles *et al.*, 1992; Seymour *et al.*, 1993). Effects of ethylene are beneficial or deleterious depending on the crop, whether treated pre- or postharvest, its physiology and its intended usage. All climacteric fruit are potentially responsive to ethylene but to differing degrees, kiwifruit, for example, being more responsive than nectarines. In general, quality factors of non-climacteric fruit are affected deleteriously by ethylene, but also to differing degrees, citrus, for example, being more affected than cherry and pineapple.

Beneficial effects of ethylene on quality of fresh fruits include promotion of red color development, degreening and stimulation of ripening, but the same attributes are detrimental if expressed via acceleration of senescence, stimulation of chlorophyll loss and excessive softening (Saltveit, 1999). Complex interactions between ethylene and quality factors may exist. For example, in citrus, positive effects include increased chlorophyll degradation and promotion of carotenoid biosynthesis (Goldschmidt *et al.*, 1993; Stewart and Wheaton, 1972), and decreased decay by inhibition of the mold rots, *Pencillium digitatum* and *P. italicum* (El-Kazzaz *et al.*, 1983). Negative effects include increased susceptibility to chilling injury (Yuen *et al.*, 1995), increased decay associated with stem-end rots (Brown and Lee, 1993; Wild *et al.*, 1976) and increased off-flavor development (McGlasson and Eaks, 1972).

Knowledge of ethylene production rates for various commodities and their responsiveness to ethylene provides information about product compatibilities for storage and transport (Thompson *et al.*, 2000). The commercial management of fruit will vary greatly depending on the commodity and whether ethylene action is to be promoted or delayed.

8.2.3   *Active ethylene concentrations*

The ethylene concentrations required to affect fruit physiology are uncertain. Threshold, half-maximal and saturating doses for ethylene-mediated responses in vegetative tissues are $0.01, 0.1$ and $10 \, \mu l \, l^{-1}$ respectively (Abeles *et al.*, 1992). In general, senescence of mature fruit (and vegetables) is promoted by an ethylene concentration of $0.1 \, \mu l \, l^{-1}$ (Knee *et al.*, 1985). This concentration is affected greatly, however, by species (Yang, 1985) and maturity (Peacock, 1972).

The potential impact of even lower ethylene concentrations on fruit quality has been highlighted by recent studies that show that exposure of oranges and strawberries to $<0.005 \, \mu l \, l^{-1}$ ethylene extended their storage life compared with fruit exposed to $0.1 \, \mu l \, l^{-1}$ (Wills and Kim, 1995; Wills *et al.*, 1999). The commercial significance of these findings is uncertain, but Wills *et al.* (2000) found average ethylene concentrations of $0.06 \, \mu l \, l^{-1}$ in wholesale markets and distribution centers, $0.017–0.035 \, \mu l \, l^{-1}$ in supermarkets, $0.029 \, \mu l \, l^{-1}$ in domestic refrigerators without apples and $0.20 \, \mu l \, l^{-1}$ in refrigerators with apples. It was estimated that ethylene in market areas could reduce the potential storage life of non-climacteric fruit and vegetables by 10–30%.

8.2.4   Control of ethylene production and action

Increased ethylene production is associated with development of quality characteristics of climacteric fruit but also with loss of storage potential. For some fruit, ethylene production after harvest is reduced by harvest at the pre-climacteric stage, application of low temperatures (at least for non-chilling sensitive products) and use of modified atmosphere (MA) or controlled atmosphere (CA) storage. Some fruits are harvested well before the climacteric stage and either induced to ripen by ethylene treatment, e.g. banana and pear (Hartmann *et al.*, 1975; Inaba and Nakamura, 1988) or cold storage, e.g. pear (Knee *et al.*, 1983). Ethylene production or internal ethylene concentrations are sometimes used as a maturity index and guide for storage potential of apples (Dilley, 1982).

Preharvest chemical strategies have been developed to delay ethylene production in the field. For apples, preharvest sprays of daminozide (Alar) were used by some growers to delay autocatalytic ethylene production and preharvest drop, and allow better color development to occur (Looney, 1971). Since the withdrawal of the chemical from the market, aminoethoxyvinyl glycine (AVG), marketed as ReTain™, has been developed for delay of ethylene production in apples. AVG inhibits the activity of 1-aminocyclopropane carboxylic acid (ACC) synthase (see section 8.3.1). ReTain™ delays fruit ripening, but its efficacy can be affected by cultivar and concentration (Autio and Bramlage, 1982). The labelled rate for ReTain™ of $124\,g\,ha^{-1}$ active ingredient (a.i.) is lower than that used in many early studies, and effectiveness is affected by growing region (Stover *et al.*, 2002). Early studies on preharvest application of AVG to pears (Romani *et al.*, 1983) have been followed up using ReTain™ (Clayton *et al.*, 2000). Inhibition of ripening by AVG can be overcome by cold storage or ethylene treatment in both apples and pears (Autio and Bramlage, 1982; Clayton *et al.*, 2000; Romani *et al.*, 1983).

A number of postharvest strategies can be used to prevent deleterious effects of ethylene on fruit quality. Ethylene-responsive fruit should be isolated from both non-biological sources, such as internal combustion engines, and biological sources, such as ripening and injured fruit and vegetables.

Ethylene can be removed by ventilation, absorption on carbon, oxidation by permanganate, ozone, ultraviolet irradiation or catalysts (Abeles *et al.*, 1992; Reid, 1992). Air exchange can reduce undesirable concentrations of ethylene, as well as CO_2 and offensive odors. The effectiveness of ventilation is limited however for climacteric fruit already producing ethylene as ripe fruit typically will be less responsive to exogenous ethylene. In such fruit, the threshold ethylene concentration for action will have been exceeded and the internal ethylene concentrations will be high because of the diffusion barrier of the skin. Moreover, postharvest handling entails restriction of ventilation, and accumulation of ethylene in apple CA stores can reach as high as $1000\,\mu l\,l^{-1}$. Even a commodity producing only 0.01 µl kg hr^{-1} would accumulate 0.12 µl l^{-1} in 24 hours when occupying a third of the room volume (Knee *et al.*, 1985).

Low ethylene concentrations obtained by using potassium permanganate can result in reduced softening and/or rotting in apples, avocado, banana, kiwifruit, lemon and mango (Knee *et al.*, 1985), but application of low ethylene storage to a climacteric fruit has been most studied for the apple. Successful use of this technology is based on a combination of avoiding the sources of ethylene, inhibition of ethylene production and removal of ethylene produced by the apples during storage (Blanpied *et al.*, 1985; Stow *et al.*, 2000). Avoidance is based on use of cultivars with low rates of ethylene production, use of electric forklifts and nitrogen gas for flushing out the atmosphere in the room, rather than fossil-fueled forklifts and atmosphere generators, and harvest of fruit before the climacteric. Inhibition of ethylene production by the apples is achieved by cooling fruit and applying CA rapidly to suppress the rate of fruit ripening, maintaining low O_2 concentrations in the room and using elevated CO_2. Removal of ethylene produced by the apples during storage is achieved by ethylene scrubbers. Ethylene concentrations in the storage atmosphere of <2 µl l^{-1} were necessary to retard softening and superficial scald (Knee and Hatfield, 1981), while Stow *et al.* (2000) found that internal ethylene concentrations of the fruit had to be maintained at <0.1 µl l^{-1} to delay initiation of softening.

8.2.5 *Pre- and postharvest ethylene application*

Preharvest chemical strategies have been developed to advance ripening of fruit in the field. Ethylene released from compounds such as ethephon has been tested extensively for use in agriculture, and many practical applications have been derived (Abeles *et al.*, 1992). Ethephon accelerates fruit growth and onset of ripening of tomatoes (Iwahori and Lyons, 1970), and is used extensively for processing tomatoes. In cantaloupe, however, ethephon decreases sugar accumulation, thereby limiting its application (Kasmire *et al.*, 1970). In the apple, ethephon treatment before harvest can be used to enhance red color development, advance starch hydrolysis, reduce incidence of the storage disorders bitter pit and superficial scald, but it can also increase softening (Larrigaudiere *et al.*, 1996; Stover *et al.*, 2002; Watkins *et al.*, 1982, 1989b).

Postharvest application of ethylene is commonly used for degreening of citrus or ripening before retail marketing of bananas harvested and shipped in an unripe stage. Other fruit can be treated with ethylene, including avocado, mango, melon, nectarine, papaya, peach, pear, persimmon, plum and tomato (Knee *et al.*, 1985). However, these fruit will also ripen in the absence of ethylene preconditioning. Because ethylene treatment irreversibly stimulates all aspects of ripening, tight control of products within the retail markets is essential. Moreover, such treatments entail risk of loss through physiological deterioration, disease and increased susceptibility to damage. Therefore operators may prefer to reduce their risk by providing unripened product to the consumer.

Nevertheless, ethylene preconditioning is likely to be used increasingly to provide a uniform product for processors, and to ensure uniformly ripe, ready-to-eat products for the consumer. 'Bartlett' pears, for example, have less chilling requirement than winter pears, but they ripen unevenly and fail to develop good color, texture and flavor if held at room temperature after harvest. Ethylene preconditioning ensures uniform ripeness and increases processed product yield of the pears (Puig *et al.*, 1996; Agar *et al.*, 2000).

8.3   Ethylene biosynthesis and perception

8.3.1   *Ethylene biosynthesis*

The discovery of 1-aminocyclopropane carboxylic acid (ACC), as an inter-mediate between methionine and ethylene (Adams and Yang, 1979; Lurssen *et al.*, 1979), led to a rapid description of the ethylene biosynthetic pathway (Figure 8.1). The major pathway of ethylene biosynthesis in higher plants is methionine → S-adenosylmethionine (SAM) → ACC → ethylene (Yang and Hoffman, 1984). The steps in the pathway are catalyzed by SAM synthase, ACC synthase (ACS), and ACC oxidase (ACO), respectively. While the biochemistry of these enzymes is well-described (Fluhr and Mattoo, 1996; John, 1997), and outside of the scope of this chapter, relevant aspects of these enzymes are:

1. *SAM synthase.* In addition to providing SAM for ethylene synthesis (Adams and Yang, 1979), SAM synthase has a pivotal role in cellular biochemistry providing SAM for synthesis of polyamines (Even-Chen *et al.*, 1982; see section 8.3.2) and transmethylation reactions in proteins, lipids and nucleic acids. SAM synthase genes are modulated in higher plants and SAM levels may regulate ethylene production. While several genes encoding SAM synthase have been isolated (Peleman *et al.*, 1989), no expression data are available for ripening fruit (Lelievre *et al.*, 1997a).
2. *ACS.* ACS is the main control site for ethylene biosynthesis. It requires pyridoxal phosphate for maximal activity (Yu *et al.*, 1979) and activity is inhibited by rhizobitoxine and its analogue aminoethoxyvinyl glycine (AVG) (Owens *et al.*, 1971). ACS exists in several isoforms, encoded by at least nine genes that appear to be expressed in a highly regulated manner. ACS genes are divergent in the 5′-noncoding regions, but have high coding region homology. ACS activity is unstable and has rapid turnover (Fluhr and Mattoo, 1996; Zarembinski and Theologis, 1994).
3. *ACO.* Initially ACO could only be assayed *in vivo*, by adding ACC substrate to tissue discs. This 'ethylene forming enzyme' (EFE) appeared to be membrane dependent (Yang and Hoffman, 1984) but it is now

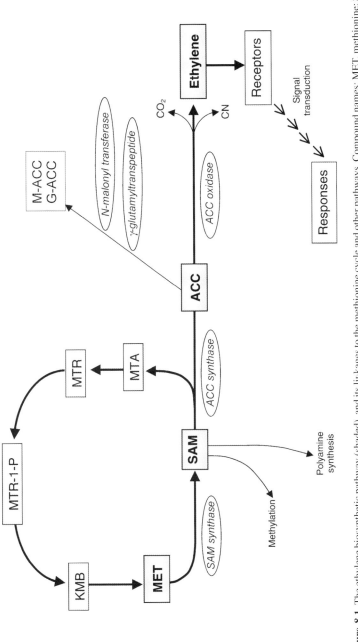

Figure 8.1 The ethylene biosynthetic pathway (shaded), and its linkages to the methionine cycle and other pathways. Compound names: MET, methionine; SAM, S-adenosylmethionine; ACC, 1-aminocyclopropane carboxylic acid; MTA, 5′-methylthioadenosine; MTR, 5′-methylthioribose; MTR-1P, 5′-methylthioribose-1-phosphate; KMB, 2-keto-4-methylthiobuytyric acid; M-ACC, 1-(malonylamino) cyclopropane-1-carboxylic acid; G-ACC, 1-(γ-L-glutamylamino)cyclopropane-1-carboxylic acid. Enzymes are shown in italics.

known to be a soluble enzyme. Sequence information indicated that ACO was a member of the dioxygenase family and *in vitro* activity requires ascorbate and Fe(II) as co-factors (Ververidis and John, 1991). In addition to O_2 as a substrate, undissociated CO_2 is a necessary co-factor for ACO (Dong *et al.*, 1992; Poneliet and Dilley, 1993). Carbon dioxide (5 to 20%) increases the V_{max}, and the K_m for ACC, O_2 and ascorbate (John, 1997). Thus CO_2 may regulate ACO activity *in vivo*. Like ACS, ACO activity is unstable and has a rapid turnover. Also like ACS, it is encoded by multiple genes that appear to be expressed in a highly regulated manner, are divergent in the 5′-noncoding regions and have high coding region homology (Fluhr and Mattoo, 1996).

There is an extensive literature showing increased transcript accumulation and activities of ACS and/or ACO during ripening of a range of climacteric fruit, including apple (Dong *et al.*, 1991; Ross *et al.*, 1992), apricot (Mbeguie-A-Mbeguie *et al.*, 1999), banana (Lopez-Gomez *et al.*, 1997; Liu *et al.*, 1999), cucumber (Shiomi *et al.*, 1998), kiwifruit (Whittaker *et al.*, 1997; Xu *et al.*, 2000), mume (Mita *et al.*, 1999), melon (Balague *et al.*, 1993; Lasserre *et al.*, 1996; Yamamoto *et al.*, 1995), passion fruit (Mita *et al.*, 1998), peach (Callahan *et al.*, 1992; Tonutti *et al.*, 1997), pear (Lelievre *et al.*, 1997b), tomato (Lincoln *et al.*, 1993; Olson *et al.*, 1991; Yip *et al.*, 1992) and winter squash (Nakajima *et al.*, 1990; Nakagawa *et al.*, 1991). Accumulation of ACS and ACO mRNA has also been shown for the non-climacteric pineapple (Cazzonelli *et al.*, 1998).

Transcript accumulation patterns for ACS and ACO are closely related to the respective enzyme activities. However, each has a multi-gene family and it is becoming increasingly clear that the enzymes are regulated differentially. Most research on isolation and characterization of multi-gene families for ACS and ACO during fruit ripening has been carried out with tomato. In this fruit, autocatalytic ethylene production is associated with increases in *LE-ACS2* and *LE-ACS4* (Kneissl and Deikman, 1996; Lincoln *et al.*, 1993; Nakatsuka *et al.*, 1997, 1998; Olson *et al.*, 1991; Rottmann *et al.*, 1991), *LE-ACO2* (Barry *et al.*, 1996; Nakatsuka *et al.*, 1997, 1998) and *ACO4* (Nakatsuka *et al.*, 1998). Differential expression of these genes appears to be associated with negative and positive feedback of ethylene biosynthesis (see section 8.3.4).

Multiple gene families for ethylene-related genes have received less attention in other fruit but in peach, *PP-ACS1*, *PP-ACO1* and *PP-ACO2* mRNAs increase with ethylene production (Mathooko *et al.*, 2001). In banana, which has a sharp increase and then decline in ethylene production during the climacteric, initiation of autocatalytic ethylene was associated with increasing *MA-ACS1* transcript abundance (Liu *et al.*, 1999). *MA-ACO1* transcript was detectable at the preclimacteric stage and accumulated during the climacteric while ACO activity and ethylene production declined, perhaps in part because availability of co-factors for the enzyme became limiting (Liu *et al.*, 1999).

8.3.2   Interaction with other pathways

Ethylene production may be influenced or regulated by interactions between its biosynthesis and other metabolic pathways (figure 8.1). In addition to its role in ethylene biosynthesis, SAM provides propylamine for polyamine biosynthesis and methyl groups in several other metabolic reactions. Polyamines act as anti-senescence compounds, whose synthesis decreases as senescence begins. This suggests that SAM metabolism is switched to ethylene synthesis at the onset of ripening (Casas *et al.*, 1990; Mattoo and White, 1991). Also, overexpression of SAM hydrolase, which degrades SAM, has been associated with inhibited ethylene production during tomato ripening (Good *et al.*, 1994). However, polyamine contents did not decline as ethylene production increased in cherimoya (Escribano *et al.*, 1994), suggesting that relationships between polyamines and ethylene may be affected by species and metabolic stage. The methionine or Yang cycle (Miyazaki and Yang, 1987) is an important interlinking pathway with ethylene biosynthesis. In addition to ACC, ACS produces 5'-methylthioadenosine (MTA), which is recycled to methionine; otherwise the low concentrations of methionine in plants and would limit the rate of ethylene synthesis and other metabolic pathways (Wang *et al.*, 1982).

ACC can be conjugated with malonate or glutathione to form 1-(malonylamino)cyclopropane-1-carboxylic acid (M-ACC) and 1-(γ-L-glutamylamino) cyclopropane-1-carboxylic acid (G-ACC), respectively (Amrhein *et al.*, 1981; Martin *et al.*, 1995; Peiser and Yang, 1998). Activity of ACC *N*-malonyltransferase that catalyzes malonylation of ACC with malonyl CoA, purified from tomato fruit, is developmentally regulated and stimulated by ethylene (Liu *et al.*, 1985c; Martin and Saftner, 1995). M-ACC accumulates in ripening apple and tomato fruit (Liu *et al.*, 1985c; Mansour *et al.*, 1986) and may be involved in autoinhibition of ethylene production in citrus albedo tissue (Liu *et al.*, 1985b). However, M-ACC is considered an inactive end product that is sequestered in the vacuole (Bouzayen *et al.*, 1989) and may serve as an ethylene source only under non-physiological conditions (Jiao *et al.*, 1986). The regulation of the levels of these conjugates, and to what extent they influence ethylene production, remains uncertain.

Other possible interactions with ethylene biosynthesis during ripening include the jasmonates (Saniewski *et al.*, 1987a, b). Endogenous jasmonate concentrations increase transiently prior to autocatalytic ethylene biosynthesis during the onset of ripening of both apple and tomato fruit, but decline subsequently (Fan *et al.*, 1998).

8.3.3   Ethylene perception

Early research indicated that biological activity of ethylene required the presence of an unsaturated bond adjacent to a terminal carbon atom in the ligand, and that it bound to a metal containing receptor site (Burg and Burg, 1967). It was proposed that CO_2, one of the first chemicals to be considered as an inhibitor of ethylene perception, displaced ethylene from these receptor sites. Oxygen appeared to promote ethylene binding to the receptor site.

Subsequently, silver was shown to negate the effects of ethylene in plant tissues (Beyer, 1976), probably at the receptor level (Veen, 1986). Application of silver, as its thiosulphate salt, inhibits fruit ripening (Saltveit et al., 1978; Tucker and Brady, 1987). Another compound, 2,5-norbornadiene, which counteracts ethylene responses competitively, inhibited ripening in apples (Blankenship and Sisler, 1989). Diazocyclopentadiene (DACP), a weak inhibitor of ethylene responses, produces much more active compounds when irradiated. These compounds have not been identified and appear to be unstable. Nevertheless, DACP irradiation products effectively inhibited ripening of apple, avocado, banana, kiwifruit, persimmon and tomato fruit (Blankenship and Sisler, 1993; Sisler and Blankenship, 1993a,b; Sisler and Lallu, 1994; Sisler and Serek, 1997; Tian et al., 1997a), although they stimulated ethylene production of the non-climacteric strawberry fruit (Tian et al., 1997b).

Each of these presumed inhibitors of ethylene perception, however, has limitations that affect commercial acceptance. Silver is a heavy metal that cannot be used on food and is environmentally harmful. As with other cyclic olefins, 2,5-norbornadiene has the disadvantages of requiring continuous exposure, a high concentration and possessing a strong odour (Sisler and Yang, 1984). DACP is explosive at high concentrations.

Cyclopropenes have been much more promising as antagonists of ethylene responses (Sisler et al., 1996). Of the compounds tested, 1-methylcyclopropene (1-MCP) has received most attention because it is more stable than cyclopropene and 1000 times more active than 3,3-dimethylcyclopropene. Under normal conditions, it is thought that ethylene binds to a metal in the receptor but subsequent dissociation from each receptor is required for an active complex to form. 1-Methylcyclopropene may bind to the ethylene receptor, competing with ethylene for the available binding sites (Sisler and Serek, 1997). Whereas ethylene diffuses rapidly from the binding site after ethylene treatment, these compounds remain bound for long periods, and formation of an active complex is prevented. The applied research with 1-MCP and its potential commercialization is reviewed in section 8.4.5.

Rapid advances in research on ethylene perception have occurred in *Arabidopsis*, where a family of genes encoding ethylene receptors, related to histidine kinases, has been identified (Bleecker, 1999; Chang and Shockey, 1999; Kieber, 1997). This has been extended to investigate fruit ripening, especially that of the tomato. The ethylene response gene (*ETR1*) was the first ethylene receptor gene characterized (Chang et al., 1993). Transgenic tomato plants expressing a mutant *Arabidopsis ETR1* gene show delayed fruit ripening (Wilkinson et al., 1995). The tomato ethylene-receptor gene family homologous to *ETR1* consists of at least five members that are expressed in virtually all tomato plant tissues (Lashbrook *et al.*, 1998; Tieman and Klee, 1999; Zhou *et al.*, 1996). Two of them are structurally divergent from the other *ETR1* homologues. *LeETR1*, *LeETR2*, *LeETR4* and *LeETR5* genes are expressed at constant levels during fruit ripening (Lashbrook *et al.*, 1998; Tieman and Klee, 1999; Zhou *et al.*, 1996). However, *LeETR4* is expressed at very high levels throughout ripening, indicating that it may be important in ethylene perception provided that it truly encodes a functional ethylene receptor (Tieman and Klee, 1999). *LeETR5* lacks the histidine within the kinase domain that is predicted to be phosphorylated, and mutation studies suggest that histidine kinase activity is not necessary for an ethylene response (Tieman and Klee, 1999).

Unlike the other *ETR* homologs, *NR* (*LeETR3*) is regulated by ethylene. *NR* was isolated from the ethylene-insensitive *Never ripe* (*Nr*) tomato mutant (Wilkinson *et al.*, 1995), which is insensitive to ethylene (Lanahan *et al.*, 1994), and provides a useful ethylene receptor mutant for studying ethylene signal transduction. *NR* mRNA accumulates during normal ripening, or in ethylene-treated mature green, wild-type fruit (Hackett *et al.*, 2000; Lashbrook *et al.*, 1998; Wilkinson *et al.*, 1995), but not in the *Nr* mutant (Payton *et al.*, 1996) or in 1-MCP-treated fruit (Nakatsuka *et al.*, 1998). Anti-sense inhibition of the *NR* gene restored ripening of the *Nr* tomato mutant, confirming that receptors are inhibited in the mutant, and that wild-type *NR* gene product is not required for normal ripening (Hackett *et al.*, 2000). Transcripts encoded by *NR* are absent in the *rin*, *nor* and *Cnr* mutants, but inducible by exogenous ethylene, indicating that these mutations act upstream of ethylene perception in the regulation of ripening (Lanahan *et al.*, 1994; Thompson *et al.*, 1999).

Kieber *et al.* (1993) also identified an *Arabidopsis* gene product, *CTR1*, whose action is downstream of *ETR1*. *CTR1*, which is constitutively suppressed and negatively regulates the ethylene response, encodes a putative serine/threonine kinase related to the MAP kinase family. An *ER50* clone, identical to this kinase (Wang and Li, 1997), and with strong homology to *Arabidopsis CTR1* gene, has been isolated from tomatoes by Zegzouti *et al.* (1999). Unlike *CTR1*, however, mRNA encoding *ER50* was up-regulated both by ethylene and during ripening. Transgenic tomato plants under- and overexpressing the gene are being assessed to address the apparent paradox of a negative regulator being upregulated during the onset of ripening (Zegzouti *et al.*, 1999).

Ethylene receptor homologue genes have been isolated in other fruit tissues (Lee *et al.*, 1998; Martinez *et al.*, 2001; Sato-Nara *et al.*, 1999). In muskmelon, expression of mRNA encoding *Cm-ERS1* and *Cm-ETR1* genes, similar to *Arabidopsis ERS1* and *ETR1* genes, respectively, were consistent with patterns of ripening within the fruit (Sato-Nara *et al.*, 1999). As shown in the tomato, these genes were involved in other developmental processes and clarification of their roles will only come with understanding of factors regulating their expression. Recent research on the *responsiveness to antagonist1* (*ran1*) mutation, which suggests that copper is required not only for ethylene binding but also for the signaling function of ethylene receptors (Woeste and Kieber, 2000), has not yet been extended to fruit.

8.3.4   Negative and positive feedback inhibition

McMurchie *et al.* (1972) developed a concept of two systems to describe regulation of ethylene production that has proved useful in understanding ethylene metabolism, and has been supported experimentally. System 1 is the ethylene-production system common to non-climacteric tissues and preclimacteric fruit. System 2 is the autocatalytic ethylene-production system active during the climacteric. Exogenous ethylene or propylene causes autoinhibition of ethylene production in system 1 tissues and autocatalytic production in system 2 tissues.

Autoinhibition of ethylene production has been demonstrated in tissues such as preclimacteric avocados, bananas and sycamore fig (McMurchie *et al.*, 1972; Vendrell and McGlasson, 1971; Zauberman and Fuchs, 1973; Zeroni *et al.*, 1976), immature tomato locule tissue (Atta-Aly *et al.*, 2000b), and wounded flavedo tissues of citrus (Riov and Yang, 1982). However, as fruit mature, they become progressively more responsive to ethylene, requiring lower exogenous concentrations to shorten the length of the preclimacteric period and to trigger autocatalytic ethylene production (Knee *et al.*, 1987; McMurchie *et al.*, 1972; McGlasson *et al.*, 1972; Peacock, 1972). This process involves a transition from system 1 to system 2 ethylene production and a change from feedback inhibition to stimulation of ethylene production.

Autoinhibition of ethylene production involves suppression of activity of ACS and ACO (Riov and Yang, 1982) or only ACS (Atta-Aly *et al.*, 2000b). In immature fruit, short-term ethylene exposure stimulates ACO transcript levels (Dong *et al.*, 1992; Kneissl and Deikman, 1996) and activity (Liu *et al.*, 1985a; Starrett and Laties, 1991), but not ACS activity or ethylene production. The shift from autoinhibition to autocatalysis is associated with stimulation of ACS and ACO activities (Atta-Aly *et al.*, 2000a; Riov and Yang, 1982.). Silver, an antagonist of ethylene action, enhances ethylene production in immature tomato fruit (Atta-Aly *et al.*, 1987), but appears to block both positive and negative feedback mechanisms (Atta-Aly *et al.*, 2000a). Inhibitors of ethylene action block autocatalytic ethylene production in system 2 tissues (see section 8.3.3).

The transition from system 1 to system 2 ethylene is probably controlled by differential regulation of multigene families of ACS and ACO. Nakatsuka *et al.* (1998), for example, concluded that *LE-ACS1A, LE-ACS3* and *LE-ACS6* mediate low system 1 ethylene production rates, together with *LE-ACO1* and *LE-ACO4*. *LE-ACS6* is negatively regulated in preclimacteric fruit, while transcripts for *LE-ACS1A* and *LE-ACS3* accumulate irrespective of the mode of feedback regulation. System 2 ethylene appears to be regulated by *LE-ASC2* and

LE-ASC4 by a positive feedback mechanism, with accompanying *LE-ACO1* and *LE-ACO4* mRNAs. Nakatsuka *et al.* (1998) also proposed that the transition from system 1 to system 2 ethylene production may be controlled by accumulation of *NR* protein. A similar model to explain the cascade of ACS and ACO gene expression during the transition from system 1 to system 2 ethylene has been suggested by Lelievre *et al.* (1997a).

8.3.5   Ethylene-dependent and independent events

Fruit ripening involves a complex array of physiological and biochemical changes. Jeffery *et al.* (1984) distinguished classes of biochemical change in ripening tomato. Ethylene-independent changes in enzymes associated with citrate and malate metabolism, loss of chlorophyll (which was enhanced by, but not intimately dependent on, ethylene) and the formation of lycopene, PG and invertase that seemed ethylene-dependent. Subsequent study of gene expression changes have shown an extensive number of genes that are ethylene-dependent or -independent during ripening. These studies have included analysis of the effects of ethylene perception inhibitors such as silver and 1-MCP, ethylene treatments and use of transgenic fruit (Davies *et al.*, 1988; Gray *et al.*, 1994; Hadfield *et al.*, 2000; Itai *et al.*, 2000; Zegzouti *et al.*, 1999). Categories of ethylene-regulated genes in climacteric fruit include those involved in transcriptional and post-transcriptional regulation, signal transduction, stress-related proteins and primary metabolism (Zegzouti *et al.*, 1999). Ethylene also induces gene activity in non-climacteric fruit (Alonso and Granell, 1995; Alonso *et al.*, 1995), although the involvement of ethylene in ripening of non-climacteric fruit remains unclear (Lelievre *et al.*, 1997a). It is becoming evident, however, that both ethylene-dependent and -independent pathways coexist in both climacteric and climacteric fruit (Lelievre *et al.*, 1997a; Hadfield *et al.*, 2000).

8.3.6   Transgenic approaches for reducing ethylene production

The tomato has been a useful fruit for ripening research with the availability of extensive chromosome maps and a variety of ripening mutants, its commercial importance, and good transformation systems (Gray *et al.*, 1994). Several direct and indirect approaches have been taken to manipulate ethylene production by genetic modification. Expression of antisense *LE-ACS2* resulted in ethylene production that was less than 0.5% of the wild-type (Oeller *et al.*, 1991; Theologis *et al.*, 1993). Fruit showed no accumulation of *ACS2* and *ACS4*, and fail to ripen in the absence of exogenous ethylene.

Effects of expression of antisense ACO in tomatoes were inconsistent. These fruit had about 3% of normal ethylene production (Hamilton *et al.*, 1990; Murray *et al.*, 1993; Picton *et al.*, 1993). While lycopene accumulation and loss of acidity were delayed in transgenic fruit, softening rates were similar to wild-type.

However, retardation of fruit softening at the over-ripe phase was associated with improved resistance to cracking and damage compared with control fruit. Inhibition of ripening was observed both on and off the plant but transgenic fruit ripened faster on the plant. Inhibition of ripening was greater in fruit harvested before than after the onset of color change. Ripening was only partially restored by exogenous ethylene treatment in ACO anti-sense tomatoes (Murray *et al.*, 1993; Picton *et al.*, 1993), in contrast to complete restoration in ACS antisense tomatoes (Oeller *et al.*, 1991).

Ethylene production was also suppressed in tomatoes expressing ACC deaminase from soil bacteria. The fruits had lower ACC accumulation and ethylene production, and delayed ripening (Klee, 1993; Klee *et al.*, 1991). As shown for ACO transgenics, this transgenic fruit ripened faster on than off the plant. Good *et al.* (1994) expressed an S-adenosylmethionine (SAM) hydrolase gene from bacteriophage T3 in tomato; this reduced SAM and ACC contents, and ethylene production by the fruit. Bolitho *et al.* (1997) demonstrated that expression of a heterologous ACO gene from apple in tomato reduced ethylene.

Transgenic approaches to modifying ethylene production have been extended to the cantaloupe melon (Ayub *et al.*, 1996; Guis *et al.*, 1997). Fruit expressing an anti-sense ACO transgene produced less than 1% ethylene; chlorophyll loss, titratable acidity loss, carotenoid accumulation in the rind, volatile production and flesh softening were suppressed relative to the wild-type. Flesh carotenoid increased before the climacteric in wild-type fruit and therefore was not affected by the transgene. Both types of fruit accumulated soluble solids, mainly sucrose, at the same rate but levels in the transgenic fruit were higher because they did not abscise and therefore remained on the plant longer. Antisense fruit softened significantly but did not yellow when harvested, but treatment with exogenous ethylene restored ripening (Bauchot *et al.*, 1998).

The research so far indicates that transgenic fruit could make a major contribution in commercial postharvest handling. However, fruit species are likely to show variable responses, dependent on the sequence of ripening events and their sensitivities to ethylene.

8.4   Effects of postharvest treatments on ethylene biosynthesis and perception

8.4.1   Storage temperature

Low temperature storage is the primary means of reducing metabolic rates and maintaining the quality of harvested fruit. However, application of this technology is restricted for chilling sensitive fruit, which are usually of tropical and subtropical origin and develop chilling injury if stored between −1.5 and 10 to 15°C. Chilling insensitive fruit, usually of temperate origin, can be stored for prolonged periods at temperatures approaching freezing, although chilling-related disorders may occur after prolonged storage. The interaction between ethylene biosynthesis and responses of fruit to chilling temperatures is not clear. The variation in responses among fruits perhaps indicates the difficulty in separating ripening-related ethylene production from that related to stress (chilling).

Ethylene treatment enhances chilling injury in avocado (Chaplin *et al.*, 1983) and orange (McCollum and McDonald, 1991; Porat *et al.*, 1999), decreases injury in muskmelon (Lipton *et al.*, 1979) and does not affect it in lemon (McDonald *et al.*, 1985). Susceptibility to chilling injury decreases with ripening in tomato (Autio and Bramlage, 1986); this could result from the action of endogenous ethylene causing a change in metabolic response to chilling during ripening.

Low temperatures can inhibit or stimulate ethylene production in chilling-sensitive fruit and the effects may be apparent while the fruit remains in the cold or after removal to warmer temperatures. The commonly observed stimulation of ethylene after chilling may be an indicator of injury. Fruit quality can be adversely affected; for example chilling-induced ethylene production in citrus fruit is associated with skin degreening (Cooper *et al.*, 1969).

In fruit at low temperatures ACC can accumulate (Cabrera and Saltveit, 1990; Lipton and Wang, 1987). In cucumbers, ACC accumulation, ACS activity and ethylene production remained low at chilling temperatures, but increased dramatically when fruit were moved to warmer temperatures (Wang and Adams, 1982). When tomatoes were kept at chilling temperatures, ethylene production was inhibited, but mRNA encoding ACO (*pTOM 13*) accumulated to levels much greater than during normal ripening (Watkins *et al.*, 1990). In avocado, ACO transcript accumulated and then declined during storage at 3°C, while they accumulated steadily at 7°C (Dopico *et al.*, 1993).

Ethylene production will not occur if irreversible injury has occurred. In cucumbers, ACS activity and ACC accumulation were inhibited if the low temperature exposure was prolonged (Wang and Adams, 1982). In nectarines and peaches, development of a chilling disorder known as woolliness was associated with inhibited ACO activity and ethylene production (Dong *et al.*, 2001a; Zhou *et al.*, 2001a). Intermittent warming of fruit during storage, which alleviated woolliness development, induced mRNA transcripts for ACS and ACO and maintained ethylene production (Zhou *et al.*, 2001b).

Transgenic melons with antisense ACO constructs did not develop chilling injury at low temperatures or after warming, but treating these fruit with ethylene prior to storage restored chilling sensitivity (Ben-Amor *et al.*, 1999). In contrast, 1-MCP application inhibited ACO transcripts and ethylene production, increased woolliness development in nectarines (Dong *et al.*, 2001a) and increased chilling injury symptoms in oranges (Porat *et al.*, 1999), suggesting that chilling effects on ethylene metabolism are specific for a particular fruit or tissue type. Information on differential expression of ACS and ACO gene families may contribute towards a fuller understanding of relationships between chilling injury and ethylene metabolism. Wong *et al.* (1999) have isolated ACS genes that are induced (*CS-ACS1*) or repressed (*CS-ACS2*) by chilling in citrus fruit, but further studies are needed.

Some fruit require exposure to chilling temperatures to induce ripening. The classic example is seen in some winter European pear cultivars where proper ripening will not occur if the chilling period is too short and mealiness will develop if chilling periods are too long (Blankenship and Richardson, 1985; Knee, 1973). Exposure to ethylene, or its analogs, can overcome the winter pear requirement for chilling (Blankenship and Richardson, 1985). The reason for the chilling requirement for pear ripening is not known but the underlying metabolic effects of chilling on ethylene biosynthesis are being increasingly understood.

ACC accumulation and ethylene production increased during cold storage of 'Conference' pears, but ACS activity remained low (Knee, 1987). However, inhibitor studies suggested that mRNA for ACS was formed at low temperature. Ripening capacity of 'Bartlett' pear was associated with increased ACC accumulation, and activities of ACS and ACO, but was not fully expressed without chilling or ethylene treatments (Agar *et al.*, 1999, 2000). In 'Passe Crassane' pear fruit, chilling increased transcript accumulation for genes encoding ACS and ACO, strongly increased ACO activity and, to a lesser extent, ACS activity (Lelievre *et al.*, 1997b). Autocatalytic ethylene production after storage was associated with decreased ACS mRNA and increased ACO mRNA levels, but higher activities of both enzymes. Treatment of fruit with propylene or 1-MCP indicated that ACS gene expression is regulated by ethylene alone during or after chilling, while ACO gene expression is induced separately by either ethylene or chilling. Pear strains that are unresponsive to chilling may be unable to produce ACC or perceive the cold stimuli (Chen *et al.*, 1997; Satoh *et al.*, 2000).

Low storage temperatures can slow down the onset of ethylene production in apples (Knee *et al.*, 1983) and lower internal ethylene concentrations were associated with lower ACS and ACO mRNA transcript levels in fruit at $0°C$ compared with $4°C$ (Cregoe *et al.*, 1993). However, chilling temperatures can also enhance ethylene production in some cultivars of apples (Knee *et al.*, 1983). ACC accumulation and ethylene production of 'Granny Smith' apples at $20°C$ are markedly increased by exposure to $0°C$ for at least eight days but this does not occur in 'Royal Gala' or 'Delicious' (Jobling *et al.*, 1991; Larrigaudiere *et al.*, 1997). Cold treatment appears to reduce resistance to ripening in 'Granny Smith' and is related to *de novo* synthesis of ACS and ACO (Larrigaudiere and Vendrell, 1993; Lelievre *et al.*, 1995). The ripening physiology of this cultivar may be more analogous to that of winter pears, although implications for quality maintenance are unclear.

Treatments that prevent chilling injury in fruit have been associated with maintenance of the polyamines, spermidine and spermine (Kramer and Wang,

1989; Wang and Ji, 1989). Increases in putrescine contents at chilling temperatures appear related to ripening and not to development of chilling injury (Escribano and Merodio, 1994; Wang and Ji, 1989). The relationships between these polyamines and ethylene biosynthesis have not been studied.

8.4.2   Storage atmospheres

Decreased O_2 and/or elevated CO_2 levels during controlled atmosphere (CA) or modified atmosphere (MA) storage conditions generally increase the storability of most horticultural crops (Kader *et al.*, 1989). Tolerances to atmospheres among crops can vary widely, the optimal concentrations being a function of the species, cultivar, harvest maturity, length of storage and storage temperature. In addition to 'normal' atmosphere regimes, short-term stress levels of these gases are sometimes employed for additional quality benefits, reduction of physiological and pathological disorders, or quarantine purposes (Fernández-Trujillo *et al.*, 2001; Nicolas *et al.*, 1989; Wang and Dilley, 2000). These treatments, which may be applied immediately after harvest or intermittently during storage, are suitable only if the tissue recovers without damage after removal from storage. Although all postharvest storage represents a stress on the product, (Kays, 1997), ethylene biosynthesis under short-term stress treatments may be affected differently than under normal atmosphere storage regimens. Regardless of whether atmospheres are within normal ranges, or at stress levels, complex interactions can exist between the gases and physiological state of the tissue, and overall the mechanisms of action of decreased O_2 and elevated CO_2 on ethylene metabolism are still unclear. Both ethylene dependent and independent ripening events are affected by these gases (Kanellis *et al.*, 1991; Rothan *et al.*, 1997).

For both non-climacteric and climacteric fruit, it is reasonable to assume that atmosphere effects are mediated through suppressed metabolic changes that result in decreased respiration. However, MA/CA conditions also suppress ethylene biosynthesis and action in climacteric fruit and therefore their benefits also may not be solely due to effects of these atmospheres on respiration (Banks *et al.*, 1984). The situation with non-climacteric fruit is less clear as recent research indicates that an important benefit of the high CO_2 levels used to maintain quality of the non-climacteric strawberry also might be to suppress ethylene production by the fruit (Kim and Wills, 1998).

Ethylene production is reduced by as much as 50% in a 3% O_2 atmosphere (Burg and Thimann, 1959). Burg and Burg (1967) suggested that low O_2 impeded ethylene binding but no experimental evidence is available. The primary effects of low O_2 are likely to be directly on ethylene biosynthesis since ACO has an absolute O_2 requirement for activity (Adams and Yang, 1979; John, 1997). The K_m for O_2 ranges from 0.3 to 6.2%, decreasing as the ACC content increases (Yip *et al.*, 1988).

Elevated CO_2 can enhance, reduce or have no effect, on ethylene biosynthesis (Mathooko, 1996). Differences in fruit responses to CO_2 may reflect positive effects of the gas on inhibiting ethylene-related metabolism versus stress-induced effects on ethylene production. These effects are also complicated by at least three other factors:

1. It can be difficult to dissect the direct effects of elevated CO_2 from those of low O_2 on ethylene biosynthesis.
2. Solubilization of CO_2 produces H^+ and HCO_3^- which may effect the pH of the cytoplasm and enzyme activities (Bown, 1985), but *in vivo* effects of CO_2 will be affected by the buffering capacity of the tissue involved.
3. As discussed in section 8.3.1, maximal activity of ACO *in vitro* requires CO_2 and it is uncertain to what extent enzyme activity is regulated *in vivo* by the gas. A fuller understanding will require description of molecular events surrounding feedback activation of ACO by CO_2 and the subcellular location of the enzyme.

Burg and Burg (1967) proposed that CO_2 displaced ethylene from ethylene receptor sites but experimental support for this concept has been difficult to obtain (Sisler and Wood, 1988). Recent research, showing that the effects of CO_2 on ethylene production were similar in control and 1-MCP-treated fruit, suggests that CO_2 effects are separable from those associated with ethylene perception (de Wild *et al.*, 1999).

Elevated CO_2 could operate through an effect of mass action, since CO_2 is the product of the oxidative deamination of ACC to ethylene (Miyazaki and Yang, 1987). However, the gradual recovery of ACS and ACO activities, and tissue capacity to produce ethylene after CO_2 has been removed from the storage atmosphere, argues against a simple mass action effect (Mathooko, 1996). Early research, based either on implications derived from measurements of ACC accumulation or by direct measurement of enzyme activities, suggested that elevated CO_2 reduced ethylene production by inhibiting ACS and/or ACO activities (Bufler, 1984; Cheverry *et al.*, 1988; Kubo *et al.*, 1990; Mathooko *et al.*, 1995b). Because ACS turnover is rapid, loss of enzyme activity could result from degradation, inhibition of synthesis, inactivation or a combination of these processes (Mathooko *et al.*, 1995b). Stimulation of ethylene production by stress levels of CO_2 are associated with accumulation of ACC and increased ACS and ACO activities (Mathooko *et al.*, 1995a). Induction of stress ethylene requires continuous presence of the gas (Mathooko, 1996).

In apples, low O_2 inhibited only the ACC to ethylene step (Li *et al.*, 1983), while high CO_2 also inhibited formation of ACC (Chaves and Tomas, 1984; Li *et al.*, 1983). The fruit have lower internal ethylene concentrations and ACC accumulation in a 2.5% O_2 atmosphere compared with air (Lau *et al.*, 1984). Patterns of delayed and decreased ethylene production of preclimacteric apple fruit in 2% O_2, 2% O_2 and 5% CO_2, or air plus 5% CO_2, compared with air alone, were associated with reduced expression of genes encoding ACS and ACO, and ACS activity (Gorny and Kader, 1996a). ACO activity in air plus 5% CO_2, however, was no different from that in air. The primary effects of CO_2 appeared to be on ACS transcription and on accumulation of active ACO protein. Fungistatic levels of 20% CO_2 or 0.25% O_2 on preclimacteric and climacteric apples also indicated that ACS transcript abundance and enzyme activity was a key step in the inhibition of ethylene production (Gorny and Kader, 1996b, 1997). The effect of these gases on ACO transcript accumulation, activity and protein was smaller in climacteric than non-climacteric fruit.

Exposure of tomatoes to 20% CO_2 for three days before transfer to air reduced accumulation of transcripts for ACS and ACO, and of mRNAs encoding both presumed ethylene dependent and independent ripening events (Rothan et al., 1997). In contrast, 20% CO_2 decreased enzyme activity, accumulation of ACS transcripts and ACC concentration in peach fruit, but had little effect on ACO transcript accumulation and ACO activity (Mathooko et al., 2001). Post-transcriptional modification, synthesis of novel proteins, protein phosphorylation and dephosphorylation in the signal transduction pathway, and calcium influx from the extracellular space could all be involved in CO_2-induced ethylene biosynthesis (Mathooko et al., 1998, 1999).

Stress levels of CO_2 caused polyamine accumulation, especially putrescine and spermidine, in cucumber fruit, concurrent with stimulated ethylene production but the pathways were not competitive (Mathooko et al., 1995a). In contrast, Munoz et al. (1999) proposed that inhibited ethylene production in CO_2-treated cherimoya fruit at both chilling and shelf-life temperatures, resulted from increased flux from SAM towards polyamine synthesis. Under elevated CO_2, spermidine and spermine increased more than putrescine and it was been suggested that these polyamines could have improved fruit quality. Further research on the effects of elevated CO_2 on intercellular pH and polyamine levels in relation to ethylene biosynthesis is required.

Recovery of ethylene biosynthesis is often slow after removal from CA; this residual effect may help maintain fruit quality during subsequent marketing. However, fruit ripening after removal from storage can be very rapid; it seems that the capacity for ethylene synthesis continues to develop during CA storage of banana (McGlasson and Wills, 1972).

8.4.3   Heat treatments

Postharvest heat treatment of fruit can be used for disinfestation, disease control and maintenance of fruit quality by delaying ripening or decreasing incidence of disorders such as chilling injury (Lurie, 1998; Paull and Chen, 2000). Successful application of the technology relies on the ability of the treated commodity to recover from heat stress and, therefore, the temperatures and exposure periods used for a given commodity are critical. For climacteric fruit subsequent recovery of ripening seems to depend on the ability to produce ethylene. Typically, ethylene production is inhibited by temperatures above 35°C (Klein, 1989; Picton and Grierson, 1988). This inhibition is associated with accumulation of ACC (Atta Aly, 1992; Yu *et al.*, 1980), and decreased expression of ripening related genes, including that encoding ACO (Lurie *et al.*, 1996; Picton and Grierson, 1988). Heat treatments also depress the activity of ACS (Biggs *et al.*, 1988; Ketsa *et al.*, 1999) and ACO (Ketsa *et al.*, 1999; Klein and Lurie, 1990; Paull and Chen, 2000).

Lurie *et al.* (1996) found that expression of ACO mRNA in tomato recovered after removal of heat but the corresponding protein for the enzyme did not always follow transcript accumulation closely. Although heat appears to inhibit ethylene perception, as indicated by inability of fruit during heating to respond to exogenous ethylene (Maxie *et al.*, 1974; Seymour *et al.*, 1987; Yang *et al.*, 1990), no information on effects of heat treatment on ethylene receptors is available. While disruption of ethylene synthesis may be responsible for the inhibition of ripening (Lurie *et al.*, 1996), Paull and Chen (2000) suggest that this is a symptom of injury. Heat treatment reduced the chilling-induced increase in ethylene production in mandarin fruit (Martinez-Tellez and Lafuente, 1997) but relatively little information about non-climacteric ethylene metabolism is available.

8.4.4   1-Methylcyclopropene

The mode of action of 1-MCP is described in section 8.3.3. 1-Methycyclopropene is structurally related to ethylene, has a non-toxic mode of action, and is applied at very low dose levels, with low measurable residues in food commodities. The USA Environmental Protection Agency has classified 1-MCP as a plant growth regulator.

The pending commercial availability of 1-MCP has led to a surge of research activity on its effects on fruit ripening. Significant effects of the chemical have been demonstrated for a range of climacteric and non-climacteric fruit (table 8.2). Some general observations and interpretations about the effects of 1-MCP on fruit systems can be made:

1. *The 1-MCP concentrations* required to saturate binding sites, and the extent and longevity of 1-MCP action, are influenced greatly by species, organ, tissue and mode of ethylene biosynthesis induction. A 'time × concentration' effect is apparent and the longer the exposure, the lower the required concentration (Sisler and Serek, 1997). Although 1-MCP binding is essentially irreversible, inhibition of ethylene action may be overcome by the production of new receptors (Sisler *et al.*, 1996).

2. *Ethylene production* is usually inhibited by 1-MCP application. However, it can be enhanced by 1-MCP in some fruit, such as the banana,

Table 8.2 Physiological processes or disorders that are delayed or decreased, increased, or unaffected in climacteric and non-climacteric fruit treated with 1-MCP

Fruit type	Decreased or delayed	Increased	Unaffected	References
Climacteric				
Apple	Ethylene production, respiration, softening, loss of titratable acidity, superficial scald, soft scald, volatile alcohol and ester formation, coreflush, browncore, greasiness, and senescent breakdown	Soluble solids	Soluble solids, starch index, carbon dioxide injury, loss of titratable acidity	Fan *et al.*, 1999a, b; Fan and Mattheis, 1999a, b; Rupasinghe *et al.*, 2000; Watkins *et al.*, 2000, unpublished data
Apricot	Ethylene production, respiration, softening, decay, loss of titratable acidity, color change, production of volatile alcohols and esters			Chahine *et al.*, 1999; Dong *et al.*, 2001c; Fan *et al.*, 2000
Avocado	Ethylene production, softening, endo-1,4-β-glucanase and polygalacturonase activities, color change			Feng *et al.*, 2000
Banana	Ethylene production, chlorophyll loss, color change, softening, 'green life', production of alcohols and related esters	Uneven and blotchy color development, ethylene production, ratio between alcohols and related esters		Golding *et al.*, 1998, 1999; Harris *et al.*, 2000; Jiang *et al.*, 1999a, b,; Macnish *et al.*, 2000; Sisler and Serek, 1997; Sisler *et al.*, 1996, 1999
Nectarine	Ethylene production, softening	Flesh woolliness and reddening, lower expressible juice	Respiration	Dong *et al.*, 2001a
Pear	Ethylene production			de Wild *et al.*, 1999; Lelievre *et al.*, 1997

Table 8.2 (continued)

Fruit type	Decreased or delayed	Increased	Unaffected	References
Climacteric				
Peach	Ethylene production, respiration, softening, loss of titratable acidity, ACC synthase and ACC oxidase activity	Flesh browning at 5°C		Fan et al., 2001; Mathooko et al., 2001
Plum	Ethylene production, respiration, softening, color change, loss of titratable acidity, aroma development, flesh browning, decay	Flesh browning	Color change	Abdi et al., 1998; Dong et al., 2001b, c
Tomato	Ethylene production, lycopene accumulation, ACC synthase and ACC oxidase activity			Nakatsuka et al., 1997, 1998; Sisler and Serek, 1997; Sisler et al., 1996, 1999
Non-climacteric				
Grapefruit	De-greening	Ethylene production	Decay	Mullins et al., 2000
Orange	De-greening, mold rot incidence	Chilling injury, volatile off-flavors, stem-end rot incidence	Weight loss, softening	Porat et al., 1999
Pineapple	Chilling injury (internal browning), decline in ascorbic acid and soluble solids content, yellowing	Ethylene production		Selvarajah et al., 2001
Strawberry	Decay	Ethylene production	Respiration, decay	Ku et al., 1999; Tian et al., 2000

indicating that 1-MCP blocks normal feedback regulation of ethylene biosynthesis (Golding *et al.*, 1998). In the grapefruit, ethylene production is enhanced even while the fruit remained green and Mullins *et al.* (2000) proposed that bound 1-MCP represses signalling pathways and shuts down the feedback mechanism that modifies ethylene biosynthesis. Tian *et al.* (2000) observed that exogenous ethylene stimulated respiration of strawberry fruit harvested at an early stage of maturity but not at a late stage. The rise in respiration in early harvested fruit was prevented by prior exposure to 1-MCP.

3. *Ripening processes* characteristic of most climacteric and some non-climacteric fruit, such as softening, yellowing, respiration and loss of titratable acidity, are usually delayed or inhibited by 1-MCP application. Soluble solid contents are sometimes unaffected by 1-MCP application and effects of 1-MCP may be affected by storage conditions. Loss of green color in bananas can be irregular, resulting in blotchy color development (see table 8.2).

4. *Cultivar* greatly influences responses of fruit to 1-MCP, as shown for apple (Fan *et al.*, 1999a; Rupasinghe *et al.*, 2000; Watkins *et al.*, 2000). In plums, 1-MCP caused significant delays of ripening in 'suppressed climacteric fruit' (Abdi *et al.*, 1998; Dong *et al.*, 2001b), while in climacteric plums the inhibition was greater (Abdi *et al.*, 1998; Dong *et al.*, 2001c). For these plums, pulses or continuous exposure to low doses of 1-MCP may be required to control ripening (Abdi *et al.*, 1998). Cultivar effects could complicate commercial adoption of 1-MCP technology.

5. *Maturity and ripeness* can affect the responses to 1-MCP, later harvested fruit usually being less responsive to 1-MCP, e.g. apricot and banana (Chahine *et al.*, 1999; Jiang *et al.*, 1999a,b), as might be expected for a compound that acts by inhibiting ethylene action. Maturity effects may limit commercial application of 1-MCP in apple cultivars such as 'McIntosh', which can have high ethylene production at harvest (Watkins *et al.*, 2000) and for 'Williams' bananas that normally have mixed maturity of fruit within bunches (Harris *et al.*, 2000).

6. *The effect of treatment temperature* may vary by commodity. 1-MCP was less effective in controlling banana ripening when applied at 2°C compared with 20°C (Macnish *et al.*, 2000), suggesting that 1-MCP binding is poorer at low temperature, perhaps due to conformational changes in a membrane-located ethylene receptor protein. Treatment at low temperature may result in relatively greater accumulation and/or non-specific binding of 1-MCP molecules in plant tissues.

7. *Volatile production* is inhibited by 1-MCP in apples, apricots and bananas (Fan and Mattheis, 1999a,b; Golding *et al.*, 1999; Fan *et al.*, 2000a; Rupasinghe *et al.*, 2000) and in bananas, it increased alcohol contents

relative to their related esters (Golding *et al.*, 1999). These results are consistent with the view that volatile production is regulated by ethylene. The commercial impact of reduced aroma is likely to vary by fruit type, being more critical for species or cultivars that are purchased on the basis of aroma. Off-flavor development was increased by 1-MCP treatment of oranges (Porat *et al.*, 1999) and higher titratable acidity in treated apricots could result in sourness (Fan *et al.*, 2000). Consumer studies on acceptability of 1-MCP treated fruit will be required.

8. *Physiological disorders* of fruit can be *reduced*: superficial scald, soft scald, coreflush, greasiness and senescent breakdown in apples (Fan *et al.*, 1999b; Rupasinghe *et al.*, 2000; Watkins *et al.*, 2000) and internal browning in pineapples and plums (Dong *et al.*, 2001b; Selvarajah *et al.*, 2001); *increased*: woolliness in nectarines and breakdown in plums (Dong *et al.*, 2001a,c); or *unaffected*: CO_2 injury in apples (Watkins, unpublished data). 1-Methylcyclopropene affected disorder development of apricots differently if treated before or after storage (Dong *et al.*, 2001c). Beneficial or detrimental effects of 1-MCP presumably depend on whether ethylene production, and associated ripening and senescence, are required for disorder development, e.g. scald and senescent breakdown, or whether normal ripening is required to prevent disorder development, e.g. woolliness in nectarines.

9. *Decay incidence* can be *increased*: oranges (Porat *et al.*, 1999); *inhibited*: apricots (Dong *et al.*, 2001b); or *unaffected*: grapefruit (Mullins *et al.*, 2000) by 1-MCP. However, research to date on 1-MCP effects on pathogenicity has been very limited, particularly when one considers that reduced ethylene sensitivity can be beneficial against some pathogens but deleterious to resistance against other pathogens (Hoffman *et al.*, 1999). In citrus, for example, mold rots are inhibited by ethylene, while stem-end rots are enhanced by ethylene (Porat *et al.*, 1999), and Porat *et al.* (1999) suggested that small amounts of endogenous ethylene may be necessary to maintain basic levels of resistance to environmental and pathological stress. Decay was inhibited by 1-MCP at concentrations of $5-15 \, nl \, l^{-1}$, but increased by $50 \, nl \, l^{-1}$ (Ku *et al.*, 1999).

The advent of 1-MCP as a commercial tool has tremendous potential to help fruit industries maintain fruit quality. However, the effects of 1-MCP described thus far indicate that much remains to be learned before commercial success can be realized. Also, assumptions that effects of inhibiting endogenous ethylene production on fruit quality will be the same as reducing ethylene levels in the atmosphere should be taken cautiously. For example, the effects of 1-MCP on the quality of oranges (Porat *et al.*, 1999) are different from those obtained with ethylene absorbers (McGlasson and Eaks, 1972; Wild *et al.*, 1976).

8.5   Summary

Management of ethylene production and exposure is an essential component of meeting consumer demand for high quality fruit with appropriate appearance, texture and taste characteristics. The action of ethylene, both exogenous and endogenous, is used to maximize development of these characteristics but these effects can conflict with the deleterious effects of ethylene on product storage and shelf life, especially for climacteric fruits. The effects of ethylene on quality of fruits are well documented. Current harvesting and handling systems were often developed empirically but can now be understood in relation to ethylene physiology. Further understanding of ethylene biosynthesis, perception and action will be required to develop more refined technologies to improve storage potential and better maintain quality.

Research of the last decade or so has greatly improved progress towards these goals. The identification of intermediates involved in ethylene biosynthesis and its associated pathways, together with research on ethylene perception and signal transduction, gene expression and use of molecular techniques to down-regulate specific genes, has provided insight into the role of ethylene during fruit ripening. Better understanding about the effects of existing storage technologies on fruit quality is being obtained at the molecular level, but more research is required on the involvement of differential gene expression and other regulatory processes related to ethylene production and perception during fruit ripening. New strategies towards controlling ethylene production to commercially control ripening and maintain fruit quality are emerging. These are principally the use of transgenic approaches and use of the inhibitor of ethylene perception, 1-MCP. Both strategies, however, are revealing the diversity and complexity of fruit systems, confounding our ability to modify fruit ripening without attendant negative effects. Therefore, while they offer significant advances towards the goals of maintaining fruit quality, the difficulties of their commercial implementation should not be underestimated.

References

Abbas, M.F. and Ibrahim, M.A. (1996) The role of ethylene in the regulation of fruit ripening in the Hillawi date palm (*Phoenix dactylifera*). *Journal of the Science of Food and Agriculture*, **72**, 306-308.

Abdi, N., Holford, P., McGlasson, W.B. and Mizrahi, Y. (1997) Ripening behaviour and responses to propylene in four cultivars of Japanese type plums. *Postharvest Biology and Technology*, **12**, 21-34.

Abdi, N., McGlasson, W.B., Holford, P., Williams, M. and Mizrahi, Y. (1998) Responses of climacteric and suppressed-climacteric plums to treatment with propylene and 1-methylcyclopropene. *Postharvest Biology and Technology*, **14**, 29-39.

Abeles, F.B., Morgan, P.W. and Saltveit, M.E. (1992) *Ethylene in Plant Biology*, Academic Press, San Diego, California, p. 414.

Adams, D.O. and Yang, S.F. (1979) Ethylene biosynthesis: identification of 1-aminocyclopropane-1-carboxylic acid as an intermediate in the conversion of methionine to ethylene. *Proceedings of the National Academy of Sciences of the USA*, **76**, 170-174.

Agar, I.T., Biasi, W.V. and Mitcham, E.J. (1999) Exogenous ethylene accelerates ripening responses in Bartlett pears regardless of maturity or growing region. *Postharvest Biology and Technology*, **17**, 67-78.

Agar, I.T., Biasi, W.V. and Mitcham, E.J. (2000) Temperature and exposure time during ethylene conditioning affect ripening of Bartlett pears. *Journal of Agricultural and Food Chemistry*, **48**, 165-170.

Aina, J.O. and Oladunjoye, O.O. (1993) Respiration, pectolytic activity and textural changes in ripening African mango (*Irvingia gabonensis*) fruits. *Journal of the Science of Food and Agriculture*, **63**, 451-454.

Akamine, E.T. and Goo, T. (1978) Respiration and ethylene production in mammee apple (*Mammea americana* L.). *Journal of the American Society for Horticultural Science*, **103**, 308-310.

Akamine, E.T. and Goo, T. (1979) Respiration and ethylene production in fruits of species and cultivars of *Psidium* and species of *Eugenia*. *Journal of the American Society for Horticultural Science*, **104**, 632-635.

Alonso, J.M. and Granell, A. (1995) A putative vacuolar processing protease is regulated by ethylene and also during fruit ripening in citrus fruit. *Plant Physiology*, **109**, 541-547.

Alonso, J.M., Chamarro, J. and Granell, A. (1995) Evidence for the involvement of ethylene in the expression of specific RNAs during maturation of the orange, a non-climacteric fruit. *Plant Molecular Biology*, **29**, 385-390.

Amrhein, N., Schneebeck, D., Skoripka, H., Tophof, S. and Stockigt, J. (1981) Identification of a major metabolite of the ethylene precursor 1-aminocyclopropane-1-carboxylic acid in higher plants. *Naturwissen*, **68**, 619-620.

Andrews, J. (1995) The climacteric respiration rise in attached and detached tomato fruit. *Postharvest Biology and Technology*, **6**, 287-292.

Antunes, M.D.C., Pateraki, I., Kanellis, A.K. and Sfakiotakis, E.M. (2000) Differential effects of low-temperature inhibition on the propylene induced autocatalysis of ethylene production, respiration and ripening of 'Hayward' kiwifruit. *Journal of Horticultural Science and Biotechnology*, **75**, 575-580.

Atta-Aly, M.A. (1992) Effects of high temperature on ethylene biosynthesis of tomato fruits. *Postharvest Biology and Technology*, **2**, 19-24.

Atta-Aly, M.A., Saltveit, M.E. and Hobson, G.E. (1987) Effect of silver ions on ethylene biosynthesis by tomato fruit tissue. *Plant Physiology*, **83**, 44-48.

Atta-Aly, M.A., Brecht, J.K. and Huber, D.J. (2000a) Ripening of tomato locule gel tissue in response to ethylene. *Postharvest Biology and Technology*, **18**, 239-244.

Atta-Aly, M.A., Brecht, J.K. and Huber, D.J. (2000b) Ethylene feedback mechanisms in tomato and strawberry fruit tissues in relation to fruit ripening and climacteric patterns. *Postharvest Biology and Technology*, **20**, 151-162.

Autio, W.R. and Bramlage, W.J. (1982) Effects of AVG on maturation, ripening, and storage of apples. *Journal of the American Society for Horticultural Science*, **107**, 1074-1077.

Autio, W.R. and Bramlage, W.J. (1986) Chilling sensitivity of tomato fruit in relation to ripening and senescence. *Journal of the American Society for Horticultural Science*, **111**, 201-204.

Ayub, R., Guis, M., Amor, B., Gillot, L., Roustan, J.P., Latche, A., Bouzayen, M. and Pech, J.C. (1996) Expression of ACC oxidase antisense gene inhibits ripening of cantaloupe melon fruits. *Nature Biotechnology*, **14**, 862-866.

Balague, C., Watson, C.F., Turner, A.J., Rouge, P., Picton, S., Pech, J.-C. and Grierson, D. (1993) Isolation of a ripening and wound-induced cDNA from *Cucumis melo* L. encoding a protein with homology to the ethylene-forming enzyme. *European Journal of Biochemistry*, **212**, 27-34.

Banks, N.H., Elyatem, S. and Hammat, T. (1984) The oxygen affinity of ethylene production by slices of apple fruit tissue. *Acta Horticulturae*, **157**, 257-260.

Barry, C.S., Blume, B., Bouzayen, M., Cooper, W., Hamilton, A.J. and Grierson, D. (1996) Differential expression of the 1-amino-cyclopropane-1-carboxylate oxidase gene family of tomato. *Plant Journal*, **9**, 525-535.

Bauchot, A.D., Mottram, D.S., Dodson, A.T. and John, P. (1998) Effect of aminocyclopropane-1-carboxylic acid oxidase antisense gene on the formation of volatile esters in cantaloupe Charentais melon (cv. Vedrandais). *Journal of Agricultural and Food Chemistry*, **46**, 4787-4792.

Ben-Amor, B., Latche, A., Bouzayen, M., Pech, J.C. and Romojaro, F. (1999) Inhibition of ethylene biosynthesis by antisense ACC oxidase RNA prevents chilling injury in Charentais cantaloupe melons. *Plant Cell and Environment*, **22**, 1579-1586.

Ben-Arie, R., Segal, N. and Guelfat-Reich, S. (1984) The maturation and ripening of the 'Wonderful' pomegranate. *Journal of the American Society for Horticultural Science*, **109**, 898-902.

Beyer, E.M. (1976) A potent inhibitor of ethylene action in plants. *Plant Physiology*, **58**, 268-271.

Biale, J.B. (1960) Respiration of fruits, in *Encyclopedia of Plant Physiology*, vol. 12 (ed. W. Ruhland), Springer, Berlin, pp. 536-592.

Biale, J.B. (1976) Recent advances in postharvest physiology of tropical and subtropical fruits. *Acta Horticulturae*, **57**, 179-187.

Biale, J.B. and Barcus, D.E. (1970) Respiratory patterns of tropical fruits of the Amazon Basin. *Tropical Science*, **12**, 93-104.

Biale, J.B. and Young, R.E. (1981) Respiration and ripening in fruits–retrospect and prospect, in *Recent Advances in the Biochemistry of Fruit and Vegetables* (eds J. Friend and M.J.C. Rhodes), Academic Press, London, pp. 1-39.

Biggs, M.S., Woodson, W.R. and Handa, A.K. (1988) Biochemical basis of high-temperature inhibition of ethylene biosynthesis in ripening tomato fruits. *Physiologia Plantarum*, **72**, 572-578.

Blankenship, S.M. and Richardson, D.G. (1985) Development of ethylene biosynthesis and ethylene-induced ripening in 'd'Anjou' pears during the cold requirement for ripening. *Journal of the American Society for Horticultural Science*, **110**, 520-523.

Blankenship, S.M. and Sisler, E.C. (1989) 2,5-Norbornadiene retards apple softening. *HortScience*, **24**, 313-314.

Blankenship, S.M. and Sisler, E.C. (1993) Response of apples to diazocyclopentadiene inhibition of ethylene binding. *Postharvest Biology and Technology*, **3**, 95-101.

Blanpied, G.D. (1972) A study of ethylene in apple, red raspberry and cherry. *Plant Physiology*, **48**, 627-630.

Blanpied, G.D., Bartsch, J.A. and Turk, J.R. (1985) A commercial development programme for low ethylene controlled-atmosphere storage of apples (eds J.A. Roberts and G.A. Tucker), in *Proceedings of the Easter School, Agricultural Science, University of Nottingham*, Butterworths, London, pp. 393-404.

Bleecker, A.B. (1999) Ethylene perception and signaling: an evolutionary perspective. *Trends in Plant Science*, **4**, 269-274.

Bolitho, K.M., Lay-Yee, M., Knighton, M.L. and Ross, G.S. (1997) Antisense apple ACC oxidase RNA reduces ethylene production in transgenic tomato fruit. *Plant Science*, **122**, 91-99.

Bouzayen, M., Latche, A., Pech, J.C. and Marigo, G. (1989) Carrier-mediated uptake of 1-(malonylamino)cyclopropane-1-carboxylic acid in vacuoles isolated from *Catharanthus roseus* cells. *Plant Physiology*, **91**, 1317-1322.

Bown, A.W. (1985) CO_2 and intracellular pH. *Plant Cell and Environment*, **8**, 459-465.

Brady, C.J. (1987) Fruit ripening. *Annual Reviews of Plant Physiology*, **38**, 155-178.

Brecht, J.K. and Kader, A.A. (1984) Ethylene production by fruit of some slow-ripening nectarine genotypes. *Journal of the American Society for Horticultural Science*, **109**, 763-767.

Brovelli, E.A., Brecht, J.K., Sherman, W.B. and Sims, C.-A. (1999) Nonmelting-flesh trait in peaches is not related to low ethylene production rates. *HortScience*, **34**, 313-315.

Brown, G.E. and Lee, H.S. (1993) Interactions of ethylene with citrus stem-end rot caused by *Diplodia natalensis*. *Phytopathology*, **83**, 1204-1208.

Brown, B.I., Wong, L.S., George, A.P. and Nissen, R.J. (1988) Comparative studies on the postharvest physiology of fruit from different species of *Annona* (custard apple). *Journal of Horticultural Science*, **63**, 521-528.

Bufler, G. (1984) Ethylene-enhanced 1-aminocyclopropane-1-carboxylic acid synthase activity in ripening apples. *Plant Physiology*, **75**, 192-195.

Burdon, J.N. and Sexton, R. (1990a) The role of ethylene in the shedding of red raspberry fruit. *Annals of Botany*, **66**, 111-120.

Burdon, J.N. and Sexton, R. (1990b) Fruit abscission and ethylene production of red raspberry cultivars. *Scientia Horticulturae*, **43**, 95-102.

Burdon, J.N. and Sexton, R. (1993) Fruit abscission and ethylene production of four blackberry cultivars (*Rubus* spp.). *Annals of Applied Biology*, **123**, 121-132.

Burg, S.P. and Burg, E.A. (1967) Molecular requirements for the biological activity of ethylene. *Plant Physiology*, **42**, 144-152.

Burg, S.P. and Thimann, K.V. (1959) The physiology of ethylene formation in apples. *Proceedings of the National Academy of Sciences USA*, **45**, 335-344.

Cabrera, R.M. and Saltveit, M.E. (1990) Physiological response to chilling temperatures of intermittently warmed cucumber fruit. *Journal of the American Society for Horticultural Science*, **115**, 256-261.

Callahan, A.M., Morgens, P.H., Wright, P. and Nichols, K.E. (1992) Comparison of Pch313 (pTOM13 homolog) RNA accumulation during fruit softening and wounding of two phenotypically different peach cultivars. *Plant Physiology*, **100**, 482-488.

Casas, J.L., Acosta, M., del Rio, J.A. and Sabater, F. (1990) Ethylene evolution during ripening of detached tomato fruit: its relation with polyamine metabolism. *Plant Growth and Regulation*, **9**, 89-96.

Cazzonelli, C.I., Cacallaro, A.S. and Botella, J.R. (1998) Cloning and characterisation of ripening-induced ethylene biosynthesis genes from non-climacteric pineapple (*Ananas comosus*) fruits. *Australian Journal of Plant Physiology*, **25**, 513-518.

Chahine, H., Gouble, B., Audergon, J.M., Souty, M., Albagnac, G., Jacquemin, G., Reich, M. and Hughes, M. (1999) Effect of ethylene on certain parameters of apricot fruit (*Prunus armeniaca*, L.) during maturation and postharvest evolution. *Acta Horticulturae*, **488**, 577-584.

Chang, C. and Shockey, J.A. (1999) The ethylene-response pathway: signal perception to gene regulation. *Current Opinion in Plant Biology*, **2**, 352-358.

Chang, C., Kwok, S.F., Bleecker, A.B. and Meyerowitz, E.M. (1993) *Arabidopsis* ethylene-response gene *ETR1*: similarity of product to two-component regulators. *Science*, **262**, 539-544.

Chaplin, G.R., Wills, R.B.H. and Graham, D. (1983) Induction of chilling injury in stored avocados with exogenous ethylene. *HortScience*, **18**, 952-953.

Chaves, A.R. and Tomas, J.O. (1984) Effect of a brief CO_2 exposure on ethylene production. *Plant Physiology*, **76**, 88-91.

Chen, P.M., Varga, D.M. and Facteau, T.J. (1997) Promotion of ripening of 'Gebhard' red 'd'Anjou' pears by treatment with ethylene. *Postharvest Biology and Technology*, **12**, 213-220.

Cheverry, J.L., Sy, M.O., Pouliquen, J. and Marcellin, P. (1988) Regulation by CO_2 of 1-amino-cyclopropane-1-carboxylic acid conversion to ethylene in climacteric fruits. *Physiologia Plantarum*, **72**, 535-540.

Chu, C.L. (1988) Internal ethylene concentration of 'McIntosh', 'Northern Spy', 'Empire', 'Mutsu', and 'IdaRed' apples during the harvest season. *Journal of the American Society for Horticultural Science*, **113**, 226-229.

Clayton, M., Biasi, W.V., Southwick, S.M. and Mitcham, E.J. (2000) ReTain™ affects maturity and ripening of 'Bartlett' pear. *HortScience*, **35**, 1294-1299.

Cooper, W.C., Rasmussen, G.K. and Waldon, E.S. (1969) Ethylene evolution stimulated by chilling in *Citrus* and *Persea* sp. *Plant Physiology*, **44**, 1194-1196.

Cregoe, B.A., Ross, G.S. and Watkins, C.B. (1993) Changes in protein and mRNA expression during cold storage of 'Cox's Orange Pippin' apple fruit. *Acta Horticulturae*, **326**, 315-323.

Davies, K.M., Hobson, G.E. and Grierson, D. (1988) Silver ions inhibit the ethylene-stimulated production of ripening-related mRNAs in tomato. *Plant Cell and Environment*, **11**, 729-738.

de Wild, H.P.J., Woltering, E.J. and Peppelenbos, H.W. (1999) Carbon dioxide and 1-MCP inhibit ethylene production and respiration of pear fruit by different mechanisms. *Journal of Experimental Botany*, **50**, 837-844.

Dilley, D.R. (1982) Ethylene and the post harvest physiology of perishables. *Agriculture and Forestry Bulletin (University of Alberta)*, **5**, 19-28.

Dong, J.G., Kim., W.T., Yip, W.K., Thompson, G.A., Li, L., Bennett, A.B. and Yang, S.F. (1991) Cloning of a cDNA encoding 1-aminocyclopropane-1-carboxylate synthase and expression of its mRNA in ripening apple fruit. *Planta*, **185**, 38-45.

Dong, J.G., Fernandez-Maculet, J.C. and Yang, S.F. (1992) Purification and characterization of 1-aminocyclopropane-1-carboxylate oxidase from ripe apple fruit. *Proceedings of the National Academy of Sciences of the USA*, **89**, 9789-9793.

Dong, L., Zhou, H-W., Sonego, L, Lers, A. and Lurie, S. (2001a) Ethylene involvement in the cold storage disorder of 'Flavortop' nectarine. *Postharvest Biology and Technology*, **23** (in press).

Dong, L., Zhou, H-W., Sonego, L, Lers, A. and Lurie, S. (2001b) Ripening of 'Red Rosa' plums: effect of ethylene and 1-methylcyclopropene. *Australian Journal of Plant Physiology*, **28** (in press).

Dong, L., Zhou, H-W. and Lurie, S. (2001c) Effect of 1-methylcyclopropene on ripening of 'Canino' apricots and 'Royal Zee' plums. *Postharvest Biology and Technology*, **23** (in press).

Dopico, B., Lowe, A.L., Wilson, I.D., Merodio, C. and Grierson, D. (1993) Cloning and characterization of avocado fruit mRNAs and their expression during ripening and low-temperature storage. *Plant Molecular Biology*, **21**, 437-449.

Downs, C.G., Brady, C.J., Campbell, J.M. and McGlasson, W.B. (1991) Normal ripening cultivars of *Pyrus serotina* are either climacteric or non-climacteric. *Scientia Horticulturae*, **48**, 213-221.

Elkashif, M.E., Huber, D.J. and Brecht, J.K. (1989) Respiration and ethylene production in harvested watermelon fruit: evidence for nonclimacteric respiratory behavior. *Journal of the American Society for Horticultural Science*, **114**, 81-85.

El-Kazzaz, M.K., Chordas, A. and Kader, A.A. (1983) Physiological and compositional changes in orange fruit in relation to modification of their susceptibility to *Penicillium italicum* by ethylene treatments. *Journal of the American Society for Horticultural Science*, **108**, 618-621.

Escribano, M.I. and Merodio, C. (1994) The relevance of polyamine levels in cherimoya (*Annona chermola* Mill.) fruit ripening. *Journal of Plant Physiology*, **143**, 207-212.

Even-Chen, Z., Mattoo, A.K. and Goren, R. (1982) Inhibition of ethylene biosynthesis by aminoethoxyvinylglycine and by polyamines shunts label from 3,4-(^{14}C)methionine into spermidine in aged orange peel discs. *Plant Physiology*, **69**, 385-388.

Fan, X. and Mattheis, J.P. (1999a) Methyl jasmonate promotes apple fruit degreening independently of ethylene action. *HortScience*, **34**, 310-312.

Fan, X. and Mattheis, J.P. (1999b) Impact of 1-methylcyclopropene and methyl jasmonate on apple volatile production. *Journal of Agricultural and Food Chemistry*, **47**, 2847-2853.

Fan, X., Mattheis, J.P. and Fellman, J.K. (1998) A role for jasmonates in climacteric fruit ripening. *Planta*, **204**, 444-449.

Fan, X., Blankenship, S.M. and Mattheis, J.P. (1999a) 1-Methylcyclopropene inhibits apple ripening. *Journal of the American Society for Horticultural Science*, **124**, 690-695.

Fan, X., Mattheis, J.P. and Blankenship, S. (1999b) Development of apple superficial scald, soft scald, core flush, and greasiness is reduced by MCP. *Journal of Agricultural and Food Chemistry*, **47**, 3063-3068.

Fan, X., Argenta, L. and Mattheis, J.P. (2000) Inhibition of ethylene action by 1-methylcyclopropene prolongs storage life of apricots. *Postharvest Biology and Technology*, **20**, 135-142.

Fan, X., Argenta, L. and Mattheis, J.P. (2001) Interactive effects of 1-MCP and temperature on 'Elberta' peach quality. *HortScience*, **36** (in press).

Feng, X., Apelbaum, A., Sisler, E.C. and Goren, R. (2000) Control of ethylene responses in avocado fruit with 1-methylcyclopropene. *Postharvest Biology and Technology*, **20**, 143-150.

Fernández-Trujillo, J.P., Nock, J.F. and Watkins, C.B. (2001) Superficial scald, carbon dioxide injury, and changes of fermentation products and organic acids in 'Cortland' and 'Law Rome' apple fruit after high carbon dioxide stress treatment. *Journal of the American Society for Horticultural Science*, **126**, 235-241.

Fluhr, R. and Mattoo, A.K. (1996) Ethylene–biosynthesis and perception. *Critical Reviews in Plant Science*, **15**, 479-523.

Golding, J.B., Shearer, D., Wyllie, S.G. and McGlasson, W.B. (1998) Application of 1-MCP and propylene to identify ethylene-dependent ripening processes in mature banana fruit. *Postharvest Biology and Technology*, **14**, 87-98.

Golding, J.B., Shearer, D., McGlasson, W.B. and Wyllie, S.G. (1999) Relationships between respiration, ethylene, and aroma production in ripening banana. *Journal of Agricultural and Food Chemistry*, **47**, 1646-1651.

Goldschmidt, E.E., Huberman, M. and Goren, R. (1993) Probing the role of endogenous ethylene in the degreening of citrus fruit with ethylene antagonists. *Plant Growth Regulation*, **12**, 325-329.

Good, X., Kellogg, J.A., Wagoner, W., Langhoff, D., Matsumura, W. and Bestwick, R.K. (1994) Reduced ethylene synthesis by transgenic tomatoes expressing S-adenosylmethionine hydrolase. *Plant Molecular Biology*, **26**, 781-790.

Gorny, J.R. and Kader, A.A. (1996a) Regulation of ethylene biosynthesis in climacteric apple fruit by elevated CO_2 and reduced O_2 atmospheres. *Postharvest Biology and Technology*, **9**, 311-323.

Gorny, J.R. and Kader, A.A. (1996b) Controlled-atmosphere suppression of ACC synthase and ACC oxidase in 'Golden Delicious' apples during long-term cold storage. *Journal of the American Society for Horticultural Science*, **121**, 751-755.

Gorny, J.R. and Kader, A.A. (1997) Low oxygen and elevated carbon dioxide atmospheres inhibit ethylene biosynthesis in preclimacteric and climacteric apple fruit. *Journal of the American Society for Horticultural Science*, **122**, 542-546.

Gray, J.E., Picton, S., Giovannoni, J.J. and Grierson, D. (1994) The use of transgenic and naturally occurring mutants to understand and manipulate tomato fruit ripening. *Plant Cell and Environment*, **17**, 557-571.

Grierson, D. and Tucker, G.A. (1983) Timing of ethylene and polygalacturonase synthesis in relation to the control of tomato fruit ripening. *Planta*, **157**, 174-179.

Guis, M., Botondi, R., Ben-Amor, M., Ayub, R., Bouzayen, M., Pech, J.C. and Latche, A. (1997) Ripening-associated biochemical traits of Cantaloupe Charentais melons expressing an antisense ACC oxidase transgene. *Journal of the American Society for Horticultural Science*, **122**, 748-751.

Hackett, R.M., Ho, C-W., Lin, Z., Foote, H.C.C., Fray, R.G. and Grierson, D. (2000) Antisense inhibition of the *Nr* gene restored normal ripening to the tomato *Never-ripe* mutant, consistent with the ethylene receptor-inhibition model. *Plant Physiology*, **124**, 1079-1085.

Hadfield, K.A., Rose, J.K.C. and Bennett, A.B. (1995) The respiratory climacteric is present in Charentais (*Cucumis melo* cv. Reticulatus F1 Alpha) melons ripened on or off the plant. *Journal of Experimental Botany*, **46**, 1923-1925.

Hadfield, K.A., Dang, T., Guis, M., Pech, J.C., Bouzayen, M. and Bennett, A.B. (2000) Characterization of ripening-regulated cDNAs and their expression in ethylene-suppressed charentais melon fruit. *Plant Physiology*, **122**, 977-983.

Hamilton, A.J., Lycett, G.W. and Grierson, D. (1990) Antisense gene that inhibits synthesis of the hormone ethylene in transgenic plants. *Nature*, **346**, 284-287.

Harris, D.R., Seberry, J.A., Willis, R.B.H. and Spohr, L.J. (2000) Effect of fruit maturity on efficiency of 1-methylcyclopropene to delay the ripening of bananas. *Postharvest Biology and Technology*, **20**, 303-308.

Hartmann, C., Boulay, M. and Moulet, A.M. (1975) Malic enzyme, ripening and need of cold by the Passe-Crassane pear. *Comptes Rendues des Hebdominales Seances de l'Academie des Sciences, Series D Sciences Naturelles*, **281**, 135-137.

Hoffman, T., Schmidt, J.S., Zheng, X. and Bent, A.F. (1999) Isolation of ethylene-insensitive soybean mutants that are altered in pathogen susceptibility and gene-for-gene disease resistace. *Plant Physiology*, **119**, 935-949.

Hoffman, N.E. and Yang, S.F. (1980) Changes of 1-aminocyclopropane-1-carboxylic acid content in ripening fruits in relation to their ethylene production rates. *Journal of the American Society for Horticultural Science*, **105**, 492-495.

Inaba, A. and Nakamura, R. (1988) Numerical expression for estimating the minimum ethylene exposure time necessary to induce ripening in banana fruit. *Journal of the American Society for Horticultural Science*, **113**, 561-564.

Ismael, A.A. and Kender, W.T. (1969) Evidence of a respiratory climacteric in highbush and lowbush blueberry fruit. *HortScience*, **4**, 342-344.

Itai, A., Tanabe, K., Tamura, F. and Tanaka, T. (2000) Isolation of cDNA clones corresponding to genes expressed during fruit ripening in Japanese pear (*Pyrus pyrifolia* Nakai): involvement of the ethylene signal transduction pathway in their expression. *Journal of Experimental Botany*, **51**, 1163-1166.

Iwahori, S. and Lyons, J.M. (1970) Maturation and quality of tomatoes with preharvest treatments of 2-chloroethylphosphonic acid. *Proceedings of the American Society for Horticultural Science*, **95**, 88-91.

Jeffery, D., Smith, C., Goodenough, P., Prosser, I. and Grierson, D. (1984) Ethylene-independent and ethylene-dependent biochemical changes in ripening tomatoes (*Lycopersicon esculentum*). *Plant Physiology*, **74**, 32-38.

Jiang, Y. and Fu, J. (2000) Ethylene regulation of fruit ripening: molecular aspects. *Plant Growth Regulation*, **30**, 193-200.

Jiang, Y., Joyce, D.C. and Macnish, A.J. (1999a) Extension of the shelf life of banana fruit by 1-methyl-cyclopropene in combination with polyethylene bags. *Postharvest Biology and Technology*, **16**, 187-193.

Jiang, Y., Joyce, D.C. and Macnish, A.J. (1999b) Responses of banana fruit to treatment with 1-methylcyclopropene. *Plant Growth Regulation*, **28**, 77-82.

Jiao, X.Z., Philosoph-Hadas, S., Su, L.-Y. and Yang, S.F. (1986) The conversion of 1-(malonyl-amino)cyclopropane-1-carboxylic acid to 1-aminocyclopropane-1-carboxylic acid in plant tissues. *Plant Physiology*, **81**, 637-641.

Jobling, J., McGlasson, W.B. and Dilley, D.R. (1991) Induction of ethylene synthesizing competency in Granny Smith apples by exposure to low temperature in air. *Postharvest Biology and Technology*, **1**, 111-118.

John, P. (1997) Ethylene biosynthesis: the role of 1-aminocyclopropane-1-carboxylate (ACC) oxidase, and its possible evolutionary origin. *Physiologia Plantarum*, **100**, 583-592.

Kader, A.A., Li, Y. and Chordas, A. (1982) Postharvest respiration, ethylene production, and compositional changes in Chinese jujube fruits. *HortScience*, **17**, 678-679.

Kader, A.A., Zagory, D. and Kerbel, E.L. (1989) Modified atmosphere packaging of fruits and vegetables. *Critical Reviews in Food Science and Nutrition*, **28**, 1-30.

Kanellis, A.K., Solomos, T. and Roubelakis-Angelakis, K.A. (1991) Suppression of cellulase and polygalacturonase and induction of alcohol dehydrogenase isoenzymes in avocado fruit mesocarp subjected to low oxygen stress. *Plant Physiology*, **96**, 269-274.

Kasmire, R-F., Rappaport, L. and May, D. (1970) Effects of 2-chloroethylphosphonic acid on ripening of cantaloupes. *Proceedings of the American Society for Horticultural Science*, **95**, 134-137.

Kays, S.J. (1997) *Postharvest Physiology of Perishable Plant Products*, Exon Press, Athens, Georgia.

Kays, S.J. and Hayes, M.J. (1978) Induction of ripening in the fruits of *Momordica charactia* L. by ethylene. *Tropical Agriculture*, **55**, 167-172.

Ketsa, S. and Daengkanit, T. (1998) Physiological changes during postharvest ripening of durian fruit (*Durio zibethinus* Murray). *Journal of Horticultural Science and Biotechnology*, **73**, 575-577.

Ketsa, S., Chidtragool, S., Klein, J.D. and Lurie, S. (1999) Ethylene synthesis in mango fruit following heat treatment. *Postharvest Biology and Technology*, **15**, 65-72.

Kidd, F. and West, F. (1925) The course of respiratory activity throughout the life of an apple. *Annual Report of the Food Investigation Board, London, 1924*, pp. 27-32.

Kieber, J.J. (1997) The ethylene signal transduction pathway in *Arabidopsis*. *Journal of Experimental Botany*, **48**, 211-218.

Kieber, J.J., Rothenberg, M., Roman, G., Feldman, K.A. and Ecker, J.R. (1993) *CTR1*, a negative regulator of the ethylene response pathway in *Arabidopsis*, encodes a member of the raf family of protein kinases. *Cell*, **72**, 427-441.

Kim, G.H. and Wills, R.B.H. (1998) Interaction of enhanced carbon dioxide and reduced ethylene on the storage life of strawberries. *Journal of Horticultural Science and Biotechnology*, **73**, 181-184.

Klee, H.-J. (1993) Ripening physiology of fruit from transgenic tomato (*Lycopersicon esculentum*) plants with reduced ethylene synthesis. *Plant Physiology*, **102**, 911-916.

Klee, H.J., Hayford, M.B., Kretzmer, K.A., Barry, G.F. and Kishore, G.M. (1991) Control of ethylene synthesis by expression of a bacterial enzyme in transgenic tomato plants. *Plant Cell*, **3**, 1187-1193.

Klein, J.D. (1989) Ethylene biosynthesis in heat treated apples, in *Biochemical and Physiological Aspects of Ethylene Production in Lower and Higher Plants* (eds H. Clijster, M. de Proft, R. Marcelle and M. van Pouche), Kluwer, Dordrecht, pp. 184-190.

Klein, J.D. and Lurie, S. (1990) Prestorage heat treatment as a means of improving poststorage quality of apples. *Journal of the American Society for Horticultural Science*, **115**, 265-269.

Knee, M. (1973) Effects of storage treatments upon the ripening of Conference pears. *Journal of the Science of Food and Agriculture*, **24**, 1137-1145.

Knee, M. (1987) Development of ethylene biosynthesis in pear fruits at $-1°$C. *Journal of Experimental Botany*, **38**, 1724-1733.

Knee, M. (1995) Do tomatoes on the plant behave as climacteric fruit? *Physiologia Plantarum*, **95**, 211-216.

Knee, M. and Hatfield, S.G.S. (1981) Benefits of ethylene removal during apple storage. *Annals of Applied Biology*, **98**, 157-165.

Knee, M., Looney, N.E., Hatfield, S.G.S. and Smith, S.M. (1983) Initiation of rapid ethylene synthesis by apple and pear fruits in relation to storage temperature. *Journal of Experimental Botany*, **34**, 1207-1212.

Knee, M., Proctor, F.J. and Dover, C.J. (1985) The technology of ethylene control: use and removal in post-harvest handling of horticultural commodities. *Annals of Applied Biology*, **107**, 581-595.

Knee, M., Hatfield, S.G.S. and Bramlage, W.J. (1987) Response of developing apple fruits to ethylene treatment. *Journal of Experimental Botany*, **38**, 972-979.

Kneissel, M.L. and Deikman, J. (1996) The tomato E8 gene influences ethylene biosynthesis in fruit but not in flowers. *Plant Physiology*, **112**, 537-547.

Kramer, G.F. and Wang, C.Y. (1989) Correlation of reduced chilling injury with increased spermine and spermidine levels in zucchini squash. *Physiologia Plantarum*, **76**, 479-484.

Ku, V.V.V., Wills, R.B.H. and Ben-Yehosua, S. (1999) 1-Methylcyclopropene can differentially affect the postharvest life of strawberries exposed to ethylene. *HortScience*, **34**, 119-120.

Kubo, Y., Inaba, A. and Nakamura, R. (1990) Respiration and C_2H_4 production in various harvested crops held in CO_2-enriched atmospheres. *Journal of the American Society for Horticultural Science*, **115**, 975-978.

Lakshminarayana, S. and Estrella, I.B. (1978) Postharvest respiratory behavior of tuna (prickly pear) fruit (*Opuntia robusta* Mill.). *Journal of Horticultural Science*, **53**, 327-330.

Lam, P.F. and Wan, C.K. (1983) Climacteric nature of the carambola (*Averrhoa carambola* L.) fruit. *Pertanika*, **6**, 44-47.

Lanahan, M.B., Yen, H.C., Giovannoni, J.J. and Klee, H.J. (1994) The Never ripe mutation blocks ethylene perception in tomato. *Plant Physiology*, **6**, 521-530.

Larrigaudiere, C. and Vendrell, M. (1993) Cold-induced climacteric rise of ethylene metabolism in Granny Smith apples. *Current Plant Science, Biotechnology and Agriculture*, **16**, 136-141.

Larrigaudiere, C., Pinto, E. and Vendrell, M. (1996) Differential effects of ethephon and seniphos on color development of 'Starking Delicious' apple. *Journal of the American Society for Horticultural Science*, **121**, 746-750.

Larrigaudiere, C., Graell, J., Salas, J. and Vendrell, M. (1997) Cultivar differences in the influences of a short period of cold storage on ethylene biosynthesis in apples. *Postharvest Biology and Technology*, **10**, 21-27.

Lashbrook, C.C., Tieman, D.M. and Klee, H.J. (1998) Differential regulation of the tomato *ETR* gene family throughout plant development. *Plant Journal*, **15**, 243-252.

Lasserre, E., Bouquin, T., Hernandez, J.A., Bull, J., Pech, J.C. and Balague, C. (1996) Structure and expression of three genes encoding ACC oxidase homologs from melon (*Cucumis melo* L.). *Molecular and General Genetics*, **251**, 81-90.

Lau, O.L., Liu, Y. and Yang, S.F. (1984) Influence of storage atmospheres and procedures on 1-aminocyclopropane-1-carboxylic acid concentration in relation to flesh firmness in 'Golden Delicious' apple. *HortScience*, **19**, 425-426.

Lee, S.A., Ross, G.S. and Gardner, R.C. (1998) An apple (*Malus domestica* L. Borkh cv Granny Smith) homologue of the ethylene receptor gene *ETR1* (Accession No. AF032448). *Plant Physiology*, **117**, 1126.

Lelievre, J.M., Tichit, L., Fillion, L., Larrigaudiere, C., Vendrell, M. and Pech, J.C. (1995) Cold-induced accumulation of 1-aminocyclopropane 1-carboxylate oxidase protein in Granny Smith apples. *Postharvest Biology and Technology*, **5**, 11-17.

Lelievre, J.M., Latche, A., Jones, B., Bouzayen, M. and Pech, J.C. (1997a) Ethylene and fruit ripening. *Physiologia Plantarum*, **101**, 727-739.

Lelievre, J.M., Tichit, L., Dao, P., Fillion, L., Nam-Young, W., Pech, J.C. and Latche, A. (1997b) Effects of chilling on the expression of ethylene biosynthetic genes in Passe-Crassane pear (*Pyrus communis* L.) fruits. *Plant Molecular Biology*, **35**, 847-855.

Li, Z., Liu, Y., Dong, J., Xu, R. and Zhu, M. (1983) Effect of low oxygen and high carbon dioxide on the levels of ethylene and 1-aminocyclopropane-1-carboxylic acid in ripening apple fruits. *Journal of Plant Growth Regulation*, **2**, 81-87.

Lincoln, J.E., Campbell, A.D., Oetiker, J., Rottmann, W.H., Oeller, P.W., Shen, N.F. and Theologis, A. (1993) LE-ACS4, a fruit ripening and wound-induced 1-aminocyclopropane-1-carboxylate synthase gene of tomato (*Lycopersicon esculentum*). *Journal of Biological Chemistry*, **268**, 19422-19430.

Lipe, J.A. (1978) Ethylene in fruits of blackberry and rabbiteye blueberry. *Journal of the American Society for Horticultural Science*, **103**, 76-77.

Lipton, W.J. and Wang, C.Y. (1987) Chilling exposures and ethylene treatment change the level of ACC in 'Honey Dew' melons. *Journal of the American Society for Horticultural Science*, **112**, 109-112.

Lipton, W.J., Aharoni, Y. and Elliston, E. (1979) Rates of CO_2 and ethylene production and of ripening of harvested and packed 'Honey Dew' muskmelons at a chilling temperature after pretreatment with ethylene. *Journal of the American Society for Horticultural Science*, **104**, 846-849.

Liu, Y., Hoffman, N.E. and Yang, S.F. (1985a) Promotion by ethylene of the capability to convert 1-aminocyclopropane-1-carboxylic acid to ethylene in preclimacteric tomato and cantaloupe fruits. *Plant Physiology*, **77**, 407-411.

Liu, Y., Hoffman, N.E. and Yang, S.F. (1985b) Ethylene-promoted malonylation of 1-aminocyclopropane-1-carboxylic acid participates in autoinhibition of ethylene synthesis in grapefruit flavedo discs. *Planta*, **164**, 565-568.

Liu, Y., Su, L.Y. and Yang, S.F. (1985c) Ethylene promotes the capability to malonylate 1-amino-cyclopropane-1-carboxylic acid and D-amino acids in preclimacteric tomato fruits. *Plant Physiology*, **77**, 891-895.

Liu, X., Shiomi, S., Nakatsuka, A., Kubo, Y., Nakamura, R. and Inaba, A. (1999) Characterization of ethylene biosynthesis associated with ripening in banana fruit. *Plant Physiology*, **121**, 1257-1265.

Looney, N.E. (1971) Interaction of ethylene, auxin, and succinic acid-2,2-dimethylhydrazide in apple fruit ripening control. *Journal of the American Society for Horticultural Science*, **96**, 350-353.

Lopez-Gomez, R., Campbell, A, Dong, J.G., Yang, S.F. and Gomez-Lim, M.A. (1997) Ethylene biosynthesis in banana fruit: isolation of a genomic clone to ACC oxidase and expression studies. *Plant Science*, **123**, 123-131.

Lurie, S. (1998) Postharvest heat treatments. *Postharvest Biology and Technology*, **14**, 257-269.

Lurie, S., Handros, A., Fallek, E. and Shapira, R. (1996) Reversible inhibition of tomato fruit gene expression at high temperature. *Plant Physiology*, **110**, 1207-1214.

Lurssen, K., Naumann, K. and Schroder, R. (1979) 1-Aminocyclopropane-1-carboxylic acid, an intermediate of the ethylene biosynthesis in higher plants. *Zeitschrift für Pflanzenphysiologie*, **92**, 285-294.

Lyons, J.M., McGlasson, W.B. and Pratt, H.K. (1962) Ethylene production, respiration and internal gas concentrations in cantaloupe fruits at various stages of maturity. *Plant Physiology*, **37**, 31-36.

Macnish, A.J., Joyce, D.C., Hofman, P.J., Simons, D.H. and Reid, M.S. (2000) 1-Methylcyclopropene treatment efficacy in preventing ethylene perception in banana fruit and grevillea and waxflower flowers. *Australian Journal of Experimental Agriculture*, **40**, 471-481.

Mansour, R., Latche, A., Vaillant, V., Pech, J.C. and Reid, M.S. (1986) Metabolism of 1-amino-cyclopropane-1-carboxylic acid in ripening apple fruits. *Physiologia Plantarum*, **66**, 495-502.

Marei, N. and Crane, J.C. (1971) Growth and respiratory response of fig (*Ficus carica*, L. cv. Mission) fruits to ethylene. *Plant Physiology*, **48**, 249-254.

Martin, M.N. and Saftner, R.A. (1995) Purification and characterization of 1-aminocyclopropane-1-carboxylic acid *N*-malonyltransferase from tomato fruit. *Plant Physiology*, **108**, 1241-1249.

Martin, M.N., Cohen, J.D. and Saftner, R.A. (1995) A new 1-aminocyclopropane carboxylic acid-conjugating activity in tomato fruit. *Plant Physiology*, **109**, 917-926.

Martinez, P.G., Gomez, R.L. and Gomez-Lim, M.A. (2001) Identification of an ETR 1-homolgue from mango fruit expressing during fruit ripening and wounding. *Journal of Plant Physiology*, **158**, 101-108.

Martinez-Tellez, M.A. and Lafuente, M.T. (1997) Effect of high temperature conditioning on ethylene, phenylalanine ammonia-lyase, peroxidase and polyphenol oxidase activities in flavedo of chilled 'Fortune' mandarin fruit. *Journal of Plant Physiology*, **150**, 674-678.

Mathooko, F.M. (1996) Regulation of ethylene biosynthesis in higher plants by carbon dioxide. *Postharvest Biology and Technology*, **9**, 247-264.

Mathooko, F.M., Kubo, Y., Inaba, A. and Nakamura, R. (1995a) Induction of ethylene biosynthesis and polyamine accumulation in cucumber fruit in response to carbon dioxide stress. *Postharvest Biology and Technology*, **5**, 51-65.

Mathooko, F.M., Kubo, Y., Inaba, A. and Nakamura, R. (1995b) Characterization of the regulation of ethylene biosynthesis in tomato fruit by carbon dioxide and diazocyclopentadiene. *Postharvest Biology and Technology*, **5**, 221-233.

Mathooko, F.M., Inaba, A. and Nakamura, R. (1998) Characterization of carbon dioxide stress-induced ethylene biosynthesis in cucumber (*Cucumis sativus* L.) fruit. *Plant and Cell Physiology*, **39**, 285-293.

Mathooko, F.M., Mwaniki, M.W., Nakatsuka, A., Shiomi, S., Kubo, Y., Inaba, A. and Nakamura, R. (1999) Expression characteristics of *CS-ACS1*, *CS-ACS2* and *CS-ACS3*, three members of the 1-aminocyclopropane-1-carboxylate synthase gene family in cucumber (*Cucumis sativus* L.) fruit under carbon dioxide stress. *Plant and Cell Physiology*, **40**, 164-172.

Mathooko, F.M., Tsunashima, Y., Owino, W.Z.O., Kubo, Y. and Inaba, A. (2001) Regulation of genes encloding ethylene biosynthetic enzymes in peach (*Prunus persica* L.) fruit by carbon dioxide and 1-methylcyclopropene. *Postharvest Biology and Technology*, **21**, 265-281.

Mattoo, A.K. and Suttle, C.S. (1991) *The Plant Hormone Ethylene*. CRC Press, Boca Raton, Florida, 337pp.

Mattoo, A.K. and White, B. (1991) Regulation of ethylene biosynthesis, in *The Plant Hormone Ethylene*. (eds A.K. Mattoo and C.S. Suttle), CRC Press, Boca Raton, Florida, pp. 21-42.

Maxie, E.C., Catlin, P.B. and Hartman, H.T. (1960) Respiration and ripening of olive fruits. *Proceedings of the American Society for Horticultural Science*, **75**, 275-291.

Maxie, E.C., Mitchell, F.G., Sommer, N.F., Snyder, R.G. and Rae, H.L. (1974) Effects of elevated temperatures on ripening of 'Bartlett' pears, *Pyrus communis* L. *Journal of the American Society for Horticultural Science*, **99**, 344-349.

Mbeguie-A-Mbeguie, D., Chahine, H., Gomez, R.M., Gouble, B., Reich, M., Audergon, J.M., Souty, M., Albagnac, G. and Fils-Lycaon, B. (1999) Molecular cloning and expression of a cDNA encoding 1-aminocyclopropane-1-carboxylate (ACC) oxidase from apricot fruit (*Prunus armeniaca*). *Physiologia Plantarum*, **105**, 294-303.

McCollum, T.G. and McDonald, R.E. (1991) Electrolyte leakage, respiration, and ethylene production as indices of chilling injury in grapefruit. *HortScience*, **26**, 1191-1192.

McDonald, R.E., Hatton, T.T., and Cubbedge, R.-H. (1985) Chilling injury and decay of lemons as affected by ethylene, low temperature, and optimal storage. *HortScience*, **20**, 92-93.

McGlasson, W.B. (1985) Ethylene and fruit ripening. *HortScience*, **20**, 51-54.

McGlasson, W.B. and Eaks, I.L. (1972) A role for ethylene in the development of wastage and off-flavors in stored 'Valencia' oranges. *HortScience*, **7**, 80-81.

McGlasson, W.B. and Wills, R.B.H. (1972) Effects of oxygen and carbon dioxide on respiration, storage life, and organic acids of green bananas. *Australian Journal of Biological Science*, **25**, 35-42.

McMurchie, E.J., McGlasson, W.B. and Eaks, I.L. (1972) Treatment of fruit with propylene gives information about the biogenesis of ethylene. *Nature*, **237**, 235-236.

Meakin, P.J. and Roberts, J.A. (1990) Dehiscence of fruit in oilseed rape (*Brassica napus* L.). II. The role of cell wall degrading enzymes and ethylene. *Journal of Experimental Botany*, **41**, 1003-1011.

Mita, S., Kawamura, S., Yamawaki, K., Nakamura, K. and Hyodo, H. (1998) Differential expression of genes involved in the biosynthesis and perception of ethylene during ripening of passion fruit (*Passiflora edulis* Sims). *Plant and Cell Physiology*, **39**, 1209-1217.

Mita, S., Kirita, C., Kato, M., and Hyodo, H. (1999) Expression of ACC synthase is enhanced earlier than that of ACC oxidase during fruit ripening of mume (*Prunus mume*). *Physiologia Plantarum*, **107**, 319-328.

Miyazaki, J.H. and Yang, S.F. (1987) Metabolism of 5-methylthioribose to methionine. *Plant Physiology*, **84**, 277-281.

Mizano, S. and Pratt, H.K. (1973) Relations of respiration and ethylene production to maturity in the watermelon. *Journal of the American Society for Horticultural Science*, **98**, 614-617.

Mullins, E.D., McCollum, T.G. and McDonald, R.E. (2000) Consequences on ethylene metabolism of inactivating the ethylene receptor sites in diseased non-climacteric fruit. *Postharvest Biology and Technology*, **19**, 155-164.

Munoz, M.T., Aguado, P., Ortega, N., Escribano, M.I. and Merodio, C. (1999) Regulation of ethylene and polyamine synthesis by elevated carbon dioxide in cherimoya fruit stored at ripening and chilling temperatures. *Australian Journal of Plant Physiology*, **26**, 201-209.

Murray, A.J., Hobson, G.E., Schuch, W. and Bird, C.R. (1993) Reduced ethylene synthesis in EFE antisense tomatoes has differential effects on fruit ripening processes. *Postharvest Biology and Technology*, **2**, 301-313.

Nakagawa, N., Mori, H., Yamazaki, K. and Imaseki, H. (1991) Cloning of complementary DNA for auxin-induced 1-aminocyclopropane-1-carboxylate synthase and differential expression of the gene by auxin and wounding. *Plant and Cell Physiology*, **32**, 1153-1163.

Nakajima, N., Mori, H., Yamazaki, K. and Imaseki, H. (1990) Molecular cloning and sequence of a complementary DNA encoding 1-aminocyclopropane-1-caboxylate synthase induced by tissue wounding. *Plant and Cell Physiology*, **31**, 1016-1021.

Nakatsuka, A., Shiomi, S., Kubo, Y. and Inaba, A. (1997) Expression and internal feedback regulation of ACC synthase and ACC oxidase gene in ripening tomato fruit. *Plant and Cell Physiology*, **38**, 1103-1110.

Nakatsuka, A., Murachi, S., Okunishi, H., Shiomi, S., Nakano, R., Kubo, Y. and Inaba, A. (1998) Differential expression and internal feedback regulation of 1-aminocyclopropane-1-carboxylate synthase, 1-aminocyclopropane-1-carboxylate oxidase, and ethylene receptor genes in tomato fruit during development and ripening. *Plant Physiology*, **118**, 1295-1305.

Nam, Y.W., Tichit, L., Leperlier, M., Cuerq, B., Marty, I. and Lelievre, J.M. (1999) Isolation and characterization of mRNAs differentially expressed during ripening of wild strawberry (*Fragaria vesca* L.) fruits. *Plant Molecular Biology*, **39**, 629-636.

Nerd, A. and Mizrahi, Y. (1999) The effect of ripening stage on fruit quality after storage of yellow pitaya. *Postharvest Biology and Technology*, **15**, 99-105.

Nicolas, J., Rothan, C. and Duprat, F. (1989) Softening of kiwifruit in storage. Effects of intermittent high CO_2 treatments. *Acta Horticulturae*, **258**, 185-192.

Oeller, P.W., Wong, L.M., Taylor, L.P., Pike, D.A. and Theologis, A. (1991) Reversible inhibition of tomato fruit senescence by antisense RNA. *Science*, **254**, 437-439.

O'Hare, T.J. (1995) Postharvest physiology and storage of rambutan. *Postharvest Biology and Technology*, **6**, 189-199.

Olson, D.C., White, J.A., Edelman, L., Harkins, R.N. and Kende, H. (1991) Differential expression of two genes for 1-aminocyclopropane-1-carboxylate synthase in tomato fruits. *Proceedings of the National Academy of Sciences of the USA*, **88**, 5340-5344.

Owens, L.D., Lieberman, M. and Kunishi, A. (1971) Inhibition of ethylene production by rhizobitoxin. *Plant Physiology*, **48**, 1-4.

Paull, R.E. and Chen, N.J. (2000) Heat treatment and fruit ripening. *Postharvest Biology and Technology*, **21**, 21-37.

Payton, S., Fray, R.G., Brown, S. and Grierson, D. (1996) Ethylene receptor expression is regulated during fruit ripening, flower senescence and abscission. *Plant Molecular Biology*, **31**, 1227-1231.

Peacock, B.C. (1972) Role of ethylene in the initiation of fruit ripening. *Queensland Journal of Agriculture and Animal Sciences*, **29**, 137-145.

Pech, J.C., Balague, C., Latche, A. and Bouzayen, M. (1994) Postharvest physiology of climacteric fruits: recent developments in the biosynthesis and action of ethylene. *Sciences des Aliments*, **14**, 3-15.

Peiser, G. and Yang, S.F. (1998) Evidence for 1-(malonylamino) cyclopropane-1-carboxylic acid being the major conjugate of aminocyclopropane-1-carboxylic acid in tomato fruit. *Plant Physiology*, **116**, 1527-1532.

Peleman, J., Saito, H., Cottyn, B., Engler, G., Seurinck, J., Van Montagu, M. and Inze, D. (1989) Structure and expression analyses of the S-adenosylmethionine synthetase gene family in *Arabidopsis thaliana*. *Gene*, **84**, 359-369.

Penarrubia, L., Aguilar, M., Margossian, L. and Fischer, R.L. (1992) An antisense gene stimulates ethylene hormone production during tomato fruit ripening. *Plant Cell*, **4**, 681-687.

Perkins-Veazie, P. (2000) Growth and ripening of strawberry fruit. *Horticulture Reviews*, **17**, 267-297.

Perkins-Veazie, P. and Nonnecke, G. (1992) Physiological changes during ripening of raspberry fruit. *HortScience*, **27**, 331-333.

Perkins-Veazie, P., Clark, J.R., Huber, D.J. and Baldwin, E.A. (2000) Ripening physiology in 'Navaho' thornless blackberries: color, respiration, ethylene production, softening, and compositional changes. *Journal of the American Society for Horticultural Science*, **125**, 357-363.

Picton, S. and Grierson, D. (1988) Inhibition of expression of tomato-ripening genes at high temperature. *Plant Cell and Environment*, **11**, 265-272.

Picton, S., Barton, S.L., Bouzayen, M., Hamilton, A.J. and Grierson, D. (1993) Altered fruit ripening and leaf senescence in tomatoes expressing an antisense ethylene-forming enzyme transgene. *Plant Journal*, **3**, 469-481.

Poneleit, L.S. and Dilley, D.R. (1993) Carbon dioxide activation of 1-aminocyclopropane-1-carboxylate (ACC) oxidase in ethylene biosynthesis. *Postharvest Biology and Technology*, **3**, 191-193.

Porat, R., Weiss, B., Cohen, L., Daus, A., Goren, R. and Droby, S. (1999) Effects of ethylene and 1-methylcyclopropene on the postharvest qualities of 'Shamouti' oranges. *Postharvest Biology and Technology*, **15**, 155-163.

Pratt, H.K. and Goeschl, J.D. (1968) The role of ethylene in fruit ripening, in *Biochemistry and Physiology of Plant Growth Substances* (eds F. Wightman and G. Setterfield), Runge, Ottawa, pp. 1295-1302.

Pratt, H.K. and Mendoza, D.B. (1980) Fruit development and ripening of the star apple (*Chrysophyllum cainito* L.). *HortScience*, **15**, 721-722.

Pratt, H.K. and Reid, M.S. (1974) Chinese gooseberry: seasonal patterns in fruit growth and maturation, ripening, respiration and the role of ethylene. *Journal of the Science of Food and Agriculture*, **25**, 747-753.

Pratt, H.K. and Reid, M.S. (1976) The tamarillo: fruit growth and maturation, ripening, respiration, and the role of ethylene. *Journal of the Science of Food and Agriculture*, **27**, 399-404.

Pratt, H.K., Goeschl, J.D. and Martin, F.W. (1977) Fruit growth and development, ripening and role of ethylene in the 'Honey Dew' muskmelon. *Journal of the American Society for Horticultural Science*, **102**, 203-210.

Puig, L., Varga, D.M., Chen, P.M. and Mielke, E.A. (1996) Synchronizing ripening in individual 'Barlett' pears with ethylene. *HortTechnology*, **6**, 24-27.

Reid, M.S. (1992) Ethylene in postharvest technology, in *Postharvest Technology of Horticultural Crops* (ed. A.A. Kader), University of California Publication 3311, pp. 97-108.

Reid, M.S. (1975) The role of ethylene in the ripening of some unusual fruits, in *Facteurs et Regulation de la Maturation des Fruits, No. 238*. CNRS, Paris, pp. 177-182.

Riov, J. and Yang, S.F. (1982) Autoinhibition of ethylene production in citrus peel discs. Suppression of 1-aminocyclopropane-1-carboxylic acid synthesis. *Plant Physiology*, **69**, 687-690.

Rogiers, S.Y., Mohan-Kumar, G.N. and Knowles, N.R. (1998) Regulation of ethylene production and ripening by saskatoon (*Amelanchier alnifolia* Nutt.) fruit. *Canadian Journal of Botany*, **76**, 1743-1754.

Romani, R., Labavitch, J., Yamashita, T., Hess, B. and Rae, H. (1983) Preharvest AVG (aminoethoxyvinylglycine) treatment of 'Bartlett' pear fruits: effects on ripening, color change, and volatiles. *Journal of the American Society for Horticultural Science*, **108**, 1046-1049.

Ross, G.S., Knighton, M.L. and Lay-Yee, M. (1992) An ethylene-related cDNA from ripening apples. *Plant Molecular Biology*, **19**, 231-238.

Rothan, C., Duret, S., Chevalier, C. and Raymond, P. (1997) Suppression of ripening-associated gene expression in tomato fruits subjected to a high CO_2 concentration. *Plant Physiology*, **114**, 255-263.

Rottmann, W.H., Peter, G.F., Oeller, P.W., Keller, J.A., Shen, N.F., Nagy, B.P., Taylor, L.P., Campbell, A.D. and Theologis, A. (1991) 1-Aminocyclopropane-1-carboxylate synthase in tomato is encoded by a multi-gene family whose transcription is induced during fruit and floral senescence. *Journal of Molecular Biology*, **222**, 937-961.

Rupasinghe, H.P.V., Murr, D.P., Paliyath, G. and Skog, L. (2000) Inhibitory effect of 1-MCP on ripening and superficial scald development in 'McIntosh' and 'Delicious' apples. *Journal of Horticultural Science and Biotechnology*, **75**, 271-276.

Saltveit, M.E. (1977) Carbon dioxide, ethylene, and color development in ripening mature green bell peppers. *Journal of the American Society for Horticultural Science*, **102**, 523-525.

Saltveit, M.E. (1999) Effect of ethylene on quality of fresh fruits and vegetables. *Postharvest Biology and Technology*, **15**, 279-292.

Saltveit, M.E., Bradford, K.J. and Dilley, D.R. (1978) Silver ion inhibits ethylene synthesis and action in ripening fruits. *Journal of the American Society for Horticultural Science*, **103**, 472-475.

Saniewski, M., Czapski, J., Nowacki, J. and Lange, E. (1987a) The effect of methyl jasmonate on ethylene and 1-aminocyclopropane-1-carboxylic acid production in apple fruits. *Biologia Plantarum*, **29**, 199-203.

Saniewski, M., Nowacki, J. and Czapski, J. (1987b) The effect of methyl jasmonate on ethylene production and ethylene-forming activity in tomatoes. *Journal of Plant Physiology*, **129**, 175-180.

Satoh, S., Kanke, C., Yoneno, T., Yohioka, T. and Hashiba, T. (2000) Characterization of pear (*Pyrus communis* L.) strains unresponsive to cold-induced ripening in relation to the production and action of ethylene. *Journal of the Japanese Society for Horticultural Science*, **69**, 176-183.

Sato-Nara, K., Yuhashi, K., Higashi, K., Hosoya, K., Kubota, M. and Ezura, H. (1999) Stage- and tissue-specific expression of ethylene receptor homolog genes during fruit development in muskmelon. *Plant Physiology*, **120**, 321-329.

Selvarajah, S., Bauchot, A.D. and John, P.J. (2001) Internal browning in cold-stored pineapples is suppressed by a postharvest application of 1-methylcyclopropene. *Postharvest Biology and Technology* (in press).

Seymour, G.B., Joh, P. and Thompson, A.K. (1987) Inhibition of degreening in the peel of bananas ripened at tropical temperatures. II. Role of ethylene, oxygen and carbon dioxide. *Annals of Applied Biology*, **110**, 153-161.

Seymour, G.B., Taylor, J.E. and Tucker, G.A. (1993) *Biochemistry of Fruit Ripening*, Chapman & Hall, London.

Shellie, K.C. and Saltveit, M.E. (1993) The lack of a respiratory rise in muskmelon fruit ripening on the plant challenges the definition of climacteric behaviour. *Journal of Experimental Botany*, **44**, 1403-1406.

Shiomi, S., Yamamoto, M., Ono, T., Kakiuchi, K., Kakamoto, J., Nakatsuka, A., Kubo, Y., Nakamura, R., Inaba, A. and Imaseki, H. (1998) cDNA cloning of ACC synthase and ACC oxidase in cucumber fruit and their differential expression by wounding and auxin. *Journal of the Japanese Society for Horticultural Science*, **67**, 685-692.

Sisler, E.C. and Blankenship, S.M. (1993a) Diazocyclopentadiene (DACP), a light sensitive reagent for the ethylene receptor in plants. *Plant Growth Regulation*, **12**, 125-132.

Sisler, E.C. and Blankenship, S.M. (1993b) Effect of diazocyclopentadiene on tomato ripening. *Plant Growth Regulation*, **12**, 155-160.

Sisler, E.C. and Lallu, N. (1994) Effect of diazocyclopentadiene (DACP) on tomato fruits harvested at different ripening stages. *Postharvest Biology and Technology*, **4**, 245-254.

Sisler, E.C. and Serek, M. (1997) Inhibitors of ethylene responses in plants at the receptor level: recent developments. *Physiologia Plantarum*, **100**, 577-582.

Sisler, E.C. and Wood, C. (1988) Interaction of ethylene and CO_2. *Physiologia Plantarum*, **73**, 440-444.

Sisler, E.C. and Yang, S.F. (1984) Anti-ethylene effects of *cis*-2-butene and cyclic olefins. *Phytochemistry*, **23**, 2765-2768.

Sisler, E.C., Serek, M. and Dupille, E. (1996) Comparison of cyclopropene, 1-methylcyclopropene and 3,3-dimethylcyclopropene as ethylene antagonists in plants. *Plant Growth Regulation*, **18**, 169-174.

Sisler, E.C., Serek, M., Dupille, E. and Goren, R. (1999) Inhibition of ethylene responses by 1-methylcyclopropene and 3-methylcyclopropene. *Plant Growth Regulation*, **27**, 105-111.

Solomos, T. (1988) Respiration in senescing plant organs: its nature, regulation, and physiological significance, in *Senescence and Aging in Plants* (eds L.D. Nooden and A.C. Leopold), Academic Press, San Diego, pp. 111-145.

Starrett, D.A. and Laties, G.G. (1991) The effect of ethylene and propylene pulses on respiration, ripening advancement, ethylene-forming enzyme, and 1-aminocyclopropane-1-carboxylic acid synthase activity in avocado fruit. *Plant Physiology*, **95**, 921-927.

Stewart, I. and Wheaton, T.A. (1972) Carotenoids in citrus: their accumulation induced by ethylene. *Journal of Agricultural Food Chemistry*, **20**, 448-449.

Stover, E., Fargione, M.J., Watkins, C.B. and Iungerman, K.A. (2002) Interactions of ReTain^{TM} (AVG) with ethephon and summer pruning on preharvest drop and fruit quality of Marshall 'McIntosh' apples. *HortScience*, **37** (in press).

Stow, J.R., Dover, C.J. and Genge, P.M. (2000) Control of ethylene biosynthesis and softening in 'Cox's Orange Pippin' apples during low-ethylene, low oxygen storage. *Postharvest Biology and Technology*, **18**, 215-225.

Sunako, T., Sakuraba, W., Senda, M., Akada, S., Ishikawa, R., Niizeki, M. and Harada, T. (1999) An allele of the ripening-specific 1-aminocyclopropane-1-carboxylic acid synthase gene (ACS1) in apple fruit with a long storage life. *Plant Physiology*, **119**, 1297-1304.

Theologis, A., Oeller, P.W., Wong, L.M., Rottman, W.H. and Gantz, D.M. (1993) Use of a tomato mutant constructed with reverse genetics to study fruit ripening, a complex developmental process. *Developmental Genetics*, **14**, 282-295.

Thompson, A.J., Tor, M., Barry, C.S., Vrebalov, J., Orfila, C., Jarvis, M.C., Giovannoni, J.J., Grierson, D. and Seymour, G.B. (1999) Molecular and genetic characterization of a novel pleiotropic tomato-ripening mutant. *Plant Physiology*, **120**, 383-389.

Thompson, J.F., Brecht, P.E., Hinsch, T. and Kader, A.A. (2000) *Marine Container Transport of Chilled Perishable Produce*, University of California Agriculture and Natural Resources Publication 21595.

Tian, M., Sheng, C.C. and Li, Y. (1987) Changes of ethylene synthesis, activity of polyphenol oxidase, permeability of plasma membrane of Dutch pear at low temperature. *Acta Botanic Sinica*, **29**, 614-619.

Tian, M.S., Bowen, J.H., Bauchot, A.D., Gong, Y.P. and Lallu, N. (1997a) Recovery of ethylene biosynthesis in diazocyclopentadiene (DACP)-treated tomato fruit. *Plant Growth Regulation*, **22**, 73-78.

Tian, M.S., Gong, Y. and Bauchot, A.D. (1997b) Ethylene biosynthesis and respiration in strawberry fruit treated with diazocyclopentadiene and IAA. *Plant Growth Regulation*, **23**, 195-200.

Tian, M.S., Prakash, S., Elgar, H.J., Young, H., Burmeister, D.M. and Ross, G.S. (2000) Responses of strawberry fruit to 1-methylcyclopropene (1-MCP) and ethylene. *Plant Growth Regulation*, **32**, 83-90.

Tieman, D.M. and Klee, H.J. (1999) Differential expression of two novel members of the tomato ethylene-receptor family. *Plant Physiology*, **120**, 165-172.

Tigchelaar, E.-C., McGlasson, W.B. and Buescher, R.W. (1978) Genetic regulation of tomato fruit ripening. *HortScience*, **13**, 508-513.

Tonutti, P., Bonghi, C., Ruperti, B., Tornielli, G.B. and Ramina, A. (1997) Ethylene evolution and 1-aminocyclopropane-1-carboxylase oxidase gene expression during early development and ripening of peach fruit. *Journal of the American Society for Horticultural Science*, **122**, 642-647.

Trinchero, G.D., Sozzi, G.O., Cerri, A.M., Vilella, F. and Fraschina, A.A. (1999) Ripening-related changes in ethylene production, respiration rate and cell-wall enzyme activity in goldenberry (*Physalis peruviana* L.), a solanaceous species. *Postharvest Biology and Technology*, **16**, 139-145.

Tucker, G. and Brady, C. (1987) Silver ions interrupt tomato fruit ripening. *Journal of Plant Physiology*, **127**, 165-169.

Veen, H. (1986) A theoretical model for anti-ethylene effects of silver thiosulphate and 2,5-norbornadiene. *Acta Horticulturae*, **181**, 129-131.

Vendrell, M. and McGlasson, W.B. (1971) Inhibition of ethylene production in banana fruit tissue by ethylene treatment. *Australian Journal of Biological Science*, **24**, 885-895.

Ververidis, P. and John, P. (1991) Complete recovery *in vitro* of ethylene-forming enzyme activity. *Phytochemistry*, **30**, 725-727.

Villavicencio, L., Blankenship, S.M., Sanders, D.C. and Swallow, W.H. (1999) Ethylene and carbon dioxide production in detached fruit of selected pepper cultivars. *Journal of the American Society for Horticultural Science*, **124**, 402-406.

Walsh, C.S., Popenoe, J. and Solomos, T. (1983) Thornless blackberry is a climacteric fruit. *HortScience*, **18**, 482-483.

Wang, C.Y. and Adams, D.O. (1982) Chilling-induced ethylene production in cucumbers. *Plant Physiology*, **69**, 424-427.

Wang, C.Y. and Ji, Z.L. (1989) Effect of low-oxygen storage on chilling injury and polyamines in zucchini squash. *Scientia Horticulturae*, **39**, 1-7.

Wang, Y. and Li, N. (1997) A cDNA sequence isolated from the ripening tomato fruit encodes a putative protein kinase. *Plant Physiology*, **114**, 1135.

Wang, Z. and Dilley, D.R. (2000) Initial low oxygen stress controls superficial scald of apples. *Postharvest Biology and Technology*, **18**, 201-213.

Wang, S.Y., Adams, D.O. and Lieberman, M. (1982) Recycling of 5'-methylthioadenosine-ribose carbon atoms into methionine in tomato tissue in relation to ethylene production. *Plant Physiology*, **70**, 117-121.

Wang, Z.Y., MacRae, E.A., Wright, M.A., Bolitho, K.M., Ross, G.S., Atkinson, R.G. and Wang, Z.Y. (2000) Polygalacturonase gene expression in kiwifruit: relationship to fruit softening and ethylene production. *Plant Molecular Biology*, **42**, 317-328.

Watkins, C.B., Harman, J.E., Reid, M.S. and Padfield, C.A.S. (1982) Starch iodine pattern as a maturity index for Granny Smith apples. 2. Differences between districts and relationship to storage disorders and yield. *New Zealand Journal of Agricultural Research*, **25**, 587-592.

Watkins, C.B., Bowen, J.H. and Walker, V.J. (1989a) Assessment of ethylene production by apple cultivars in relation to commercial harvest dates. *New Zealand Journal of Crop and Horticultural Science*, **17**, 327-333.

Watkins, C.B., Hewett, E.W., Bateup, C., Gunson, A. and Triggs, C.M. (1989b) Relationships between maturity and storage disorders in 'Cox's Orange Pippin' apples as influenced by preharvest calcium or ethephon sprays. *New Zealand Journal of Crop and Horticultural Science*, **17**, 283-292.

Watkins, C.B., Picton, S. and Grierson, D. (1990) Stimulation and inhibition of expression of ripening-related mRNAs in tomatoes as influenced by chilling temperatures. *Journal of Plant Physiology*, **136**, 318-323.

Watkins, C.B., Nock, J.F. and Whitaker, B.D. (2000) Responses of early, mid and late season apple cultivars to postharvest application of 1-methylcyclopropene (1-MCP) under air and controlled atmosphere storage conditions. *Postharvest Biology and Technology*, **19**, 17-32.

Whittaker, D.J., Smith, G.S. and Gardner, R.C. (1997) Expression of ethylene biosynthetic genes in *Actinidia chinensis* fruit. *Postharvest Biology and Technology*, **34**, 45-55.

Wild, B.L., McGlasson, W.B. and Lee, T.H. (1976) Effect of reduced ethylene levels in storage atmospheres on lemon keeping quality. *HortScience*, **11**, 114-115.

Wilkinson, J.Q., Lanahan, M.B., Yen, H.C., Giovannoni, J.J. and Klee, H.J. (1995) An ethylene-inducible component of signal transduction encoded by *Never-ripe*. *Science*, **270**, 1807-1809.

Wills, R.B.H. and Kim, G.H. (1995) Effect of ethylene on postharvest life of strawberries. *Postharvest Biology and Technology*, **6**, 249-255.

Wills, R.B.H., Ku, V.V.V., Shohet, D. and Kim, G.H. (1999) Importance of low ethylene levels to delay senescence of non-climacteric fruit and vegetables. *Australian Journal of Experimental Agriculture*, **39**, 221-224.

Wills, R.B.H., Warton, M.A. and Ku, V.V.V. (2000) Ethylene levels associated with fruit and vegetables during marketing. *Australian Journal of Experimental Agriculture*, **40**, 465-470.

Woeste, K.E. and Kieber, J.J. (2000) A strong loss-of-function mutation in RAN1 results in constitutive activation of the ethylene response pathway as well as a rosette-lethal phenotype. *Plant Cell*, **12**, 443-455.

Wong, W.S., Ning, W., Xu, P.L., Kung, S.D., Yang, S.F. and Li, N. (1999) Identification of two chilling-regulated 1-aminocyclopropane-1-carboxylate synthase genes from citrus (*Citrus sinensis* Osbeck) fruit. *Plant Molecular Biology*, **41**, 587-600.

Xu, Z.C., Hyodo, H., Ikoma, Y., Yano, M. and Ogawa, K. (2000) Relation between ethylene-producing potential and gene expression of 1-aminocyclopropane-1-carboxylic acid synthase in *Actinidia chinensis* and *A. deliciosa* fruits. *Journal of the Japanese Society for Horticultural Science*, **69**, 192-194.

Yamamoto, M., Miki, T., Ishiki, Y., Fujinami, K., Yanagisawa, Y., Nakagawa, H., Ogura, N., Hirabayashi, T. and Sato, T. (1995) The synthesis of ethylene in melon fruit during the early stage of ripening. *Plant Cell Physiology*, **326**, 591-695.

Yang, S.F. (1985) Biosynthesis and action of ethylene. *HortScience*, **20**, 41-44.

Yang, S.F. and Hoffman, N.E. (1984) Ethylene biosynthesis and its regulation in higher plants. *Annual Reviews of Plant Physiology*, **35**, 155-189.

Yang, R.F., Cheng, M.T.S. and Shewfelt, R.L. (1990) The effect of high temperature and ethylene treatment on the ripening of tomatoes. *Journal of Plant Physiology*, **136**, 941-942.

Yip, W.K., Jiao, X.Z. and Yang, S.F. (1988) Dependence of *in vivo* ethylene production rate on 1-aminocyclopropane-1-carboxylic acid content and oxygen concentrations. *Plant Physiology*, **88**, 553-558.

Yip, W.K., Moore, T. and Yang, S.F. (1992) Differential accumulation of transcripts for four tomato 1-aminocyclopropane-1-carboxylate synthase homologs under various conditions. *Proceedings of the National Academy of Sciences of the USA*, **89**, 2475-2479.

Yu, Y.B., Adams, D.O. and Yang, S.F. (1979) 1-Aminocyclopropane-carboxylate synthase, a key enzyme in ethylene biosynthesis. *Archives of Biochemistry and Biophysics*, **198**, 280-286.

Yu, Y.B., Adams, D.O. and Yang, S.F. (1980) Inhibition of ethylene production by 2,4-dinitrophenol and high temperature. *Plant Physiology*, **66**, 286-290.

Yuen, C.M.C., Tridjaja, N.O., Wills, R.B.H. and Wild, B.L. (1995) Chilling injury development of 'Tahitian' lime, 'Emperor' mandarin, 'Marsh' grapefruit and 'Valencia' orange. *Journal of the Science of Food and Agriculture*, **67**, 335-339.

Zarembinski, T.I. and Theologis, A. (1994) Ethylene biosynthesis and action: a case of conservation. *Plant Molecular Biology*, **26**, 1579-1597.

Zauberman, G. and Fuchs, Y. (1973) Ripening processes in avocados stored in ethylene atmosphere in cold storage. *Journal of the American Society for Horticultural Science*, **98**, 477-480.

Zegzouti, H., Jones, B., Frasse, P., Marty, C., Maitre, B., Latche, A., Pech, J.C. and Bouzayen, M. (1999) Ethylene-regulated gene expression in tomato fruit: characterization of novel ethylene-responsive and ripening-related genes isolated by differential display. *Plant Journal*, **18**, 589-600.

Zeroni, M., Galil, J. and Ben-Yehoshua, S. (1976) Autoinhibition of ethylene formation in nonripening stages of the fruit of sycomore fig (*Ficus sycomorus* L.). *Plant Physiology*, **57**, 647-650.

Zhang, L.X. and Paull, R.E. (1990) Ripening behavior of papaya genotypes. *HortScience*, **25**, 454-455.

Zheng, Y.H., Xi, Y.F. and Ying, T.J. (1993) Studies on postharvest respiration and ethylene production of loquat fruits. *Acta Horticulturae Sinica*, **20**, 111-115.

Zhou, D., Kalaitzis, P., Mattoo, A.K. and Tucker, M.L. (1996) The mRNA for an ETR1 homologue in tomato is constitutively expressed in vegetative and reproductive tissues. *Plant Molecular Biology*, **30**, 1331-1338.

Zhou, H.W., Dong, L., Ben-Arie, R. and Lurie, S. (2001a) The role of ethylene in the prevention of chilling injury in nectarines. *Journal of Plant Physiology*, **158**, 55-61.

Zhou, H.W., Lurie, S., Ben-Arie, R., Dong, L., Burd, S., Weksler, A. and Lers, A. (2001b) Intermittent warming of peaches reduces chilling injury by enhancing ethylene production and enzymes mediated by ethylene. *Journal of Horticultural Science and Biotechnology* (In press).

9 Management of postharvest diseases

David Sugar

9.1 Introduction

Postharvest diseases affect fruit quality. The consumer will not purchase fruit which is obviously infected, while the presence of internal infections will spoil the eating experience and deter future consumption. Producers suffer economic loss because infected fruit must be discarded and they can lose confidence in their ability to store fruit. This chapter will examine the nature of postharvest diseases, the agents and conditions which lead to their development, and the diverse strategies employed by growers and packing house operators to minimize the likelihood of disease occurrence.

9.2 The nature of postharvest disease

Postharvest diseases of fruit crops, as distinguished from postharvest physiological disorders or other problems of abiotic origin, are caused mostly by fungal infection. The problem is then the presence of infected tissue, also known as decay, which is typically different from surrounding healthy tissue in color and/or texture. In most cases, infected tissue forms a discrete zone, known as a lesion, which extends radially from an infection point in a characteristic pattern determined by the interaction between the host fruit and the pathogen. In some postharvest diseases, the border between infected and apparently healthy tissue is sharply defined, in others it is more diffuse. A defined lesion border may indicate that fungal advance has been halted by defense mechanisms in the host fruit, or may merely indicate the collapse of older infected tissue, while the pathogen continues to penetrate, and newly infected tissue remains symptomless. In some diseases, lesions rarely extend beyond 1–2 cm in diameter, while in other diseases the entire fruit is uniformly infected.

The lesion may appear only as altered host tissue, or it may be covered to a varying extent by tissues of the pathogen, including fungal mycelium, spore-bearing structures and spores. The form and color pattern of the pathogen and the symptoms displayed by the host form the basis for classical diagnosis of the disease. Since postharvest diseases are caused by a wide variety of fungi, which in turn respond differentially to environmental and cultural conditions, accurate diagnosis is essential to the design of effective management schemes. Some diseases are indistinguishable from one another by their appearance. Culture of the pathogen from lesion tissue on an agar medium allows identification based on mycelial features (pattern and color) and sporulation (spore bearing structures, color and shape of spores).

Postharvest diseases are to a high degree host specific. Although some pathogens, most notably *Botrytis cinerea*, will infect fruits of various genera, most pathogens only infect hosts of distinct fruit types (e.g. pome, stone or citrus). Distinct species within a fungal genus may infect different fruits; *Penicillium expansum* is the principal cause of blue mold decay in pome fruits, while *P. italicum* and *P. digitatum* cause blue and green molds, respectively, in citrus. While all stone fruits are highly perishable, they may be ranked in order of decreasing susceptibility to postharvest decay: cherries, nectarines, peaches, plums and apricots (Eckert and Ogawa, 1988). Major postharvest diseases of pome, stone and citrus fruits are listed in table 9.1.

From the moment of harvest, fruits begin a process of deterioration. Efforts to preserve fruit quality and manage fungal decay can at best only slow this

Table 9.1 Important postharvest diseases of pome, stone, and citrus fruits and the causal fungi

Pome fruits	
Blue mold	*Penicillium expansum*
Gray mold	*Botrytis cinerea*
Mucor rot	*Mucor piriformis*
Bull's-eye rot	*Neofabraea malicorticis* anamorph: *Cryptosporiopsis curvispora*
Alternaria fruit rot	*Alternaria* spp.
Cladosporium fruit rot	*Cladosporium herbarum*
Side rot	*Phialophora malorum*
Black rot	*Physalospora obtusa*
Bitter rot	*Glomerella cingulata* anamorph: *Collectotrichum gloeosporioides*
White rot	*Botryospheria ribis*
Stone fruits	
Brown rot	*Monilinia fructicola, M. laxa*
Gray mold	*Botrytis cinerea*
Rhizopus rot	*Rhizopus stolonifer*
Alternaria rot	*Alternaria* spp.
Sour rot	*Geotrichum candidum*
Blue mold	*Penicillium* spp.
Citrus fruits	
Blue mold	*Penicillium italicum*
Green mold	*Penicillium digitatum*
Alternaria (black) rot	*Alternaria citri*
Sour rot	*Geotrichum candidum*
Stem end rot	*Phomopsis citri, Diplodia natalensis*

inexorable process. With the gradual deterioration of the fruit, natural barriers to infection by pathogenic fungi diminish. Furthermore, many of the changes in fruit development that are desirable for human enjoyment and nutrition are simultaneously advantageous to pathogenic fungi. Fruit ripening marks a critical change in the natural balance of resistance and susceptibility. Ripening is typically accompanied by changes in fruit texture, sweetness and acidity, and in the quantities of various phenolic compounds present. Texture changes in pome and stone fruits function in a similar fashion to fungal attack; enzymes break down pectic substances that maintain tissue structure. Thus fungal breakdown of fruit tissue is facilitated. Furthermore, the increase in sugars associated with maturity can enhance the food base for fungal infection. Phenolic substances that may function in disease resistance generally diminish with ripeness and while this may increase palatability, it may also facilitate fungal advance.

9.3 Opportunities for infection

Pathogenic fungi, along with non-pathogenic fungal species, yeast and bacteria, are naturally present on the surfaces of harvested fruit. In fact, it may be assumed that at least some fungal propagules capable of infection are present on nearly all individual fruit as they enter the postharvest period. Ubiquitous fungi, such as *Botrytis cinerea* and *Penicillium* spp. are produced on decaying vegetative matter in the orchard and accumulate as airborne spores on fruit surfaces during the growing season. Contamination may also occur with dust raised during the harvest operation, from contact with harvest containers, from air in the storage room, and from surfaces and solutions used to handle the fruit and move it through the packing procedures. In the latter process, pathogen spores may be redistributed from one fruit to another through contamination of the handling solution or machinery surfaces. Nevertheless, the overall probability of disease developing is quite low; most harvested fruit produced in modern fruit operations are successfully stored, marketed and consumed. In the absence of specific control measures, the likelihood of surface contaminants infecting fruit depends on time, the presence of an infection court, and on the relative susceptibility of the individual fruit. At storage temperatures appropriate to the fruit type, infection proceeds more slowly than at ambient temperatures. Thus the fruit may be consumed before infection takes place or lesions are manifest. The amount of time necessary for lesion development varies among pathogens; in pome fruit, a relatively aggressive fungus like *Botrytis cinerea* may cause visible lesions within six weeks at $-1°C$, while slower-growing *Alternaria* spp. may require three months or more, depending on the fruit condition.

The fungal propagule that is most often dispersed in the orchard and leads to fruit infection is the spore. In most cases, the fungal spore causing fruit infection is a conidium, an asexual spore produced by fungal mycelium actively growing on living or decaying vegetative tissue. In order to germinate and initiate infection, spores usually require: 1. oxygen, although most postharvest pathogens are able to infect and grow at reduced partial pressures of oxygen; 2. water, either as the moist microenvironment of a fresh fruit wound or as a film of free moisture; and 3. a source of metabolizable organic compounds, available as leakage from ruptured cells at wounds, natural secretions, or stored resources within the spore itself (Sommer *et al.*, 1992). Growth from the spore begins as one or more germ tubes, which may form specialized structures called appressoria. These adhere to the host surface through gelatinous secretions and form infection pegs to penetrate host tissues.

9.3.1   The role of wounds

Certain fungi can infect apparently sound fruit during cold storage, in the absence of wounds or injuries. *Botrytis cinerea* and *Mucor piriformis* in pome fruit, *Monilinia fructicola* in stone fruit, and *Penicillium digitatum* in citrus, can move from a decaying fruit to infect adjacent sound fruit, creating pockets or 'nests' of diseased fruit in the storage container (Ogawa *et al.*, 1995; Rosenberger, 1990; Spotts, 1990c). The original infection, in this case, would probably have originated at a wound or from infection through a natural opening prior to cold storage. Wounds are probably the most common infection court for postharvest disease. Wounding breaks the continuity of the natural barriers to infection, the waxy cuticle and subtending cell layers, which may be suberized or otherwise less penetrable than cortical tissue below. Wounds may or may not be visible to the naked eye, and thus it is difficult to conclude whether or not a given lesion originated at a fruit lenticel, through a puncture wound too small to see, or through bruising, which can disrupt the cuticle or lenticel structure without causing an actual puncture. In a study of surface wounds in pear fruit, the average diameter of visible wounds was approximately 5 mm (Spotts *et al.*, 1998b). Wounded fruit should be culled out during the sorting process in the packing house, but small wounds are difficult to observe, especially on fruit surfaces varying in color and pattern due to partial russeting or prominent lenticels. Spotts *et al.* (1998b) found that treating pear fruit with a vegetable dye that adhered to wound tissue made cullage by commercial sorters more likely. Even with pathogens that are capable of penetrating fruit without wounds, such as *Botrytis cinerea* and *Monilinia fructicola*, the probability of infection is greatly increased by wounding. With most pathogens, the likelihood of postharvest disease is generally proportional to the incidence and severity of wounding (Eckert and Ogawa, 1985).

Although wounds can create infection opportunities for pathogens that could not otherwise penetrate the fruit cuticle, fruit can react to wounds in ways that form barriers to infection. Cell walls adjacent to wounds in apple fruit increase in tannins, lignins, other phenolic substances and callose following wounding, and become more resistant to infection by *Botrytis cinerea* and *Penicillium expansum* than fresh wounds (Bostock and Stermer, 1989; Lakshiminarayana *et al.*, 1987). Accumulation of resistance factors at wound sites requires time and proceeds at a faster rate with increasing temperature. Apple fruit inoculated four or more days after wounding and storage at 5°C had a lower rate of infection than those inoculated on the same day as wounding. In wounded pear fruit stored at -1°C, tannins and callose accumulated rapidly but lignins were not detected within 30 days (Spotts *et al.*, 1998b). Incubation of wounded pears at 28°C for 1 day accelerated accumulation of these resistance-related substances, and resistance to decay by *B. cinerea* and *P. expansum* correlated positively with positive staining for these compounds. In citrus, increased phenolic substances and lignins were detected in cell walls adjacent to wounds 16 h after injury and incubation at 30°C (Brown and Barmore, 1983). At 30 h after injury, fruit tissue was macerated by polygalacturonase enzyme from *Penicillium digitatum*, but at 48 or more hours after injury, resistance to maceration was manifest. Since wounds can become resistant to infection, treatments to protect wounds from infection may not need to be persistent (Smilanick, 1994). Because of the critical role of wounds in the development of many types of postharvest disease, it is of paramount importance that disease management strategies consider careful harvest a fundamental component, with the personnel management, physical infrastructure and handling procedures necessary to achieve it.

Most postharvest fruit pathogens are capable of infecting appropriate host fruit at any time after harvest if inoculated into fresh wounds. In the absence of distinct wounds or natural openings described above, infection courts can develop with the gradual loss of integrity of physical barriers that were effective in preventing infection by surface contaminants early in the postharvest life of the fruit. Small cracks in the cuticle, physical breaks in the epidermal tissue and the declining levels of phenolic substances provide opportunities for fungal advance. In pome fruit, after four or more months of storage there is a gradual increase in the incidence of lesions caused by slow-growing ubiquitous fungi, particularly *Alternaria* spp. and *Cladosporium herbarum*. High incidence of decay caused by these fungi may be simply a symptom of fruit senescence. Market quality of such fruit would be low, even in the absence of fungal infection.

Other natural openings can also serve as infection courts for pathogenic fungi. In some fruit cultivars, the end of the peduncle or fruit stem may remain fleshy and penetrable by fungi following separation from the tree branch or spur. The stem is thus susceptible to postharvest infection by fungi, as in stem-end decay of d'Anjou pears by *Penicillium expansum* (Chen *et al.*, 1981). Citrus fruits can develop stem-end rots following infections of the stem button by species of *Diplodia*, *Phomopsis* or *Alternaria*. Infection takes place during the growing season, but remains quiescent until the stem button becomes senescent (Brown and Wilson, 1968). Senescence in the fruit stem generally proceeds more rapidly than in the cortical tissues of the fruit.

9.3.2 Lenticel infections

Fruit lenticels, which provide openings for gas exchange between fruit cells and the surrounding atmosphere, can also serve as infection courts for initiation of postharvest disease (Kidd and Beaumont, 1925; Baker and Heald, 1934). Dissection of blue mold (*Penicillium expansum*) lesions on 'Delicious' apples led to the conclusion that a large proportion were true lenticel infections (English *et al.*, 1946). However, lenticels may not be 'open' to ready infection by pathogens. Clements (1935) found that only 4.8 to 29.8% of apple lenticels were open to infiltration by dye solutions, depending on cultivar. Lenticels on pears were found to be impenetrable by dye solution unless subjected to substantial impact bruising (Sugar and Spotts, 1993b). Hypodermal cells surrounding lenticel cavities may be cutinized and/or contain relatively high tannin levels, both of which can serve to preclude fungal infection (Clements, 1935). Lenticel condition may vary with the growing environment, fruit size and growth rate, fruit maturity, and the accessibility of lenticels to fungal infection may change with fruit senescence during storage.

9.3.3 Quiescent infections

Several diseases may originate with infection of developing fruit in the orchard and remain quiescent or latent until after harvested fruit are in storage. Apples and pears may be infected, possibly via lenticels, during the growing season by spores of *Neofabraea* (formerly *Pezicula*) *malicorticis* originating from limb cankers on the tree (Spotts, 1990a). Even though fruit infections may take place as early as immediately post petal-fall, disease lesions (bull's-eye rot) typically appear after 4–5 months of storage at around 0°C. Lenticel infection of apples and pears in the orchard is favored by summer rainfall and thus is less common in production areas with dry summers, except where overhead sprinkler irrigation is used. Brown rot of stone fruits, caused by *Monilinia fructicola*, may also infect developing fruit soon after petal-fall, and remain quiescent until fruit approach maturity on the tree, when expanding lesions can develop (Northover and Cerkauskas, 1994). In quiescent diseases the fungal infection typically proceeds beyond the fruit cuticle and sometimes through the outer layer of epidermal cells. Progress is then stalled until critical changes in fruit physiology take place and allow pathogenesis to proceed (Verhoeff, 1974). A consequence of this early penetration of the outer fruit layers is that the fungus is protected from contact with most fungicides subsequently applied in the orchard, as well as from environmental factors such as dryness and solar radiation that decrease survival on the fruit surface.

A well-studied example of some of the mechanisms involved in quiescent infections is *Colletotrichum gloeosporioides* in avocado. *C. gloeosporioides* attacks many tropical and subtropical fruits (Prusky, 1996). Germinated spores on the fruit surface in the orchard are able to penetrate the cuticle, where subcuticular hyphae enter into a quiescent phase until after harvest. Resistance to further invasion of the unripe avocado is associated with an antifungal compound in the pericarp tissue that is present before infection (Prusky and Keen, 1993). Levels of the antifungal compound can also be regulated by the host via reactive oxygen species formed in response to fungal infection (Beno-Moualem and Prusky, 2000). High levels of these reactive oxygen species can promote increased levels of the antifungal compound, resulting in quiescence of the pathogen. With ripening, the antifungal compound is oxidized by the host enzyme lipoxygenase, removing the inhibition and allowing the infection to advance. Prior to ripening, lipoxygenase is regulated by levels of the non-specific inhibitor epicatechin, which is present in the peels of unripe but not of ripe avocados (Prusky and Keen, 1993). Reactive oxygen species were also found to enhance the phenylpropanoid pathway, leading to formation of epicatechin (Ardi *et al.*, 1998). Treatment of avocados with reduced pathogenicity mutants of *C. gloeosporioides* induced increased formation of the antifungal compound, conferring greater resistance in the fruit to attack by the wild-type strain (Yakoby *et al.*, 2001).

While the specific interactions described above for quiescent infections in the avocado–*C. gloeosporioides* interaction may be shared by other host–pathogen systems, many other mechanisms may come into play simultaneously or apply differentially to particular systems. Four general situations can be described to explain the conditions that cause a change from pathogen quiescence to activity: 1. pre-formed antifungal compounds that are present in unripe, but not ripe, fruit; 2. the accumulation of antifungal compounds formed in response to initial infection (phytoalexins), which diminish with fruit ripening; 3. the unripe fruit does not fulfill the nutritional requirements of the pathogen and/or the energy resources of the unripe fruit are insufficient to sustain pathogen activity; 4. pathogen enzymes are able to facilitate cuticle penetration but cannot break down internal tissues until their composition has changed with ripening (Swinburne, 1983).

9.3.4   Floral infections

Infections of flowers of Bartlett pear in California by *Botrytis cinerea* were associated with postharvest infections at the calyx end and incidence of postharvest decay at the calyx end were reduced by applications of benomyl during bloom (Sommer *et al.*, 1985). However, it is not clear whether the quiescent fungus resides in the fruit tissue or in the senescent flower parts that remain attached to the fruit. Infection of stone fruit blossoms by the brown rot fungus, *Monilinia fructicola*, can also lead to fruit infections. In this case, rapid sporulation of the fungus on infected blossoms provides inoculum for subsequent fruit infections.

Floral infection may also lead to the presence of fungi in the core area in pome fruit, with or without actual infection of fruit tissue and typically without any external evidence of disease. If fungal mycelium grows superficially within the locules of the fruit without penetration of the tissues, it is known as 'moldy core'. Despite the absence of infection, the appearance of moldy core can have a negative effect on the consumer's experience if the fruit is sliced so as to expose the mold. The mycelium may also penetrate the core flesh, but be restricted to the carpellary region. Infection of this type is known as 'dry core rot'. If infection takes place while fruit are on the tree, the production of ethylene within the fruit in response to injury (infection in this case) can cause the fruit to color and drop from the tree prematurely. Harvested fruit infected with dry core rot may be more advanced in maturity than non-infected fruit, with more rapid deterioration of quality during storage. The fungi most commonly associated with moldy core and dry core rot in apples and pears are common orchard fungi of the genera *Alternaria*, *Ulocladium*, *Stemphylium*, *Cladosporium*, *Epicoccum*, *Coniothyrium* and *Pleospora* (Cobrink *et al.*, 1985; Spotts *et al.*, 1988). A distinct 'wet' core rot develops if fungi attack the core more aggressively, causing a soft rot that may extend beyond the carpellary region. The principal fungal genus involved in this disease in apples is *Penicillium*, but *Mucor*, *Fusarium*, *Pestalotia*, *Botryosphaeria* and *Botrytis* have also been reported as associated with wet core rot (Cobrink *et al.*, 1985; Spotts *et al.*, 1988).

The key factor in susceptibility to core rots is the presence of an open sinus extending from the floral tissues into the core. In pome fruit, the likelihood of an open sinus is determined by the intrinsic anatomy of the fruit cultivar and by environmental conditions that modify fruit shape. Of seven widely grown apple cultivars in the Pacific Northwest region of the USA, Braeburn and Granny Smith had the lowest frequency of open sinuses, while Royal Gala and Fuji had the highest frequency (Spotts *et al.*, 1999). Cool temperatures during early fruit development favor relatively elongated fruit shape; increasing length:diameter ratio is inversely related with sinus opening, consequently with incidence of core rots (Spotts, 1990b; Tufts and Hansen, 1931). When fruit of 'Delicious' apple varying in length:diameter ratio were dipped in a suspension of *Penicillium expansum* spores, shorter fruit developed more core rot (Spotts *et al.*, 1988). Fungi causing moldy core and dry core rot colonize senescent floral tissues during wet weather and, when the sinus opening is present, growth can continue into the core region. Fungi causing wet core rot, in contrast, are believed to enter most commonly through the open sinus when harvested fruit are handled in water in the packing house. Hydrostatic pressure associated with immersion facilitates entry of the water, which is commonly contaminated with spores of various orchard fungi, into the core (Spotts, 1990b).

9.4   Factors influencing fruit susceptibility to postharvest disease

9.4.1   Fruit maturity

Generally, immature fruit are highly resistant to facultative pathogens. They become increasingly susceptible with maturity or ripeness, gradually or over a relatively short period of time. D'Anjou pear fruit were highly resistant to infection by *Botrytis cinerea*, *Mucor piriformis*, and *Penicillium expansum* until at or near harvestable maturity, even when spores of the fungi were inoculated into artificial fruit wounds (Spotts, 1985). In contrast, infection by *Neofabrea* (*Pezicula*) *malicorticis*, a fungus capable of quiescent infection, took place throughout the growing season. One aspect of the change from resistance to susceptibility as fruit mature is the decline in content of various phenolic compounds. In immature apple fruit, relatively high levels of *p*-coumaryl quinic and chlorogenic acids are associated with inhibition of spore germination and mycelial growth of *B. cinerea* (Ndubizu, 1976). The concentrations of these compounds decline with fruit maturity. Two of the major phenol compounds found in pear fruit, arbutin and chlorogenic acid, were found to inhibit spore germination of *P. expansum* and mycelial growth of *M. piriformis* (Boonyakiat *et al.*, 1986). However, arbutin decreased growth of *B. cinerea in vitro* while chlorogenic acid increased both germination and growth of *B. cinerea*. Postharvest disease resistance in certain high-tannin cultivars of apple was related to reduction in activity of fungal enzymes by oxidized polyphenols (Byrde, 1956). Fawcett and Spencer (1968) found that *Sclerotinia fructigena* can convert chlorogenic and quinic acids to compounds with greater antifungal activity in culture.

9.4.2   Interaction with pathogen enzymes

Several enzymes may be involved in penetration of the host's protective layers and breakdown of host tissues. Cutinases may be secreted to hydrolyze the cutin polymer overlaying the epidermis (Wang *et al.*, 2000). Once the pathogen has penetrated the cuticle and epidermal layers, either through wounding, other infection courts or enzymatic hydrolysis, further infection and the development of lesions are dependent upon maceration of the cortical tissues. Contact between adjacent fruit on the tree may result in thinner cuticles and increased frequency and size of cuticular cracks, leading to increased decay incidence

(Michailides and Morgan, 1997). Thinner cuticles are also associated with high levels of nitrogen fertilization (Crisosto *et al.*, 1997).

Cortical cells are bound in the tissue structure by pectic substances, which form the middle lamella of fruit tissue. The breakdown of pectic substances in the middle lamella is a critical step in both the natural softening of fruit during ripening and in maceration of tissues by pathogenic fungi. The middle lamella is continuous with the individual cell walls of the tissue and is composed of several pectins. Apple fruit cell walls and middle lamellae contain relatively unbranched polygalacturonide (Knee, 1977). Thus, of necessity, various unrelated pathogens share the ability to produce and secrete pectolytic enzymes in order to hydrolyze and solubilize polyuronides as a critical step in the infection process (Collmer and Keen, 1986; Stelzig, 1984). Polygalacturonase (PG) is a key pectolytic enzyme produced by various postharvest pathogens. Resistance to many types of decay in immature fruit is in part associated with the presence of compounds that inhibit PG activity. Polygalacturonase inhibition may be accomplished by the presence of phenolic compounds, by proteinaceous inhibitors, or by specific cations associated with the pectic substances.

Abu-Goukh *et al.* (1983) found that the decline in resistance of pear fruit to infection by *Penicillium expansum* and *Botrytis cinerea* with advancing maturity was highly correlated with declines in the concentration of a proteinaceous PG inhibitor. Purified pear PG inhibitor inhibited *B. cinerea* to a greater extent than *P. expansum*, which may indicate variation in the structure or abundance of PGs among pathogens and the involvement of other pectolytic enzymes in PG hydrolysis (Abu-Goukh and Labavitch, 1983). Polygalacturonase inhibitor proteins from different plant species appear to vary in the kinetics and specificity of their inhibition.

9.4.3   Fruit calcium and decay resistance

Calcium ions are reported to link pectic molecules in the middle lamella and cell walls of apple and other fruit and the loss of cell-to-cell cohesion during ripening is attributed in part to movement of calcium ions from the middle lamella (Knee and Bartley, 1981; Stow, 1989). Fruit calcium content can be increased by spraying calcium solutions onto trees in the orchard or by postharvest immersion in calcium solutions (Sugar *et al.*, 1991; Conway *et al.*, 1994). Absorption from postharvest immersion in calcium solutions can be greatly facilitated by pressure or vacuum infiltration. Calcium ions that become associated with pectic substances following calcium treatments are believed to reduce decay by steric hindrance of PG activity (Conway *et al.*, 1988). Conway *et al.* (1988) found that the calcium content of extracted cell walls of apple fruit was positively correlated with the percentage of calcium in the pressure infiltration solution in which the apples were treated. When extracted cell walls of high-calcium apples were used as a substrate for PG extracted from *P. expansum*, approximately 60% less soluble polyuronide product of the PG activity was recovered as compared with cell walls from low-calcium apples. Fruit firmness was positively correlated with the concentration of calcium in the fruit following infiltration treatments of apple (Sams and Conway, 1984) and decay incidence correlated negatively with calcium concentration following infiltration of apples and orchard spray treatment of pear (Sams and Conway, 1984; Sugar *et al.*, 1991). Calcium orchard sprays were also associated with reduction of quiescent infections of apple by *Gloeosporium* (*Neofabraea*) *perrenans* (Sharples, 1980). Fallahi *et al.* (1997) concluded that in order to affect fruit firmness significantly and reduce postharvest decay in apples, it is necessary to raise the concentration of flesh calcium to $800–1000\,\mu g\,g^{-1}$ (dry weight), while amelioration of bitter pit, a physiological disorder of apple fruit associated with low fruit calcium, required flesh calcium concentrations of only ca. $250\,\mu g\,g^{-1}$. In pear, where cropping is less consistent than in apple and calcium concentrations vary considerably, emphasis has been placed on enhancing overall calcium levels and identifying relatively low calcium fruit lots for limited storage potential (Sugar *et al.*, 1992).

Peaches that received multiple calcium sprays during the growing season had increased calcium content in the fruit flesh but severity of brown rot decay (*Monilinia fructicola*) was not reduced (Conway *et al.*, 1987). Pressure-infiltration of calcium solutions reduced brown rot severity but caused injury to the fruit surface. Postharvest dips of peaches in solutions of various calcium salts reduced brown rot incidence and severity on mist-inoculated fruit (Biggs *et al.*, 1997). Calcium in artificial growth media also reduced mycelial growth and PG activity of *M. fructicola*.

In addition to direct interference with fungal maceration enzymes, calcium may affect postharvest disease development in other ways. Ferguson (1984) regards calcium as having anti-senescence properties. As such, the multiplicity of catabolic changes that occur with fruit senescence and facilitate fungal decay may be retarded. Among them, the maintenance of the integrity of cell membranes may be of particular importance. The formation of phytoalexins is likely to depend on cellular integrity. Furthermore, as tissue structure is broken down, increased permeability of host cell membranes can provide critical nutritional supplies to the pathogen (Edney, 1983; Stelzig, 1984).

Just as enrichment of fruit calcium through sprays and postharvest treatments can enhance fruit resistance to postharvest pathogens, variation in the native calcium content among individual fruit and fruit lots can be expected to influence the probability of encountering decay problems. Various crop and orchard factors, including soil characteristics, crop load, fruit size, and tree vigor affect the availability and distribution of calcium in the tree. As these factors affect susceptibility to calcium-related fruit disorders such as bitter pit in apple, they can affect decay susceptibility and should be considered in assessing the comparative storability of fruit lots (Edney, 1983).

9.4.4   Fruit nitrogen and decay resistance

Relatively high nitrogen apple fruit tend to be larger, greener and softer than lower nitrogen fruit and they respire at a higher rate. They develop internal browning and breakdown with a greater frequency (Bramlage *et al.*, 1980). High respiration tends to be associated with more rapid senescence and loss of resistance factors than occur in fruits with low respiration (Conway, 1984). Since nitrogen fertilization tends to increase fruit size, and larger fruit tend to have lower concentrations of calcium, nitrogen effects on fruit susceptibility to postharvest decay may be due in part to indirect effects on fruit calcium (Fallahi *et al.*, 1997). Furthermore, nitrogen fertilization tends to promote vigor; indeed, sustaining vegetative growth and renewal of fruiting structures are primary objectives of nitrogen application. With greater amounts of vegetative growth, calcium content of fruit tends to diminish (a situation known as 'shoot-fruit competition') resulting in heightened fruit susceptibility to postharvest disease (Sanchez *et al.*, 1995). While the precise mechanisms are as yet unclear, and the influence on disease susceptibility may result from multiple and inter-related responses, it is apparent that excessive nitrogen fertilization can increase susceptibility of fruit to decay. The association of relatively high fruit nitrogen content and increased decay severity has been demonstrated in pear (Sugar *et al.*, 1992), apple (Edney, 1973) and stone fruits (Crisosto *et al.*, 1997).

9.5   Disease management strategies

There are many opportunities for producers to affect the probability of postharvest fruit disease. Since many of these opportunities occur during the growing season rather than in the postharvest period, it is important for growers to understand that diseases that appear during storage can be affected much earlier, and the responsibility for postharvest disease control must be shared by both the grower and the packer/storage operator, who in many cases is not the same individual or even the same commercial organization. There is no single treatment that can be applied in order to control postharvest disease adequately. Accordingly, an integrated approach is appropriate, in which multiple steps are combined additively into a postharvest disease management strategy. Though the quantitative benefits of any individual step may be relatively small, the sum of all steps can amount to significant disease control. Given the complexities of the interactions of pathogen, host and environment, integration of diverse tactics is the only way that postharvest disease can be effectively managed. Furthermore, a diverse management strategy provides resilience in the event of failure by one or more components in a given situation.

The three factors required for plant disease are summarized in the 'plant disease triangle': a pathogen, a susceptible host, and a suitable environment for infection and disease development. Factors influencing fruit susceptibility were discussed above.

9.5.1 Presence of the pathogen

The primary source of pathogen inoculum for postharvest infection is the orchard. Airborne spores accumulate on fruit surfaces during the growing season, originating from decaying vegetative matter in or near the orchard, from soil, or from existing infections on other tree parts. The quantity of spores produced in an orchard is thus influenced by the quality of disease control earlier in the growing season and by conditions on the orchard floor. Tall weeds provide high relative humidity in the soil surface, favoring development and sporulation of pathogens that are also general saprophytes, such as *Botrytis cinerea*. Fallen fruit, or those left on the orchard floor following fruit thinning can also provide critical food sources for soil-dwelling pathogens like *Mucor piriformis*, which then increase in population (Dobson *et al.*, 1989). Pathogen spores are then transported to the packing house on fruit surfaces and in soil and debris inadvertently carried on the bottom of harvest containers.

9.5.2 Storage conditions

Low temperature storage is the first line of defense against postharvest pathogens. Temperate fruits can usually be stored at temperatures low enough to retard the development of postharvest decay significantly. Pears are typically stored at -1.1 to $-0.5°C$, apples at -1 to $4°C$ and peaches at $0–2°C$ (Hardenburg *et al.*, 1986). Unfortunately, successful postharvest pathogens are capable of growing and causing disease at these temperatures, albeit much more slowly than at warmer temperatures. In addition, the demand for extended duration of fruit storage results in sufficient time for postharvest infections to occur. It is a common goal of fruit handlers to bring down the internal temperature of the fruit as soon as possible after harvest. Steps include avoiding harvesting during the hottest hours of the day, keeping filled harvest containers in shade while awaiting transport and removing field heat as quickly as possible.

Depending on the fruit type and the management objectives, field heat may be removed in different ways: immersion in cold water (hydrocooling); forced air cooling, in which stacked field containers are arranged in cold rooms with powerful fans such that air flow is directly through, rather than between, the containers; or simple placement in cold rooms (Chapter 5). Hydrocooling solutions may contain a disinfectant such as chlorine or ozone to limit accumulation of pathogen spores from the fruit surfaces (Eckert and Ogawa, 1988; Tukey, 1993). Despite use of a disinfectant, hydrocooling is generally associated with increased risk of postharvest infection and its application is limited to fruit types such as cherries or fruit lots not targeted for long term storage. Delays of only one day in cooling stone fruit may cause significant increases in decay severity (Sommer, 1982). *Rhizopus* rot of stone fruits is perhaps the most temperature-sensitive postharvest fruit disease, as the causal fungus (*Rhizopus stolonifer*) cannot grow at temperatures below ca. 5°C (Pierson, 1966).

The relative humidity of storage rooms and containers is maintained at high levels in order to maintain fruit turgor. In pome and stone fruits, humidity levels of approximately 95% are necessary to avoid fruit dehydration during long-term storage. While some studies have shown that relative humidity levels of 99–100% are necessary for germination of spores of major fungal pathogens, these levels may be present at wound sites despite lower ambient humidity (Spotts and Peters, 1981).

The probability that spores of a pathogenic fungus will come into contact with a wound can be influenced by management in the orchard and in the packing house. Apples and pears are typically removed from harvest containers by immersion dumping or flotation in water in order to avoid physical injury. A fungistatic disinfectant, sodium *O*-phenyl phenate, is often used in immersion solutions. Despite the presence of a disinfectant, large numbers of spores can accumulate in the solution, which can serve as a redistribution focus for postharvest pathogens (Blanpied and Purnasiri, 1968; Sugar and Spotts, 1993a). The likelihood of infestation during contact with the immersion solution depends on the species and population levels of the pathogens present (Spotts and Cervantes, 1986). Penetration of spore-bearing solution into wounds, and displacement of air from the wounds, can be affected by the surface tension of the solution and by the amount of hydrostatic pressure corresponding to the depth of fruit immersion (Sugar and Spotts, 1993b).

9.5.3   Fungicide treatment

Application of treatments intended to prevent fungal infection may begin in the orchard, or treatments may be applied after harvest only. Postharvest treatments may be applied as harvested fruit arrive at the packing house and/or immediately before fruit are packed into boxes for final storage or sale. Antifungal treatments are not universally effective and must complement appropriate storage practices, as they rarely influence the rate of physiological deterioration of the fruit. Fungicides are most effective when: the treated fruit possesses some intrinsic resistance to infection; environmental conditions are least favorable to pathogen activity; and pathogen populations are low (Eckert and Ogawa, 1988). Beginning around 1970, benzimidazole fungicides, including benomyl, thiabendazole, thiophanate-methyl, and carbendazim were extensively studied and applied to a broad range of fruit crops. The effectiveness of this group of fungicides in reducing postharvest diseases promoted increased fruit production because long-term storage became less risky. It also focused much academic research towards fungicide treatment, drawing attention away from other aspects of disease management. The benzimidazoles are highly effective in controlling decays caused by *Penicillium* spp. in multiple fruit crops, as well as *Botrytis cinerea*, *Monilinia* spp., and several less-important postharvest pathogens, but are not effective against *Alternaria* spp., *Mucor* spp., or *Rhizopus stolonifer* (Eckert and Ogawa, 1988).

Orchard applications of benzimidazoles can be effective in reducing quiescent infections via lenticels by *Gloeosporium* (*Neofabraea*) spp. and residues from orchard sprays may persist sufficiently to inhibit postharvest infections by susceptible pathogens (Edney, 1983; Coyier, 1970). These fungicides, especially benomyl, are capable of penetrating the skin of treated apples and pears (Ben-Arie, 1975). The fungicide iprodione was registered for postharvest treatment of stone fruits in the USA in the early 1990s but postharvest treatment was not acceptable for fruit imported to Japan. Consequently, treatment application shifted to preharvest, with sufficient penetration of the fruit cuticle for residues to persist and be of benefit in reduction of postharvest decay.

In most pome, stone and citrus fruits, postharvest fungicide treatments are predominantly applied in the packing house, where fruit turning on rolling conveyors pass through a brief spray, with excess spray solution removed as the fruit pass over rolling sponges. In many cases, the fungicide is combined with a fruit wax at the application site, and fruit may pass through a heat tunnel to dry the wax–fungicide combination on the fruit surface. Fungicide mixtures may also be applied, especially in stone fruits, to broaden the spectrum of fungal species controlled (Eckert and Ogawa, 1988). Apple and pear cultivars susceptible to the physiological disorder known as superficial scald are often treated with a drench solution of antioxidant upon arrival at the packing house from the orchard and a fungicide may be included in the antioxidant solution (Spotts and Sanderson, 1994).

Spread of certain postharvest diseases from infected fruit to adjacent fruit in packed boxes can be inhibited by the use of copper-impregnated paper fruit wraps. Pears are often wrapped in paper to prevent surface marking during transit, and paper wraps may include oil and/or an antioxidant for prevention of superficial scald in addition to copper. Potentially spreading decays by *Botrytis cinerea* and *Mucor piriformis* may be contained by these wraps (Bertrand and Saulie-Carter, 1980; Cooley and Crenshaw, 1931).

9.5.4 Fungicide resistance

Continuous application of benzimidazole fungicides in orchards and packing houses over many years has resulted in the selection for benzimidazole-resistant strains of key postharvest pathogens, including *Penicillium expansum*, *Botrytis cinerea* and *Monilinia fructicola* (Eckert *et al.*, 1994; Michailides *et al.*, 1987; Prusky *et al.*, 1985; Rosenberger and Meyer, 1979; Smith, 1988). Selection pressure is exacerbated by their penetration through the fruit wax and cuticle, facilitating persistence of residues from pre- and postharvest treatments through the storage period. Resistance to benzimidazole fungicides is attributed to a spontaneously-occurring single gene mutation in the pathogen and strains that resist one benzimidazole are cross-resistant to other fungicides in this group (Prusky *et al.*, 1985).

The development of resistance to benzimidazole fungicides is often cited as a critical factor in stimulating research into alternative control measures, and in efforts to build multi-faceted and thereby more resilient postharvest disease management strategies (Sugar *et al.*, 1994; Wisniewski and Wilson, 1992). Resistance management has become an integral part of fungicide intro- duction and deployment by chemical manufacturers (Wade and Delp, 1990). Central to most resistance management strategies is the alternate or combined use of unrelated fungicides with distinct modes of action, so that resistant strains selected by one fungicide may be suppressed by others. Avoiding use of benzimidazole fungicides in the orchard for control of foliar diseases (e.g. scab or powdery mildew in pome fruits) as well as for suppression of postharvest pathogens resulted in less selection pressure for resistant strains and consequent slower appearance of decays attributable to resistance (Eckert and Ogawa, 1988). Fungicide treatment can thus cause a shift in the relative proportions of sensitive and resistant types in populations of a pathogen. Fungicides can also selectively suppress sensitive species and allow non-sensitive species to predominate at infection sites shared with sensitive species. This has been documented in the prevalence of *Alternaria* causing apple decay through fruit stems where benomyl selectively suppressed *Penicillium expansum* (Sitton and Pierson, 1983) and where *Phialophora malorum* predominated in fruit lesions where *Cladosporium herbarum* was suppressed by thiabendazole (Sugar *et al.*, 1986).

9.5.5 Biological control

Numerous species of bacteria and yeast suppress postharvest fungal decay (Wilson and Wisniewski, 1989). These biological control (biocontrol) agents can suppress postharvest disease development by occupying infection courts and competing with pathogens for nutrients, by secreting antifungal compounds (antibiosis), by direct parasitism of the pathogen or by inducing resistance responses in the host (Droby and Chalutz, 1994). Postharvest disease control is a particularly attractive application of biocontrol, since conditions in the storage environment are largely predictable and protected from fluctuating climatic con- ditions and potentially harmful radiation (Pusey, 1994). Biocontrol agents must, of course, be able to survive and function at storage temperatures appropriate for the crop, which may be at or below $0°C$. In addition, the response of biocontrol agents to fungicides and controlled atmospheres will determine which com- binations of postharvest practices are compatible. Combinations of different biocontrol agents may improve control over either agent alone, if the mode of action of the agents varies (Janisiewicz and Bors, 1995). Several biocontrol agents have been identified as tolerant to commonly used postharvest fungicides, such as thiabendazole (Chalutz and Wilson, 1990), and thus can be used in combination with them. However, specific combinations of biocontrol agent and chemical treatment may not be compatible. For example, *Candida sake* strain CPA-1 was tolerant of benomyl, sulfur, fluzilazol, ziram, thiabendazole and diphenylamine, but viability of the yeast was reduced by captan, imazalil and ethoxyquin (Usall *et al.*, 2001).

Generally, bacteria and yeast, in part because of their large surface-to-volume ratio, are able to take up nutrients from solution more rapidly and in greater volumes than do germinating filamentous fungi (Blakeman, 1985). Filonow (1998) demonstrated that biocontrol yeasts in apple wounds took up and utilized more ^{14}C-labelled sugars than germinating conidia of *Botrytis cinerea* and suggested that competition for sugars by yeasts played a role in the biocontrol. However, some yeast species that competed for sugars with equal efficiency differed in their biocontrol effectiveness, indicating that factors other than competition for sugars must also be involved.

Many bacteria, and some fungi, that have been evaluated for biocontrol potential have been found to produce antibiotic compounds (Pusey, 1991). The bacterium *Bacillus subtilis* strain B-3 produced an antifungal peptide in culture and both cultured bacteria and cell-free filtrates of the bacterial culture controlled brown rot of peach (Gueldner *et al.*, 1988; Pusey, 1991). Although some bacteria can produce antibiotic compounds in culture, they may not do so at fruit infection sites (Droby and Chalutz, 1994). Furthermore, the role of bacterial antibiosis in biocontrol has not been demonstrated unambiguously. The bacterium *Pseudomonas cepacia* produces a powerful antifungal compound, pyrrolnitrin (Janisiewicz and Roitman, 1988). However, while the bacterium controlled *Penicillium digitatum* infection of lemons, a pyrrolnitrin-resistant mutant of the fungus was equally controlled (Smilanick and Denis-Arrue, 1992). The use of antibiotic-producing biocontrol agents may not be acceptable for human consumption and could lead to development of resistant strains of pathogens.

Direct parasitism of pathogens by biocontrol agents has not been well established. However, cells of the yeast *Pichia guilliermondii* attached to mycelium of *Botrytis cinerea* and were associated with mycelial degradation (Wisniewski *et al.*, 1991). The yeast produces enzymes capable of degrading fungal cell walls. This ability may function in biocontrol in addition to colonization of the wound site and competition for nutrients (Droby and Chalutz, 1994).

Although applications of selected biocontrol agents can significantly reduce postharvest decay, the level of control is usually less than with fungicides. Accordingly, efforts have been made to enhance the performance of biocontrol agents by combining them with low dosages of fungicides, other materials that inhibit pathogens, suppressive storage atmospheres, or other practices that favor disease resistance in the fruit.

Materials that weaken the pathogen or strengthen the host may allow biocontrol agents to work more effectively. Glycolchitosan is a soluble derivative of chitin, a polysaccharide found in fungal cell walls and also in crustacean shells. It is believed that glycolchitosan derived from crustacean shells can induce resistance responses in fruit, possibly mimicking fungal invasion. Both glycolchitosan and the yeast *Candida saitoana* reduced postharvest disease in apples and citrus and the two combined were more effective than either alone (El-Ghaouth *et al.*, 2000a). The glucose analog 2-deoxy-D-glucose can inhibit fungal growth by being taken up in place of other sugars, though it is not metabolized by the fungus (El-Ghaouth *et al.*, 1995). When combined with *C. saitoana*, disease control was greater than with either agent alone (El-Ghaouth *et al.*, 2000b). Addition of calcium salts to yeast suspensions enhanced biocontrol of *Penicillium expansum* and *Botrytis cinerea* in apple (McLaughlin *et al.*, 1990). The amino acids L-proline and L-asparagine enhanced biocontrol of *Penicillium expansum* on apple (Janisiewicz *et al.*, 1992). Several studies have found biocontrol agents highly tolerant of a wide range of controlled atmospheres potentially used in fruit storage (Benbow and Sugar, 1997; Lurie *et al.*, 1995; Roberts, 1990; Roberts, 1994). Application of yeast and bacterial biocontrol agents to wounds following heat treatment at $38°C$ for 4 days added to control of *Penicillium expansum* in apples provided by the heat treatment (Leverentz *et al.*, 2000).

Biocontrol agents can also be applied in the orchard prior to harvest. This approach is particularly appealing in that it places the biocontrol agent on the fruit surface before the period of heightened risk of wounding that takes place during harvest and transport of fruit. Only those species of biocontrol agent that are able to survive on fruit surfaces in the preharvest period, with attendant heat and drying, would be suitable for this application (Benbow and Sugar, 1999). Teixidó *et al.* (1998) found that survival and disease control were improved by selecting yeast strains that were highly osmotolerant in culture. Trials in apple (Leibinger *et al.*, 1997; Teixidó *et al.*, 1998) and pear (Benbow and Sugar, 1999) have shown that orchard biocontrol treatments are less effective than postharvest treatments. High populations of biocontrol agents are less likely to enter a wound when applied before than after wounding.

9.5.6   Storage atmospheres

Long term storage of apples and pears often involves controlled atmosphere (CA) conditions, in which oxygen levels are drastically lowered from the ambient level of 20.9% and CO_2 is allowed to rise moderately over the ambient level of 0.03%. Typical CA storage parameters for apples and pears are 2–3% O_2 with CO_2 maintained at less than 1%. While CA storage has had a dramatic effect in preservation of fruit quality, the effect on postharvest diseases has been disappointing. Infection and lesion development by wound pathogens and development of lesions following quiescent orchard infections are somewhat delayed as compared with storage in ambient atmospheres. This results from slowing of fruit deterioration and direct effects on postharvest pathogens (Sommer, 1989). Opportunistic pathogens penetrating the surfaces of senescent fruit are delayed to a greater extent, in proportion to the delay in fruit senescence. Low (2%) O_2 slightly reduced the severity of disease on pears inoculated with the relatively aggressive pathogen *Penicillium expansum* in two of three years, as compared with storage in ambient atmospheres. Low O_2 was more effective in suppressing the weaker pathogen *Phialophora malorum* (Sugar *et al.*, 1994). Storage of d'Anjou pears in 1% O_2 for eight months resulted in reduced incidence of stem-end decay as compared with pears stored in ambient atmosphere (Chen *et al.*, 1981).

Elevated CO_2 atmospheres can be suppressive to pathogenic fungi (Littlefield *et al.*, 1966) but high CO_2 can cause internal injury to stored fruit. Precise thresholds for CO_2 injury are difficult to identify because numerous factors, including O_2 level, fruit maturity, duration of storage and environmental conditions during the growing season, can all affect susceptibility (Hansen and Mellenthin, 1962). This has generally led to a conservative approach by storage managers, avoiding more than slight CO_2 accumulation in the storage atmosphere, despite potential disease control benefits. Storage of apples at CO_2 concentrations above 2.8% in ambient O_2 for two months reduced lesion development by *Botrytis cinerea*, *Penicillium expansum* and *Pezicula malicorticis* (Sitton and Patterson, 1992). *Botrytis cinerea* was the most sensitive of these pathogens and decay was reduced to zero by 12% CO_2. Pear decay by *B. cinerea* was also reduced to zero by six weeks storage in 12% CO_2 with 5% O_2 (Benbow and Sugar, 1997) but storage of pears at 12% CO_2 for six weeks can also cause internal injury (Wang and Mellenthin, 1975). Pears stored for six months at 3% CO_2 with 1.5% O_2 also showed reduced incidence of fungal decay (Drake, 1998). Decay of apples by *Gloeosporium album* was reduced by storage in 5% CO_2 with 3% O_2, and decreased production of pectolytic enzymes by the pathogen was observed (Edney, 1964).

Controlled atmosphere storage uses nitrogen enrichment to reduce oxygen levels actively. Oxygen surrounding stored fruit may also be reduced and CO_2 increased passively by modified atmosphere packaging (MAP), in which fruit respiration is responsible for changes in gas concentrations in the package atmosphere and selective permeability of the packaging materials determines sustained gas levels. Increased levels of CO_2 in MAP are associated with reduced postharvest decay (Miller and Sugar, 1997). In addition, volatile compounds produced by fruit can be trapped in the atmosphere surrounding the fruit and may play a role in suppressing decay fungi (Mattheis and Fellman, 2000).

9.5.7 Novel treatments

Various chemical, physical and biological treatments have been shown to induce resistance responses in herbaceous plants (Sticher *et al.*, 1997) but few treatments have been validated for orchard crops or harvested fruit (El-Ghaouth, 1994). Low doses of ultraviolet light-C (254 nm, UV-C) reduced the incidence of brown rot of peach (*Monilinia fructicola*) and green mold (*Penicillium digitatum*) of tangerine (Stevens *et al.*, 1997). When the fruit were treated with *Debaryomyces hansenii* two to three days after UV-C treatment, the reduction of storage rots was greater than when UV-C was used alone, but was generally not as effective as fungicide treatment. Chitosan (from crab shells) and margosan-O (from neem seed) elicited the formation of disease defense compounds in the flavedo of oranges (Fajardo *et al.*, 1998). Treatment with the yeast biocontrol agent *Candida oleophila* induced resistance to *Penicillium digitatum* in citrus fruit, whether applied to wounds or on non-wounded surfaces (Droby *et al.*, 2000). Responses measured included: increased production of ethylene and phenylalanine ammonia lyase activity; accumulation of phytoalexins in grapefruit peel; and accumulation of the antifungal enzymes chitinase and β-1-3-glucanase in grapefruit flavedo. Germination and growth of *P. digitatum* and green mold disease incidence were suppressed within a 4 cm radius of citrus wounds treated with *C. oleophila* but not in the area around untreated wounds (Fajardo *et al.*, 1998). Induction of chitinase and β-1,3-glucanase was observed. Proteins inhibitory to polygalacturonase were also detected in orange flavedo, but they were present at constitutive levels rather than induced by treatments. Increased levels of chitinase, β-1,3-glucanase and peroxidase were observed in apple fruit treated with *Aureobasidium pullulans* and infections by *Botrytis cinerea* and *Penicillium expansum* were suppressed (Ippolito *et al.*, 2000). While wounding alone induced formation of these antifungal enzymes in the host, the levels following treatment of wounds with *A. pullulans* were much higher. Postharvest application of jasmonates, which can function as signal compounds in the induction of resistance responses, reduced green mold decay in citrus (Droby *et al.*, 1999).

Fumigation of fruit in storage with naturally occurring compounds can reduce populations of pathogens on fruit surfaces and potentially reduce decay. Vapors of acetic, formic and propionic acids controlled *Monilinia fructicola*, *Penicillium expansum* and *Rhizopus stolonifer* on cherry, *P. expansum* on pome fruit and *P. digitatum* on citrus fruit (Sholberg, 1998). While some treatments caused injury to the fruit, non-injurious concentrations inhibitory to the pathogens were identified. Gaseous ozone in the storage atmosphere can reduce surface contamination by suppressing pathogen germination and can attack aerial mycelium involved in fruit-to-fruit spread of infection (Spalding, 1968). Gaseous ozone does not appear to control fungi that have already penetrated the fruit surface and may be phytotoxic to the fruit, depending on the dosage and exposure time. Ozone can destroy ethylene gas in the storage atmosphere, which may indirectly aid control of senescence-related infections. An ozone atmosphere of 1 ppm prevented sporulation of *Penicillium digitatum* on citrus fruit for 15 days. However, once removed from ozone, sporulation resumed within 48 h (Harding, 1968).

9.5.8   Integrated management of postharvest disease

Since no single treatment or management technique can completely control all types of postharvest disease, it makes sense that integration of multiple techniques into postharvest disease management programs can maximize disease control. Furthermore, management tools such as fungicide treatment or biocontrol work best when pathogen inoculum pressure is low and when fruit resistance is high enough to slow the infection process. Methods appropriate for inclusion in an integrated program and their relative importance to the program will necessarily vary among fruit types and diseases and among production regions.

Combinations of biocontrol agents with various practices in order to enhance performance of the biocontrol agents, as discussed above, constitute a current focus in the development of integrated management programs. Biocontrol agents have been combined with reduced rates of fungicides (Chand-Goyal and Spotts, 1996), other antifungal compounds (Janisiewicz and Bors, 1995), calcium salts (Janisiewicz *et al.*, 1998), prestorage heat treatment (Leverentz *et al.*, 2000) and modified atmospheres (Benbow and Sugar, 1997). Further development of this concept has seen the combination of fungicide, biocontrol and modified atmosphere (Spotts *et al.*, 1998a) and the combination of nitrogen management, calcium, harvest maturity, biocontrol and fungicide treatments (Sugar *et al.*, 1994).

Control of postharvest diseases in the near future may include more effective fungicides and biocontrol agents and improved methods for eliciting resistance responses in fruit. Despite anticipated advancements, it is likely that successful management of postharvest diseases will depend upon a succession of practices and conditions that affect all aspects of disease dynamics in order to build effective programs additively.

References

Abu-Goukh, A.A. and Labavitch, J.M. (1983) The *in vivo* role of 'Bartlett' pear fruit polygalacturonase inhibitors. *Physiological Plant Pathology*, **23**, 123-135.

Abu-Goukh, A.A., Strand, L.L. and Labavitch, J.M. (1983) Development-related changes in decay susceptibility and polygalacturonase inhibitor content of 'Bartlett' pear fruit. *Physiological Plant Pathology*, **23**, 101-109.

Ardi, R., Kobiler, I., Keen, N.T. and Prusky, D. (1998) Involvement of epicatechin biosynthesis in the resistance of avocado fruits to postharvest decay. *Physiological and Molecular Plant Pathology*, **53**, 269-285.

Baker, K.F. and Heald, F.D. (1934) An investigation of factors affecting the incidence of lenticel infection of apples by *Penicillium expansum*. *Washington Agricultural Experiment Station Bulletin*, **298**, 1-48.

Ben-Arie, R. (1975) Benzimidazole penetration, distribution, and persistence in postharvest treated pears. *Phytopathology*, **65**, 1187-1189.

Benbow, J.M. and Sugar, D. (1997) High CO_2 CA storage combined with biocontrol agents to reduce postharvest decay of pear, in *Proceedings of the 1997 Controlled Atmosphere Research Conference (CA '97)*, Postharvest Horticulture Series, 16, University of California, Davis, Department of Pomology, pp. 270-276.

Benbow, J.M. and Sugar, D. (1999) Fruit surface colonization and biological control of postharvest diseases of pear by preharvest yeast applications. *Plant Disease*, **83**, 839-844.

Beno-Moualem, D. and Prusky, D. (2000) Early events during quiescent infection development by *Colletotrichum gloeosporioides* in unripe avocado fruits. *Phytopathology*, **90**, 553-559.

Bertrand, P.F. and Saulie-Carter, J.L. (1980) Mucor rot of pears and apples. *Oregon State Agricultural Experiment Station Special Report*, **568**, 21 pp.

Biggs, A.L., El-Kholi, M.M., El-Neshawy, S. and Nickerson, R. (1997) Effects of calcium salts on growth, polygalacturonase activity, and infection of peach fruit by *Monilinia fructicola*. *Plant Disease*, **81**, 399-403.

Blakeman, J.P. (1985) Ecological succession of leaf surface microorganisms in relation to biological control, in *Biological Control on the Phylloplane* (eds C. Windels and S.E. Lindow), APS Press, St Paul, Minnesota, pp. 6-30.

Blanpied, G.D. and Purnasiri, A. (1968) *Penicillium* and *Botrytis* rots of McIntosh apples handled in water. *Plant Disease Reporter*, **53**, 825-828.

Boonyakiat, D., Spotts, R.A. and Richardson, D.G. (1986) Effects of chlorogenic acid and arbutin on growth and spore germination of decay fungi. *HortScience*, **21**, 309-310.

Bostock, R.M. and Stermer, B.A. (1989) Perspectives on wound healing in resistance to pathogens. *Annual Review of Phytopathology*, **27**, 343-371.

Bramlage, W.J., Drake, M. and Lord, W.J. (1980) The influence of mineral nutrition on the quality and storage performance of pome fruits grown in North America, in *Mineral Nutrition of Fruit Trees* (eds D. Atkinson, J.E. Jackson, R.O. Sharples and W.M. Waller), Butterworths, London, pp. 29-39.

Brown, G.E. and Barmore, C.R. (1983) Resistance of healed citrus exocarp to penetration by *Penicillium digitatum*. *Phytopathology*, **73**, 691-694.

Brown, G.E. and Wilson, W.C. (1968) Mode of entry of *Diplodia natalensis* and *Phomopsis citri* into Florida oranges. *Phytopathology*, **58**, 736-740.

Byrde, R.J.W. (1956) The varietal resistance of fruits to brown rot: II. The nature of resistance in some varieties of cider apple. *Horticultural Science*, **32**, 227-238.

Chalutz, E. and Wilson, C.L. (1990) Biocontrol of green and blue mold and sour rot of citrus by *Debaryomyces hansenii*. *Plant Disease*, **74**, 134-137.

Chand-Goyal, T. and Spotts, R.A. (1996) Postharvest biological control of blue mold of apple and brown rot of sweet cherry by natural saprophytic yeasts alone or in combination with low doses of fungicides. *Biological Control*, **6**, 253-259.

Chen, P.M., Spotts, R.A. and Mellenthin, W.M. (1981) Stem-end decay and quality of low oxygen stored d'Anjou pears. *Journal of the American Society for Horticultural Science*, **106**, 695-698.

Clements, H.F. (1935) Morphology and physiology of the pome lenticels of *Pyrus malus*. *Botanical Gazette*, **97**, 101-117.

Cobrink, J.C., Kotze, J.M. and Visagie, T.S. (1985) Colonization of apples by fungi causing core rot. *Horticultural Science*, **2**, 9-13.

Collmer, A. and Keen, N.T. (1986) The role of pectic enzymes in plant pathogenesis. *Annual Review of Phytopathology*, **24**, 383-409.

Conway, W.S. (1984) Preharvest factors affecting postharvest losses from disease, in *Postharvest Pathology of Fruits and Vegetables: Postharvest Losses in Perishable Crops* (ed. H.E. Moline), University of California Division of Natural Resources Bulletin 1914 (Publication NE-87), Oakland, California, pp. 11-16.

Conway, W.S., Greene, G.M. and Hickey, K.D. (1987) Effects of preharvest and postharvest calcium treatments of peaches on decay caused by *Monilinia fructicola*. *Plant Disease*, **71**, 1084-1086.

Conway, W.S., Gross, K.C., Boyer, C.D. and Sams, C.E. (1988) Inhibition of *Penicillium expansum* polygalacturonase activity by increased apple cell wall calcium. *Phytopathology*, **78**, 1052-1055.

Conway, W.S., Sams, C.E. and Kelman, A. (1994) Enhancing the natural resistance of plant tissues to postharvest disease through calcium applications. *HortScience*, **29**, 751-754.

Cooley, J.S. and Crenshaw, J.H. (1931) *Control of Botrytis rot of pears with chemically treated wrappers*. US Department of Agriculture Circular 177, 9 pp.

Coyier, D.L. (1970) Control of storage decay in d'Anjou pear fruit by preharvest application of benomyl. *Plant Disease Reporter*, **54**, 647-650.

Crisosto, C.H., Johnson, R.S., DeJong, T. and Day, K.R. (1997) Orchard factors affecting postharvest stone fruit quality. *HortScience*, **32**, 820-823.

Dobson, R.L., Michailides, T.J., Cervantes, L.A. and Spotts, R.A. (1989) Population dynamics of *Mucor piriformis* in pear orchard soils as related to decaying pear fruit. *Phytopathology*, **79**, 657-660.

Drake, S.R. (1998) Quality of 'Bosc' pears as influenced by elevated carbon dioxide storage. *Journal of Food Quality*, **22**, 417-425.

Droby, S. and Chalutz, E. (1994) Mode of action of biocontrol agents for postharvest diseases, in *Biological Control of Postharvest Diseases of Fruits and Vegetables–Theory and Practice* (eds C.L. Wilson and M.E. Wisniewski), CRC Press, Boca Raton, Florida, pp. 63-75.

Droby, S., Porat, R., Cohen, L., Weiss, B., Shapiro, B., Philosoph-Hadas, S. and Meir, S. (1999) Suppressing green mold decay in grapefruit with postharvest jasmonate application. *Journal of the American Society for Horticultural Science*, **124**, 184-188.

Droby, S., Porat, R., Vinokur, V., Cohen, L., Weiss, B. and Daus, A. (2000) Induction of resistance to postharvest decay of citrus fruit by the yeast biocontrol agent *Candida oleophila*. *Phytopathology*, **90**, S20 (abstract).

Eckert, J.W. and Ogawa, J.M. (1985) The chemical control of postharvest diseases: subtropical and tropical fruits. *Annual Review of Phytopathology*, **23**, 421-454.

Eckert, J.W. and Ogawa, J.M. (1988) The chemical control of postharvest diseases: deciduous fruits, berries, vegetables, and root/tuber crops. *Annual Review of Phytopathology*, **26**, 433-469.

Eckert, J.W., Sievert, J.R. and Ratnayake, M. (1994) Reduction of imazalil effectiveness against citrus green mold in California packinghouses by resistant biotypes of *Penicillium digitatum*. *Plant Disease*, **78**, 791-794.

Edney, K.L. (1964) The effect of the composition of the storage atmosphere on the development of rotting of Cox's Orange Pippin apples and the production of pectolytic enzymes by *Gloeosporium* spp. *Annals of Applied Biology*, **54**, 327-334.

Edney, K.L. (1973) Fungal disorders, in *The Biology of Apple and Pear Storage* (eds J.C. Fidler, B.G. Wilkinson, K.L. Edney and R.O. Sharples), Research Review No. 3, Commonwealth Agricultural Bureaux, London, pp. 133-172.

Edney, K.L. (1983) Top fruit, in *Post-harvest Pathology of Fruits and Vegetables* (ed. C. Dennis), Academic Press, London, pp. 43-71.

El-Ghaouth, A. (1994) Manipulation of defense systems with elicitors to control postharvest diseases, in *Biological Control of Postharvest Diseases of Fruits and Vegetables–Theory and Practice* (eds C.L. Wilson and M.E. Wisniewski), CRC Press, Boca Raton, Florida, pp. 153-167.

El-Ghaouth, A., Wilson, C.L. and Wisniewski, M.E. (1995) Sugar analogs as potential fungicides for postharvest pathogens of apple. *Plant Disease*, **79**, 254-258.

El-Ghaouth, A., Smilanick, J.L., Brown, G.E., Ippolito, A., Wisniewski, M. and Wilson, C.L. (2000a) Application of *Candida saitoana* and glycolchitosan for the control of postharvest diseases of apple and citrus fruit under semi-commercial conditions. *Plant Disease*, **84**, 243-248.

El-Ghaouth, A., Smilanick, J.L., Wisniewski, M. and Wilson, C.L. (2000b) Improved control of apple and citrus fruit decay with a combination of *Candida saitoana* and 2-deoxy-D-glucose. *Plant Disease*, **84**, 249-253.

English, W.H., Ryall, A.L. and Smith, E. (1946) Blue mold decay of Delicious apples in relation to handling practices. *US Department of Agriculture Circular*, **751**, 20 pp.

Fajardo, J.E., McCollum, T.G., McDonald, R.E. and Mayer, R.T. (1998) Differential induction of proteins in orange flavedo by biologically based elicitors and challenged by *Penicillium digitatum* Sacc. *Biological Control*, **13**, 143-151.

Fallahi, E., Conway, W.S., Hickey, K.D. and Sams, C.E. (1997) The role of calcium and nitrogen in postharvest quality and disease resistance of apples. *HortScience*, **32**, 831-835.

Fawcett, C.H. and Spencer, D.M. (1968) *Sclerotinia fructigena* infection and chlorogenic acid content in relation to antifungal compounds in apple fruits. *Annals of Applied Biology*, **61**, 245-253.

Ferguson, I.B. (1984) Calcium in plant senescence and fruit ripening. *Plant Cell and Environment*, **7**, 477-489.

Filonow, A.B. (1998) Role of competition for sugars by yeasts in the biocontrol of gray mold of apple. *Biocontrol Science and Technology*, **8**, 243-256.

Gueldner, R.C., Reilly, C.C., Pusey, P.L., Costello, C.E., Arrendale, R.F., Himmelsbach, D.S., Crumley, F.G. and Culter, H.G. (1988) Isolation and identification of iturins as antifungal peptides in biological control of peach brown rot with *Bacillus subtilis*. *Journal of Agricultural and Food Chemistry*, **36**, 366-370.

Hansen, E. and Mellenthin, W.M. (1962) Factors influencing susceptibility of pears to carbon dioxide injury. *Proceedings of the American Society for Horticultural Science*, **80**, 146-153.

Hardenburg, R.E., Watada, A.E. and Wang, C.Y. (1986) *The Commercial Storage of Fruits, Vegetables, and Florist and Nursery Crops*. Agricultural Handbook 66, Agricultural Research Service, US Department of Agriculture, Washington DC.

Harding, P.R. (1968) Effect of ozone on *Penicillium* mold decay and sporulation. *Plant Disease Reporter*, **52**, 245-247.

Ippolito, A., El Ghaouth, A., Wilson, C.L. and Wisniewski, M. (2000) Control of postharvest decay of apple fruit by *Aureobasidium pullulans* and induction of defense responses. *Postharvest Biology and Technology*, **19**, 265-272.

Janisiewicz, W.J. and Bors, B. (1995) Development of a microbial community of bacterial and yeast antagonists to control wound-invading postharvest pathogens of fruits. *Applied and Environmental Microbiology*, **61**, 3261-3267.

Janisiewicz, W.J. and Roitman, J. (1988) Biological control of blue mold and gray mold on apple and pear with *Pseudomonas cepacia*. *Phytopathology*, **78**, 1697-1700.

Janisiewicz, W.J., Usall, J. and Bors, B. (1992) Nutritional enhancement of biocontrol of blue mold on apples. *Phytopathology*, **82**, 1364-1370.

Janisiewicz, W.J., Conway, W.S., Glen, D.M. and Sams, C.E. (1998) Integrating biological control and calcium treatment for controlling postharvest decay of apples. *HortScience*, **33**, 105-109.

Kidd, M.N. and Beaumont, B.A. (1925) An experimental study of the fungal invasion of apples in storage with particular reference to invasion through the lenticels. *Annals of Applied Biology*, **12**, 14-33.

Knee, M. (1977) Cell wall degradation in senescent fruit tissue in relation to pathogen attack, in *Cell Wall Biochemistry Related to Specificity in Host–Plant Pathogen Interactions* (eds B. Solheim and J. Raa), Columbia University Press, New York, pp. 259-262.

Knee, M. and Bartley, I.M. (1981) Composition and metabolism of cell wall polysaccharides in ripening fruits, in *Recent Advances in the Biochemistry of Fruits and Vegetables* (eds J. Friend and M.J.C. Rhodes), Academic Press, London, pp. 133-148.

Lakshiminarayana, S., Sommer, N.F., Polito, V. and Fortlage, R.J. (1987) Development of resistance to infection by *Botrytis cinerea* and *Penicillium expansum* in wounds of mature apple fruits. *Phytopathology*, **77**, 1674-1678.

Leibinger, W., Breuker, B., Hahn, M. and Mendgen, K. (1997) Control of postharvest pathogens and colonization of the apple surface by antagonistic microorganisms in the field. *Plant Disease*, **87**, 1103-1110.

Leverentz, B., Janisiewicz, W.J., Conway, W.S., Saftner, R.A., Fuchs, Y., Sams, C.E. and Camp, M.J. (2000) Combining yeasts or a bacterial biocontrol agent and heat treatment to reduce postharvest decay of 'Gala' apples. *Postharvest Biology and Technology*, **21**, 87-94.

Littlefield, N.A., Wankier, B.M., Salunkhe, D.K. and McGill, J.N. (1966) Fungistatic effects of controlled atmospheres. *Applied Microbiology*, **14**, 579-581.

Lurie, S., Droby, S., Chalupowicz, L. and Chalutz, E. (1995) Efficacy of *Candida oleophila* strain 182 in preventing *Penicillium expansum* infection of nectarine fruits. *Phytoparasitica*, **23**, 231-234.

Mattheis, J. and Fellman, J.K. (2000) Impacts of modified atmosphere packaging and controlled atmospheres on aroma, flavor, and quality of horticultural commodities. *HortTechnology*, **10**, 507-510.

McLaughlin, R.J., Wisniewski, M.E., Wilson, C.L. and Chalutz, E. (1990) Effect of inoculum concentration and salt solutions on biological control of postharvest diseases of apple with *Candida* spp. *Phytopathology*, **80**, 456-461.

Michailides, T.J. and Morgan, D.M. (1997) Influence of fruit-to-fruit contact on the susceptibility of French prune to infection by *Monilinia fructicola*. *Plant Disease*, **81**, 1416-1424.

Michailides, T.J., Ogawa, J.M. and Opgenorth, D.C. (1987) Shift of *Monilinia* spp. and distribution of isolates sensitive and resistant to benomyl in California prune and apricot orchards. *Plant Disease*, **71**, 893-896.

Miller, M.M. and Sugar, D. (1997) Modified atmosphere packaging and its applications in storage, decay control, and marketing of pears, in *Proceedings of the 1997 Controlled Atmosphere Research Conference (CA '97)*, Postharvest Horticulture Series, 16, University of California, Davis, Department of Pomology, pp. 277-284.

Ndubizu, T.O.C. (1976) Relation of phenolic inhibitors to resistance of immature apple fruit. *Journal of Horticultural Science*, **51**, 311-319.

Northover, J. and Cerkauskas, R.F. (1994) Detection and significance of symptomless latent infections of *Monilinia fructicola* in plums. *Canadian Journal of Plant Pathology*, **16**, 30-36.

Ogawa, J.M., Zehr, E.I. and Biggs, A.R. (1995) Brown rot, in *Compendium of Stone Fruit Diseases* (eds J.M. Ogawa, E.I. Zehr, G.W. Bird, D.F. Ritchie, K. Uriu and J.K. Uyemoto), APS Press, St Paul, Minnesota, pp. 7-10.

Pierson, C.F. (1966) Effect of temperature on growth of *Rhizopus stolonifer* on peaches and agar. *Phytopathology*, **56**, 276-278.

Prusky, D. (1996) Quiescent infections by postharvest pathogens. *Annual Review of Phytopathology*, **34**, 413-434.

Prusky, D. and Keen, N.T. (1993) Involvement of preformed antifungal compounds in the resistance of subtropical fruits to fungal decay. *Plant Disease*, **77**, 114-119.

Prusky, D., Bazak, M. and Ben-Arie, R. (1985) Development, persistence, survival, and strategies for control of thiabendazole-resistant strains of *Penicillium expansum* on pome fruits. *Phytopathology*, **75**, 877-882.

Pusey, P.L. (1991) Antibiosis as mode of action in postharvest biological control, in *Biological Control of Postharvest Diseases of Fruits and Vegetables, Workshop Proceedings* (eds C.L. Wilson and E. Chalutz), US Department of Agriculture, Agricultural Research Service ARS-92, pp. 127-141.

Pusey, P.L. (1994) Enhancement of biocontrol agents for postharvest diseases and their integration with other control strategies, in *Biological Control of Postharvest Diseases of Fruits and Vegetables – Theory and Practice* (eds C.L. Wilson and M.E. Wisniewski), CRC Press, Boca Raton, Florida, pp. 77-88.

Roberts, R.G. (1990) Postharvest biological control of gray mold of apple by *Cryptococcus laurentii*. *Phytopathology*, **80**, 526-530.

Roberts, R.G. (1994) Integrating biological control into postharvest disease management strategies. *HortScience*, **29**, 578-762.

Rosenberger, D.A. (1990) Gray mold, in *Compendium of Apple and Pear Diseases* (eds A.L. Jones and H.S. Aldwinkle), APS Press, St Paul, Minnesota, pp. 55-56.

Rosenberger, D.A. and Meyer, F.W. (1979) Benomyl-tolerant *Penicillium expansum* in apple packing houses in Eastern New York. *Plant Disease Reporter*, **63**, 37-40.

Sams, C.E. and Conway, W.S. (1984) Effect of calcium infiltration on ethylene production, respiration rate, soluble polyuronide content, and quality of 'Golden Delicious' apple fruit. *Journal of the American Society for Horticultural Science*, **109**, 53-57.

Sanchez, E.E., Khemira, H., Sugar, D. and Righetti, T.L. (1995) Nitrogen management in orchards, in *Nitrogen Fertilization in the Environment* (ed. P. Bacon), Marcel Dekker, New York, pp. 327-380.

Sharples, R.O. (1980) The influence of orchard nutrition on the storage quality of apples and pears grown in the United Kingdom, in *Mineral Nutrition of Fruit Trees* (eds D. Atkinson, J.E. Jackson, R.O. Sharples and W.M. Waller), Butterworths, London, pp. 17-27.

Sholberg, P.L. (1998) Fumigation of fruit with short-chain organic acids to reduce the potential of postharvest decay. *Plant Disease*, **82**, 689-693.

Sitton, J.W. and Patterson, M.E. (1992) Effect of high-carbon dioxide and low-oxygen controlled atmospheres on postharvest decays of apples. *Plant Disease*, **76**, 992-995.

Sitton, J.W. and Pierson, C.F. (1983) Interaction and control of Alternaria stem decay and blue mold of d'Anjou pears. *Plant Disease*, **67**, 904-907.

Smilanick, J.L. (1994) Strategies for the isolation and testing of biocontrol agents, in *Biological Control of Postharvest Diseases of Fruits and Vegetables – Theory and Practice* (eds C.L. Wilson and M.E. Wisniewski), CRC Press, Boca Raton, Florida, pp. 25-41.

Smilanick, J.L. and Denis-Arrue, R. (1992) Control of green mold of lemons with *Pseudomonas* species. *Plant Disease*, **76**, 481-485.

Smith, C.M. (1988) History of benzimidazole use and resistance, in *Fungicide Resistance in North America* (ed. C.E. Delp), APS Press, St Paul, Minnesota, pp. 23-24.

Sommer, N.F. (1982) Postharvest handling practices and postharvest diseases of fruit. *Plant Disease*, **66**, 357-364.

Sommer, N.F. (1989) Manipulating the postharvest environment to enhance or maintain resistance. *Phytopathology*, **79**, 1377-1380.

Sommer, N.F., Buchanan, J.R., Fortlage, R.J. and Bearden, B.E. (1985) Relation of floral infection to *Botrytis* blossom-end rot in storage. *Plant Disease*, **69**, 340-343.

Sommer, N.F., Fortlage, R.J. and Edwards, D.C. (1992) Postharvest diseases of selected commodities, in *Postharvest Technology of Horticultural Crops* (ed. A.A. Kader), University of California Division of Natural Resources Publication 3311, Oakland, California, pp. 117-160.

Spalding, D.H. (1968) *Effects of Ozone Atmospheres on Spoilage of Fruits and Vegetables after Harvest*. US Department of Agriculture, Agricultural Research Service Marketing Research Report 801, 10 pp.

Spotts, R.A. (1985) Effect of preharvest pear fruit maturity on decay resistance. *Plant Disease*, **69**, 388-390.

Spotts, R.A. (1990a) Bull's-eye rot, in *Compendium of Apple and Pear Diseases* (eds A.L. Jones and H.S. Aldwinkle), APS Press, St Paul, Minnesota, p. 56.

Spotts, R.A. (1990b) Moldy core and core rot, in *Compendium of Apple and Pear Diseases* (eds A.L. Jones and H.S. Aldwinkle), APS Press, St Paul, Minnesota, pp. 29-30.

Spotts, R.A. (1990c) Mucor rot, in *Compendium of Apple and Pear Diseases* (eds A.L. Jones and H.S. Aldwinkle), APS Press, St Paul, Minnesota, pp. 57-58.

Spotts, R.A. and Cervantes, L.A. (1986) Populations, pathogenicity, and benomyl resistance of *Botrytis* spp., *Penicillium* spp., and *Mucor piriformis* in packinghouses. *Plant Disease*, **70**, 106-108.

Spotts, R.A. and Peters, B.B. (1981) The effect of relative humidity on germination of pear decay fungi and d'Anjou pear decay. *Acta Horticulturae*, **124**, 75-78.

Spotts, R.A. and Sanderson, P.G. (1994) The postharvest environment and biological control, in *Biological Control of Postharvest Diseases of Fruits and Vegetables – Theory and Practice* (eds C.L. Wilson and M.E. Wisniewski), CRC Press, Boca Raton, Florida, pp. 43-56.

Spotts, R.A., Holmes, R.J. and Washington, W.S. (1988) Factors affecting wet core rot of apples. *Australasian Plant Pathology*, **17**, 53-57.

Spotts, R.A., Cervantes, L.A., Facteau, T.J. and Chand-Goyal, T. (1998a) Control of brown rot and blue mold of sweet cherry with preharvest iprodione, postharvest *Cryptococcus infirmo-miniatus*, and modified atmosphere packaging. *Plant Disease*, **82**, 1158-1160.

Spotts, R.A., Sanderson, P.G., Lennox, C.L., Sugar, D. and Cervantes, L.A. (1998b) Wound healing and staining of mature d'Anjou pear fruit. *Postharvest Biology and Technology*, **13**, 27-36.

Spotts, R.A., Cervantes, L.A. and Mielke, E.A. (1999) Variability in postharvest decay among apple cultivars. *Plant Disease*, **83**, 1051-1054.

Stelzig, D.A. (1984) Physiology and pathology of fruits and vegetables, in *Postharvest Pathology of Fruits and Vegetables: Postharvest Losses in Perishable Crops* (ed. H.E. Moline), University of California Division of Natural Resources Bulletin 1914 (Publication NE-87), Oakland, California, pp. 36-41.

Stevens, C., Khan, V.A., Lu, J.Y., Wilson, C.L., Pusey, P.L., Igwegbe, E.C.K., Kabwe, K., Mafolo, Y., Liu, J., Chalutz, E. and Droby, S. (1997) Integration of ultraviolet (UV-C) light with yeast treatment for control of postharvest storage rots of fruits and vegetables. *Biological Control*, **10**, 98-103.

Sticher, L., Mauch-Mani, B. and Métraux, B. (1997) Systemic acquired resistance. *Annual Review of Phytopathology*, **35**, 235-270.

Stow, J. (1989) The involvement of calcium ions in maintenance of apple fruit tissue structure. *Journal of Experimental Botany*, **40**, 1053-1057.

Sugar, D. and Spotts, R.A. (1993a) Dispersal of inoculum of *Phialophora malorum* in pear orchards and inoculum redistribution in pear immersion tanks. *Plant Disease*, **77**, 47-49.

Sugar, D. and Spotts, R.A. (1993b) The importance of wounds in infection of pear fruit by *Phialophora malorum* and the role of hydrostatic pressure in spore penetration of wounds. *Phytopathology*, **83**, 1083-1086.

Sugar, D., Powers, K.A. and Basile, S.A. (1986) Interactions among fungi causing side rot of pear. *Plant Disease*, **70**, 1132-1134.

Sugar, D., Powers, K.A. and Hilton, R.J. (1991) Enhanced resistance to side rot in pears treated with calcium chloride during the growing season. *Plant Disease*, **75**, 212-214.

Sugar, D., Righetti, T.L., Sanchez, E.E. and Khemira, H. (1992) Management of nitrogen and calcium in pear trees for enhancement of fruit resistance to postharvest decay. *HortTechnology*, **2**, 382-387.

Sugar, D., Roberts, R.G., Hilton, R.J., Righetti, T.L. and Sanchez, E.E. (1994) Integration of cultural methods with yeast treatment for control of postharvest fruit decay in pear. *Plant Disease*, **78**, 791-795.

Swinburne, T.R. (1983) Quiescent infections in post-harvest diseases, in *Post-Harvest Pathology of Fruits and Vegetables* (ed. C. Dennis), Academic Press, New York, pp. 1-21.

Teixidó, N., Viñas, I., Usall, J. and Magan, N. (1998) Control of blue mold of apples by preharvest application of *Candida sake* grown in media with different water activity. *Phytopathology*, **88**, 960-964.

Tufts, W.P. and Hansen, C.J. (1931) Variation in shape of Bartlett pears. *Proceedings of the American Society for Horticultural Science*, **28**, 627-633.

Tukey, B. (1993) Overview of ozone use at Snokist growers. *Tree Fruit Postharvest Journal*, **4**, 14-15.

Usall, J., Teixidó, N., Torres, R., Ochoa de Eribe, X. and Viñas, I. (2001) Pilot tests of *Candida sake* (CPA-1) applications to control postharvest blue mold on apple fruit. *Postharvest Biology and Technology*, **21**, 147-156.

Verhoeff, K. (1974) Latent infections by fungi. *Annual Review of Phytopathology*, **12**, 99-110.

Wade, M. and Delp, C.J. (1990) The fungicide resistance action committee: an update on goals, strategies, and North American initiatives. *American Chemical Society Symposium Series*, **421**, 320-333.

Wang, C.Y. and Mellenthin, W.M. (1975) Effect of short-term high CO_2 treatments on storage of d'Anjou pears. *Journal of the American Society for Horticultural Science*, **100**, 492-495.

Wang, G.Y., Michailides, T.J., Hammock, B.D., Lee, Y.M. and Bostock, R.M. (2000) Affinity purification and characterization of a cutinase from the fungal plant pathogen *Monilinia fructicola* (Wint.) Honey. *Archives of Biochemistry and Biophysics*, **382**, 31-38.

Wilson, C.L. and Wisniewski, M.E. (1989) Biological control of postharvest diseases of fruits and vegetables: an emerging technology. *Annual Review of Phytopathology*, **27**, 425-441.

Wisniewski, M.E. and Wilson, C.L. (1992) Biological control of postharvest diseases of fruits and vegetables: recent advances. *HortScience*, **27**, 94-98.

Wisniewski, M.E., Biles, C., Droby, S., McLaughlin, R., Wilson, C.L. and Chalutz, E. (1991) Mode of action of the postharvest biocontrol yeast, *Pichia guilliermondii*. I. Characterization of attachment to *Botrytis cinerea*. *Physiological and Molecular Plant Pathology*, **39**, 245-258.

Yakoby, N., Zhou, R., Kobiler, I., Dinoor, A. and Prusky, D. (2001) Development of *Colletotrichum gloeosporioides* restriction enzymes-mediated integration mutants as biocontrol agents against anthracnose disease in avocado fruits. *Phytopathology*, **91**, 143-148.

10 Genetic control of fruit ripening

Graham B. Seymour and Kenneth Manning

10.1 Introduction

Prior to the early 1980s, most work on the ripening of fleshy fruits focused on the physiology and biochemistry of postharvest changes. The emphasis was on the climacteric rise in respiration and the role of ethylene. Ripening was known to involve the action of numerous enzymes but whether they were synthesized *de novo* at the onset of the process, or activated when required, was still a matter of debate. Furthermore, there was little information on the extent to which these events were regulated; ripening was still considered to be a degradative process with random mixing of enzymes and substrates. Since the advent of molecular biology, it has become clear that ripening is a highly complex, tightly controlled developmental phase. Its complexity is now being revealed with DNA microarray technology and genomics approaches. It is apparent that to make significant steps to enhance fruit quality we will need to identify not only the principal biochemical events leading to changes in fruit colour, texture and flavour, but also the molecular mechanisms whereby these changes are initiated and regulated. The aim of this chapter is to summarize recent developments in our understanding of the biochemistry and molecular biology of ripening and to place these in the context of future prospects for the modification of these traits by breeding or genetic manipulation.

10.2 Evolution and development of fleshy fruits

Seed dispersal is the principal function of a fruit and the angiosperms have developed an extraordinary array of dispersal mechanisms. Fruits are essentially classed as dehiscent or indehiscent, depending on whether they shed their seeds by forming a dehiscence zone or whether they rely on being eaten or the fruit body decomposing for their release. Although, the evolution of fleshy fruits is not well understood, it is likely that the most primitive fruits consisted of a single carpel that dried when it was ripe and opened along one side. In concert with flower evolution, more complex structures would have been formed with the fusion of carpels to yield many-chambered fruits. Some of these became fleshy and attractive to animals which enhanced the dispersal of their seeds. The fossil record indicates that some of these events probably occurred in the Cretaceous period (Eriksson *et al.*, 2000).

The fruits of the model plant *Arabidopsis* are dry siliques composed of two fused carpels and seed dispersal involves the formation of a dehiscence zone allowing the two valves to detach from the replum (Ferrándiz *et al.*, 1999). By contrast, tomato and grape are indehiscent and are classified as berries. Here the pericarp is entirely fleshy (Gillaspy *et al.*, 1993). Alternatively seeds can be permanently enclosed within a fruiting structure called the achene. Some of these achenes become dry fruits, including the nuts, while others form grains. The strawberry is a fleshy fruit with achene-like structures; the fleshy parts derive from the receptacle. Other fruits such as cherry and peach are intermediate between nuts and berries in that the inner pericarp becomes hard and woody while the outer pericarp is fleshy (Coombe, 1976; Thomas, 1981). Despite the many different forms of fleshy tissues their mode of ripening often shares many features in common including changes in colour, texture and flavour. The biochemistry of these events is highly complex and involves the concerted expression of numerous genes. In tomato this can be illustrated by a visit to the Expressed Sequence Tag (EST) database at the TIGR Tomato Gene Index Website (http://www.tigr.org/tdb/lgi/). Most fruits probably share common biochemical mechanisms to generate similar physical changes in their characteristics during ripening and this is shown by a comparison of genes likely to be involved in fruit softening (see below). However, individual species clearly make use of different subsets of these genes. How these genes are regulated is still not known but it is possible that the genes controlling the ripening of fleshy fruits may have been conserved during evolution and ripening-related sets of metabolic genes may be regulated in a similar way in different species. In the next sections the sheer complexity of the ripening process is illustrated and the possibility that that similar regulatory events underlie these changes is discussed along with the implications of this information for effective manipulation of crop quality.

10.3   Texture

This is probably the key quality discriminator for fresh fruits, or at least those that show any appreciable softening. If the process is too slow the product is unpalatable, while rapid softening results in a short shelf-life and susceptibility to disease. Texture is a quantitative trait involving numerous loci and therefore gene products. This is illustrated from genetic mapping of fruit firmness in tomato (Fulton *et al.*, 2000). At the biochemical level, texture changes are known to be linked to cell wall disassembly and possibly alterations in turgor pressure. Intensive research in the past 15 years has revealed a plethora of cell-wall degrading activities and wall modifications that are associated with ripening (Chapter 3). The most important advances have resulted from the use of transgenic plants to test hypotheses on the role of individual cell wall enzymes (see review by Brummell and Harpster, 2001). The results of these experiments are consistent with the hypothesis that texture changes involve a multitude of cell wall events and illustrate the complexity of the ripening process.

To provide a framework for discussing cell wall modifications we must first consider the structure of the cell wall. The primary wall of fleshy fruits (like other plant organs) is generally composed of a network of polysaccharides and proteins. In essence, cellulose microfibrils composed of β-glucan chains are interlocked with xyloglucan molecules composed of a glucan backbone with xylosyl side chains. This network is embedded in a second matrix of galacturonic acid containing polymers known as pectins. Both the primary wall and the middle lamella region of the cell wall appear to contain these polymers, although the middle lamella is rich in pectin, which is thought to act as a principal component in the glue between cells (Carpita and Gibeaut, 1993). The cell wall proteins can be of a structural nature, interlocking and possibly attached to the polysaccharide components, or possess hydrolytic activities or non-enzymic wall loosening functions, e.g. expansins. Polygalacturonase (PG) is perhaps the best studied cell wall hydrolase. This enzyme is active against unesterified galacturonic acid polymers and high levels of PG activity are present in ripening tomatoes. The increase in the activity of the enzyme shows an excellent correlation with changes in fruit texture. It was postulated that PG was likely to be a major factor involved in pectin disassembly (see Hobson and Grierson, 1993), leading to cell separation and fruit softening. The advent of molecular biology provided a way to test this hypothesis. A PG protein was purified from ripening tomato fruit and an N-terminal sequence was obtained from this protein sample. The nucleotide sequence derived from this information matched exactly that of a cDNA clone isolated from a differential screen of a cDNA library from ripening tomato fruits (Slater *et al.*, 1985; Grierson *et al.*, 1986). This sequence was then used to suppress the expression of the endogenous PG gene (Smith *et al.*, 1988; Sheehy *et al.*, 1988). Although a dramatic reduction in PG mRNA was observed, there was no marked effect on fruit softening. In a complementary study, up-regulation of a PG transgene in *rin*, a non-softening tomato mutant lacking PG mRNA, also failed to induce any change in fruit texture (Giovannoni *et al.*, 1989).

Prior to these molecular experiments, analysis of tomato cell wall preparations had shown that pectin degradation during ripening was highly restricted *in vivo*. However, pectins showed a reduction of about 50% in their weight-average molecular weight during ripening and this was likely to be due to PG activity (Seymour *et al.*, 1987a,b). On analysis of the pectins from the low PG transgenic plants it was clear that depolymerization of these molecules had been inhibited but solubilization appeared to be unaffected (Smith *et al.*, 1990). More recent studies suggest that PG acts to depolymerize covalently bound pectin and produce a water soluble fraction (see Brummell and Harpster, 2001). These data indicated that while PG is involved in pectin degradation and probably some aspects of pectin solublilzation, it is likely to be only one of the many components that contribute to the softening process. The complex nature of the cell wall and quantitative trait loci (QTL) analysis (Fulton *et al.*, 2000) indicate that numerous events including the presence of a variety of hydrolytic activities are likely to be required for normal softening. These hydrolytic activities are highly regulated both at the level of transcription and in the wall itself. Indeed, not all the factors involved are enzymic. The first convincing evidence for the role of a wall modifying protein in fruit softening has come from experiments on expansins (Brummell *et al.*, 1999). These proteins are thought to act by disrupting the hydrogen bonds between cellulose microfibrils and their interlocking xyloglucans result in cell wall loosening. Recently, six expansin genes have been discovered in tomato fruits. One of these *LeExp1* is expressed to high levels at the onset and during ripening. Expansin proteins which are closely related to *LeExp1* have been found in ripening fruits from a number of species, including strawberry, suggesting that this gene product plays a significant role in ripening. This hypothesis was tested in tomato using transgenic plants with suppression or overexpression of *LeExp1* mRNA. In those lines with reduced *LeExp1* expression, the fruit were firmer, while enhanced *LeExp1* expression resulted in softer fruits. A link between *LeExp1* expression and texture is likely to reflect a direct effect on some aspect of wall loosening caused directly by *LeExp1* and probably indirectly through modification of wall hydrolase activity (Brummell and Harpster, 2001).

The enzyme β-galactosidase also appears to play a direct role in fruit softening. One of the most apparent changes in the fine structure of the cells of ripening fruits is often a loss of pectic neutral sugars and particularly galactosyl residues. This change was first explored in detail by Ken Gross at USDA/ARS (Beltsville, MD, USA) and subsequently his laboratory and others have investigated the loss of wall galactosyl residues and their role in fruit softening (Carey *et al.*, 1995; Smith *et al.*, 1998; Smith and Gross, 2000). The galactosyl residues are likely to occur as side-chains on pectins and the enzyme responsible for their removal is β-galactosidase. A ripening-related β-galactosidase has been purified from tomato fruit and a cDNA representing this isoform has been cloned. Very recently, Ken Gross, David Smith and colleagues at USDA have shown that suppression of the mRNA for this gene early in ripening in transgenic fruits resulted in reduced galactan degradation and a 40% increase in fruit firmness (Ken Gross and David Smith, personal communication). The role of galactosyl residues in fruit texture requires further study, but they appear to be present in tissues where strength and support are required (Jones *et al.*, 1997; McCartney *et al.*, 2000). These molecules may act directly to increase wall strength or perhaps indirectly to control the pore size in the pectin matrix, thereby inhibiting the movement of hydrolytic enzymes.

Other wall hydrolases that have received attention include endo-β-1,4-glucanases (EGases), xyloglucan endotransglycosylase (XET) and, very recently, pectate lyase. EGases are expressed in a wide range of fruits, the best example being the very high levels found in avocado. They are not active against crystalline cellulose and their endogenous substrates are not well characterized, but probably include xyloglucans, and integral and peripheral regions of non-crystalline cellulose (Brummell and Harpster, 2001). Transgenic experiments in both tomato (see Brummell and Harpster, 2001) and strawberry (Woolley *et al.*, 2001) have failed to show a link between EGase gene expression and softening. However, in both cases other EGase enzymes were likely to be have been active in the transgenic fruits and thus suppression of multiple EGases may be necessary to observe phenotypic changes (Brummell and Harpster, 2001). Xyloglucan endotransglycosylase catalyses the reversible cleavage of xyloglucan molecules and has been considered as a candidate for wall loosening during fruit ripening. Xyloglucan endotransglycosylase activity has been shown to increase during ripening in tomato but reduced accumulation of a ripening-related XET mRNA failed to affect fruit softening. Again the large number of XET genes in plants may mean that suppression of multiple XET genes is required before an effect on fruit texture is apparent (see Brummell and Harpster, 2001, and references therein). Pectate lyase genes have received much less attention in fruits and only recently have these sequences been found in fruit tissues, the first reports having come from studies on banana and strawberry (Domínguez-Puigjaner *et al.*, 1997; Medina-Suarez *et al.*, 1997; Medina Escobar *et al.*, 1997). The role of PEL in fruit softening remains unclear.

Until now, work has focused on the most obvious changes such as those in the polysaccharides components that make up the majority of the cell wall. Other wall components that may be modified during ripening include the wall structural proteins. Evidence that a protein component may be involved in modulating texture in apples has recently come to light. Cox's Orange Pippin apples when grown at different sites in the UK show differences in their texture from store. Using electron energy loss spectroscopy (EELS) to image elements in the apple cell walls, it has proved possible to demonstrate a correlation between the presence of high molecular weight nitrogen components (possibly cell-wall structural proteins) and retention of texture during storage (Huxham *et al.*, 1999). Indeed spraying developing apples with nitrate-containing nutri-ent formulations had the effect of producing fruit with enhanced postharvest texture characteristics (D. Johnson, personal communication). The identity of this nitrogen component remains unknown.

This brief résumé of the cell wall changes demonstrates that these events are tightly controlled and involve the coordinated expression of a range of different genes. Enhancing fruit texture may, therefore, rely on manipulating the regula-tory genes controlling this process but the identity of these genes is not known at present. Recent work in *Arabidopsis* may, however, provide clues as to the nature of the control system. Transcription factors that may be directly or indirectly involved in the regulation of cell wall disassembly have been described in

Arabidopsis siliques. A set of genes containing the conserved regulatory domain known as the 'MADS-box' promote or inhibit the formation of the abscission zone in these dryfruits. The *SHATTERPROOF* genes promote dehiscence zone differentiation at the valve replum boundary and *FRUITFULL* is a negative regulator of *SHATTERPROOF* expression. *SHATTERPROOF* genes promote lignification of cells adjacent to the dehiscence zone, while the presence of *FRUITFULL* expression results in the loss of lignified cells adjacent to where the dehiscence zone normally forms (Liljegren *et al.*, 2000; Ferrándiz *et al.*, 2000). Whether similar genes are active in regulating cell wall properties in fleshy fruits has not been reported.

10.4   Colour

The pigmentation of ripe fruits not only acts as an inducement for consumers, but also the compounds responsible for these colours such as anthocyanins and certain carotenoids are thought to have substantial health benefits including protection from cancers (Giovannucci *et al.*, 1995); deficiencies may cause blindness and premature death (Mayne, 1996). Indeed, carotenoids, which are essential for human health, cannot be synthesized *de novo* in the body and must be acquired from the diet (Parker, 1996). Tomato fruit and its products (ketchup, juices, soups and sauces) are one of the principal sources of carotenoids in the Western diet. Alterations in fruit colour are normally reliant on the up-regulation of enzymes in specific biochemical pathways, particularly carotenoid or anthocyanin biosynthesis. Therefore, enhancing fruit colour might be possible simply by altering the expression of one or a few key genes in the relevant biosynthetic pathway. This approach has been investigated using transgenic tomato fruits. Fruit of the cultivated tomato develop a red colour as the result of a marked increase in carotenoid biosynthesis in the flesh during ripening, including the synthesis of high levels of lycopene. This process occurs in the plastids, which undergo conversion from chloroplasts to chromoplasts.

The biosynthesis of the major carotenoid pigments in tomato follows the scheme in figure 10.1. Genes encoding enzymes for steps in this pathway have been cloned and their function confirmed by generating transgenic plants where the gene was silenced, e.g. phytoene synthase-1 (*Psy1*) (Fray and Grierson, 1993) or by using molecular genetic approaches to isolate genes by map-based cloning (Ronen *et al.*, 2000). Analysis of the transcriptional events during carotenoid synthesis demonstrates that *Psy*, and *Pds* that encodes phytoene desaturase, are up-regulated while mRNA of *CrtL-b* and *CrtL-e*, which encode lycopene β-cyclase and lycopene ε-cyclase, respectively, are down-regulated. The result is a 500-fold accumulation of lycopene. Experiments to increase carotenoid levels in fruits have included the up-regulation of endogenous genes, e.g. *Psy1* (Fray *et al.*, 1995) and the use of a bacterial

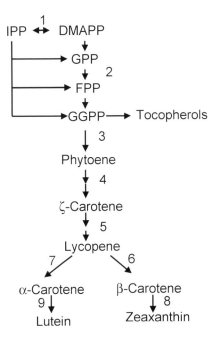

Figure 10.1 Summary of the biosynthetic pathway for carotenoids. Numbers indicate enzyme responsible for the conversion. 1 = isopentenyl diphosphate isomerase; 2 = geranylgeranyl diphosphate synthase; 3 = phytoene synthase; 4 = phytoene desaturase; 5 = ζ-carotene desaturase; 6 = β-cyclase; 7 = β- and ε-cyclase; 8 = β-hydroxylase; and 9 = β- and ε-hydroxylase.

gene to modulate the existing carotenoid biosynthetic pathway (Römer *et al.*, 2000). In the case of *Psy1*, expression of a full length transgene resulted in silencing of the endogenous gene in some cases and, where the gene was over-expressed at a high level, dwarfism resulted due to redirection of geranyl-geranyl pyrphosphate (GGPP) from the gibberellin pathway (Fray and Grierson, 1993; Fray *et al.*, 1995). The expression of *Erwinia uredovora* phytoene desaturase (*Crt1*) in transgenic tomato fruits (Römer *et al.*, 2000) resulted in a threefold elevation of the level of β-carotene, but the total carotenoid content was not enhanced. The increase in β-carotene was at the expense of lycopene and a reduction in the total level of carotenoids. These data suggest feedback inhibition within the pathway (Römer *et al.*, 2000) and indicate that an overall increase in carotenoid biosynthesis may require modulation of the components that regulate the pathway.

Alterations in the 'normal' pattern of carotenoid accumulation are apparent in several naturally occurring tomato mutants and the study of the genetic basis of these mutations should provide an insight into the regulation of carotenoid biosynthesis. In the recessive mutant *high-pigment* (*hp*) the same pigments are present, but the level of carotenoids is twice that in normal fruits because plastid number is increased (Yen *et al.*, 1997; Bramley, 1997). The *hp* gene has been mapped to chromosome 2, but it is not allelic with any known carotenoid biosynthetic genes and its identity remains unknown (Yen *et al.*, 1997). The *hp-2* mutant also has deeply pigmented fruits when compared with the wild-type, but it is non-allelic with *hp*. The *hp-2* gene has recently been cloned and has been identified as a homologue of a gene which causes a de-etiolated phenotype of the *det1* mutant in *Arabidopsis*. The function of this nuclear located protein is, however, unclear (Mustilli *et al.*, 1999).

The second major class of fruit pigments are the anthocyanins. These are responsible for the bright and vivid red/orange colours of a range of fruits including citrus, grape, apple, pear, plum, cherry, apricot, peach and soft fruits such as strawberry, blackcurrants, raspberry and gooseberry. Anthocyanins are present in fruits of several species of the Rosaceae family. The anthocyanins are only one of several categories of plant flavonoids that comprise more than 5000 compounds. The flavonoids are structurally complex. The basic C_6–C_3–C_6 flavone skeleton, in which three carbon atoms bridge the phenyl rings, can be substituted with side groups and further derivatized by O-glycosides that additionally can be acylated (see review by Robards and Antolovitch, 1997). The coloured pigments are anthocyanidin aglycones which are formed by the hydrolysis of flavonoid glycosides in acidic conditions. Fruit colour not only depends upon the chemistry of individual pigments and their concentration but also upon the pH environment and the presence of other (phenolic) substances that may subtly alter the hue, a phenomenon known as co-pigmentation (Yoshitama *et al.*, 1992). Anthocyanins have limited commercial value as food colourants because of their instability. However, there may be opportunities to engineer anthocyanins with improved stability by increasing their degree of acylation (Hong and Wrolstad, 1990). Flavonoids have been studied in relation to a number of important functions in plants including disease resistance and ultraviolet ray protection. These compounds are receiving increasing attention on account of their antioxidant (Rice-Evans *et al.*, 1997), anti-ulcer (Martin *et al.*, 1993) and potential anti-carcinogenic (Hertog *et al.*, 1992) medicinal properties and are seen as beneficial in the diet.

The biochemical pathway of anthocyanin biosynthesis is well established and much is known about its genetic regulation (Holton and Cornish, 1995). Most of this information has come from the investigation of flower colour, but recent studies of ripening genes indicate that a similar pathway operates in fruits. In grape berries, the enzymes chalcone synthase (CHS), chalcone isomerase (CHI), flavanone-3-hydroxylase (F3H), dihydroflavonol-4-reductase (DFR), leucoanthocyanidin dioxygenase (LDOX) and UDP glucose–flavonoid 3-o-glucosyl transferase (UFGT) are co-ordinately up-regulated at the onset of anthocyanin synthesis (Boss *et al.*, 1996). A gene encoding DFR is differentially expressed in ripening strawberry fruit (Moyano *et al.*, 1998) as are several of the other genes in this pathway, including a family of F3H genes (Manning, 1998).

The formation of the principal anthocyanins cyanidin- and pelargonidin-3-glucoside in strawberry fruit may require different F3H isoforms to hydroxylate the B-ring. In common with anthocyanin biosynthesis in flowers, the pathway in fruits is likely to involve regulatory genes that coordinate the expression of a set of biosynthetic genes. Although the whole pathway is up-regulated, it may be that it is the last step, in which anthocyanins are stabilized by glycosylation via UFGT, which determines colour intensity.

In common with flavour (discussed below), colour is a subtle quality attribute that also conveys freshness to the consumer. For strawberries, there is considerable genetic variation in fruit colour, which ranges from white (a complete absence of pigment) to very dark red. The optimal colour of fresh strawberries, however, is intermediate in this range with a bright orange/red colour being preferred. Colour intensity is therefore less important than colour quality. Genetic modification may be a good way of introducing novel colours, as has been done with flowers (Davies *et al.*, 1998) but the desirability of this approach will depend on consumer attitudes to such fruits. Furthermore, modifying the anthocyanin pathway may have important consequences for flavour as many flavonoids contribute to the organoleptic properties of fruits. Indeed, these phenolic compounds possess astringent properties (Ozawa *et al.*, 1987) that make fruits unpalatable when unripe. It may be no coincidence that the disappearance of astringency during ripening is accompanied by the *de novo* synthesis of anthocyanins.

10.5   Flavour

The importance of fruits as a food commodity owes much to their flavour characteristics. Fruits produce a vast array of chemicals that are detected by the oral and nasal senses and these give each fruit a sensory fingerprint that is unique. The human perception of flavour is, however, a complex area involving a myriad of factors and issues pertinent to this topic have been discussed elsewhere (Kays and Wang, 2000; Baldwin *et al.*, 2000).

Flavour is generally accepted to be a combination of aroma and taste sensations. The principal compounds in fruits detected by taste are organic acids and sugars that produce sour and sweet sensations, respectively. Fruits differ qualitatively and quantitatively in the composition of individual sugars and acids, but the overall palatability of the fruit will depend on the balance between these. Certain compounds may contribute to other aspects of taste, producing hot or bitter flavours. It is aroma, however, that really distinguishes different fruits and gives rise to the subtleties of flavour that enable the stage of ripeness, the freshness and even the cultivar to be discriminated. This is possible because of the extreme sensitivity of the olfactory organ and the vast array of volatile compounds produced in fruits. In strawberry, for example, as many as 360 volatile compounds have been detected (Latrasse, 1991).

In any fruit there will only be a few key character impact compounds. In strawberries these are the esters methyl butanoate, ethyl butanoate, methyl hexanoate, hexyl acetate and ethyl hexanoate. The compounds 2,5-dimethyl-4-hydroxy-3(2H)-furanone (DMHF; Pickenhagen et al., 1981) and its methyl derivative 2,5-dimethyl-4-methoxy-3(2H)-furanone (DMMF; Pyysalo et al., 1979) have a strawberry-like aroma but no single compound has been found to encapsulate the taste of the ripe fruit. The same is true for tomato where the principal flavour compounds contributing to ripe tomato flavour are cis-3-hexanal, cis-3-hexanol, hexanal, 3-methylbutanal, 6-methyl-5-hepten-2-one, 1-pentan-3-one, trans-2-hexanal, methyl salicylate, 2-isobutylthiazole and β-ionone (Buttery, 1993). In contrast, the compound 3-methylbutyl acetate is the principal contributor to the flavour of banana (Berger, 1991). Flavour manipulation in different fruits will therefore require different approaches. Since flavour is so complex, just a few examples of the key pathways of flavour biogenesis will be discussed below.

While the chemical nature of the volatiles of some fruits is well characterized, the biosynthesis of these compounds and the genetic controls of the biochemical pathways leading to their formation are much less well understood. Fruit aromas can be classified into the main chemical groups: esters, aldehydes, ketones, alcohols, terpenes, furanones and sulfur compounds. Of these, esters are one of the most important classes and their formation is catalysed by the enzyme alcohol acyltransferase (AAT) (figure 10.2). The enzyme has been reported in banana (Harada et al., 1985), apple (Fellman et al., 1991) and strawberry (Pérez et al., 1996). In strawberry there was a dramatic increase in AAT activity at the onset of ripening when the fruit acquires the ability to synthesize aroma compounds. A gene encoding a ripening enhanced and fruit-specific AAT from strawberry has been described by Aharoni et al. (2000). The recombinant protein encoded by the AAT gene catalyses the formation of esters from a wide range of alcohols suggesting it has a role in the biogenesis of fruity aroma volatiles in strawberry and so is likely to be a good target for manipulation. However, differences in the specificity of the enzyme for different substrates cannot account for the changes in ester composition during fruit development or the differences

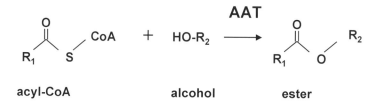

Figure 10.2 Formation of esters by alcohol acyltransferase (AAT).

between cultivars. Interestingly, a good correlation was found between AAT activity and flavour quality in strawberries (Pérez *et al.*, 1996). It is not known if more than one form of AAT is present in strawberry fruit. Ester composition in fruits could also be modified by esterases. It is probable that some of the esters formed in apple, for example, are hydrolysed by esterases (Knee and Hatfield, 1981).

The availability of ester precursors is likely to be an important factor determining the types of esters formed in fruits. The enzymes alcohol dehydrogenase (ADH), pyruvate decarboxylase (PDC) and pyruvate dehydrogenase (PDH) may play a key role in providing ethanol and acetyl CoA substrates for ester biosynthesis (figure 10.3). The activity of ADH is strongly up-regulated during the ripening of strawberry fruit (Mitchell and Jelenkovic, 1995) suggesting this enzyme has an important role in flavour development and that it could be a good candidate for manipulating flavour. Several isoforms of ADH are present in strawberry and these exhibit broad substrate specificities implying they are involved in the interconversions of a number of flavour aldehydes and alcohols. The metabolism of natural volatiles by intact strawberry fruit supports such a role for ADH (Hamilton-Kemp *et al.*, 1996). The presence of ADH is often associated with anaerobic metabolism (Kennedy *et al.*, 1992) and carbon dioxide can cause some strawberry cultivars to accumulate acetaldehyde and ethanol (Fernández-Trujillo *et al.*, 1999) that may contribute to 'off-flavours' (Ke *et al.*,

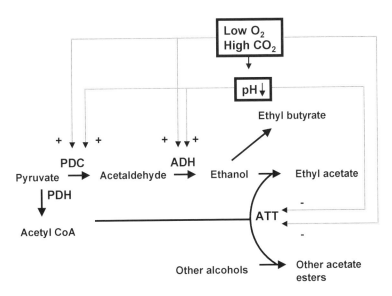

Figure 10.3 Interrelationships between the enzymes alcohol dehydrogenase (ADH), pyruvate decarboxylase (PDC) and pyruvate dehydrogenase (PDH) in the biosynthesis of esters affecting flavour (redrawn from Ke *et al.*, 1994).

1991). The product of *adh2* is present in young fruit 15 days prior to ripening but decreases before increasing late in ripening. When gene expression was modified in transgenic tomatoes the activity of ADH was positively correlated with the content of hexanol and *cis*-3-hexenol in the fruit although the levels of aldehydes remained unaltered (Speirs *et al.*, 1998). Both of these alcohols contribute to the 'ripe fruit' flavour of tomato and this is an example of how flavour might be altered by using genetic modification to change the balance between aldehydes and alcohols.

The physiological function of PDC has also been examined in relation to anaerobiosis. Along with ADH, PDC activity increases under the non-physiological conditions of low O_2 and high CO_2 that are used to extend the postharvest storage of a range of fruits. In tomato fruit, the activity of PDC is not linked to the increase in ADH activity and ethanol accumulation during normal ripening (Chen and Chase, 1993). However, PDC gene expression is enhanced during the normal ripening of grape (Or *et al.*, 2000) and strawberry (Manning, 1998; Aharoni *et al.*, 2000). Two different PDC genes appear to be expressed in strawberry suggesting the enzyme has a role in supplying substrates for the biosynthesis of aroma compounds. If pyruvate production was non-limiting then overexpression of PDC might be expected to increase the proportion of ethyl esters formed in this pathway, whereas down-regulation of this enzyme might have the opposite effect and lower the proportion of ethyl esters in favour of the formation of acetate esters.

Volatile compounds may arise from more complex precursors including lipids, amino acids, carotenoids and terpenes. Lipid metabolism is likely to be the source of a range of aldehydes, alcohols and esters with important aroma characteristics. The enzyme lipoxygenase is the first step in a pathway oxidizing the lipids linolenic (18:2) and linoleic (18:3) acid. These substrates are cleaved by the enzyme hydroperoxide lyase (HPL) to form an aldehyde and an oxoacid (Vick and Zimmermann, 1980). Thus hexenal and hexanal are produced from the action of HPL on oxidized linoleic and linolenic acids, respectively. These aldehydes have a very low odour threshold and are important contributors to the background 'green-note' flavour of many fruits. Hydroperoxide lyases have been characterized from a number of fruits including: pear (Kim and Grosch, 1981), green bell pepper (Shibata *et al.*, 1995), olive (Salas and Sánchez, 1999), tomato (Suurmeijer *et al.*, 2000) and strawberry (Pérez *et al.*, 1999). The activity of the enzyme from strawberry, which utilized the 13-hydroperoxide of linolenic acid as the preferred substrate to produce (3Z)-hexenal, increased markedly during the white stage of fruit development. In conjunction with a sharp increase in hexanal this was taken to be evidence of a sequential pathway for the formation of green odour compounds in strawberry (Pérez *et al.*, 1999).

Genetic modification of HPL, a key step diverting lipids into the flavour pathway, has not been reported but down-regulation of two lipoxygenase isoforms in transgenic tomato fruit failed to alter volatile composition significantly (Griffiths *et al.*, 1999). These authors suggested that a third lipoxygenase isoform could be important for volatile production. Convincing evidence that some aroma volatiles originate from lipid metabolism comes from a study in which the lipid composition of transgenic tomatoes was altered by ectopically expressing a Δ-9 desaturase gene from yeast (Wang *et al.*, 1996). This resulted in marked changes in the volatile profile, with increases in several C-6 compounds, notably *cis*-3-hexenol, 1-hexanol, hexanal and *cis*-3-hexenal, but also other compounds not associated with fatty acid precursors, such as 6-methyl-5-hepten-2-one and 2-isobutylthiazole. These data reinforce the idea that the biogenesis of volatile flavour compounds is extremely complex and interwoven such that the product of an enzyme in one metabolic pathway may be the substrate for another enzyme in a different pathway.

The production of volatiles via the lipid pathway presumably needs continued lipid biosynthesis since these compounds are lost from the fruit. A ripening enhanced acyl carrier protein (ACP), an essential component of the fatty acid synthase complex, in strawberry may be required to maintain the production of aroma volatiles (Manning, 1998) as there is no evidence of major changes in lipid composition in this fruit. If the ripening enhanced form of ACP is related to flavour generation, then up-regulating this gene is predicted to increase the biogenesis of lipid-derived flavour volatiles.

Two main strategies are possible for modifying fruit flavour. Overall flavour intensity could be enhanced by up-regulating all of the flavour pathways together. Identifying key ripening regulatory genes having such an effect is an important goal in current research. Alternatively, subgroups of compounds within one or more of the pathways described above might be altered to enhance particular flavour notes or introduce novel fruity characteristics. This will not be straightforward because, apart from the considerations of consumer acceptability, attempts to modify the abundance of one or a group of aroma compounds may produce changes in other volatiles that would be undesirable. An example of how this might occur is found in mutant plants deficient in ADH in which the activity of HPL in the lipoxygenase pathway is altered (Bate *et al.*, 1998). The engineering of fruit flavour will also need to take into account the effects of cultural and environmental factors (Baldwin *et al.*, 2000). Predicting how altering volatile profiles affects our perception of flavour represents an additional interesting and complex challenge for the future.

10.6 Hormonal regulation

Some of the earliest attempts to modify fruit ripening have involved the treatments that affect either the synthesis or the perception of the hormone ethylene, although the scientific basis of ethylene-mediated ripening was unknown at the time. Ethylene has since been the subject of intensive investigation mainly because of its potent effects specifically on the ripening phase of fruit development. The 1-aminocyclopropane carboxylic acid (ACC)-dependent ethylene biosynthetic pathway, which is now thoroughly characterized (Barry *et al.*, 1997), has been the subject of numerous transgenic studies to alter the course of ripening in a range of fruits. These have focused on inhibiting ethylene formation either by reducing the availability of the ethylene precursor ACC, by down-regulating the enzyme ACC synthase or expressing an enzyme that removes ACC, or by inhibiting the enzyme ACC oxidase that converts ACC to ethylene. Lowering the amount of ethylene produced or delaying its production have been successful strategies in a number of instances (Grierson *et al.*, 1992). An excellent example of the effectiveness of this approach is the inhibition of ethylene biosynthesis in melon in which ripening of the fruit was blocked by the antisense down-regulation of ACC oxidase (Ayub *et al.*, 1996). The ripening of the melon fruits on the plant was delayed and this resulted in fruits with enhanced storage-life. Flavour was also improved as the fruits could be left on the plant for longer. Ripening of these fruits could be induced by the application of ethylene, thereby enabling the effects of the transgene to be reversed.

One of the difficulties in manipulating ripening by regulating ethylene biosynthesis is that once ripening is initiated the fruit have a similar shelf-life to that of normal fruits, as in the case of melon above. Almost complete suppression of ethylene production, however, leads to fruits that never ripen fully. The ethylene signal transduction pathway (McGrath and Ecker, 1998) controls the downstream events that in fruits are manifested as changes in texture, colour and flavour. This pathway provides further opportunities for regulating ripening. The *etr1-1* gene from *Arabidopsis*, which encodes a receptor that confers dominant ethylene insensitivity, results in delayed fruit ripening and flower senescence when expressed in tomato and petunia plants (Wilkinson *et al.*, 1997). The tomato ripening mutant, *Nr*, has a defective ethylene receptor that results in dramatic pleiotropic effects on all aspects of ripening and also affects other ethylene-dependant responses, such as leaf epinasty and the senescence and abscission of flowers (Lanahan *et al.*, 1994; Wilkinson *et al.*, 1995). Ethylene receptors belong to a small multigene family (Bleecker, 1999) that may be differentially regulated during fruit development (Tieman and Klee, 1999; Sato-Nara *et al.*, 1999). The precise function of each member of this family has yet to established, but manipulating their expression independently offers the possibility of fine tuning developmental processes in fruits. Manipulating the expression of genes encoding components of the ethylene signal transduction pathway down-stream of the receptors may enable different aspects of ripening physiology to be regulated.

The ripening of non-climacteric fruits such as strawberry does not appear to be amenable to modification by regulating ethylene biosynthesis per se, as these fruits produce low levels of ethylene and do not respond to exogenous applications of ethylene or its precursor ACC (Given *et al.*, 1988). It is possibile that non-climacteric fruits have a defective signal transduction component that makes them insensitive to this hormone, either because they cannot detect ethylene or cannot fully transduce the signal. However, the ripening of strawberry fruit (Given *et al.*, 1988; Manning, 1994) and grape berries (Davies *et al.*, 1997) is inhibited by auxin. In comparison with ethylene, much less is known about the role of auxin in fruit development but there may be opportunities to manipulate the level of this hormone within fruit tissues by metabolic means. For example, a ripening related gene from strawberry putatively encoding UDP glucose–flavonoid 3-*O*-glucosyl transferase (UFGT) is postulated to be involved in auxin conjugation, a possible mechanism for regulating the level of free auxin in the fruit (Manning, 1998). In strawberry the movement of auxin from the achenes to the fleshy receptacle is essential for fruit growth. Cessation of this process as the fruit matures is hypothesized to lower the level of free auxin in the receptacle and de-repress ripening (Given *et al.*, 1988). Auxin transport from cell to cell is believed to be mediated by influx and efflux carriers (Bennett *et al.*, 1998). Modifying the expression of auxin transporter genes may offer an alternative approach and alter ripening behaviour by regulating the concentration of auxin in the fruit.

10.7 Ripening-regulatory genes: current progress and future prospects

As illustrated in the preceding paragraphs, ripening is the end result of a complex series of events brought to a climax by maturation of the seeds and the marked changes in colour, texture and flavour. These events are controlled by the presence of phytohormones such as ethylene and auxin. Recent work, however, suggests that another level of control is also operating, where developmental signals prime the fruit for ripening and allow it to become responsive to the presence of various hormones. Knowledge of these developmental cues has come to light through the study of pleiotropic fruit ripening mutations in tomato. Here, single gene mutations such as *rin* and *nor* suppress nearly all aspects of normal ripening. The phenotypes cannot simply be reversed by the addition of exogenous ethylene. Developmental regulation of genes such as *E8* and ACC synthase is disrupted by *rin* and *nor* so that they are not expressed (Giovannoni *et al.*, 1998). Recently both the *rin* and *nor* genes have been isolated from tomato by genetic map-based cloning. The genes are both transcription factors and it will be interesting to see if orthologues of *rin* and *nor* can be found in all fleshy fruits (J.J. Giovannoni, personal communication).

We have recently described a new pleiotropic ripening mutation, colourless non-ripening (*Cnr*). This dominant mutation produces fruits where carotenoid biosynthesis in the pericarp is completely abolished and the tissue remains firm, although, curiously, becomes extremely mealy, showing a loss of cell-to-cell adhesion (Thompson *et al.*, 1999) (figure 10.4). We have mapped the gene to chromosome 2 and are using a genetic map-based approach to isolate the *Cnr* gene. The aim is to be able to place *nor*, *rin* and *Cnr* in a framework that describes the molecular regulation of ripening. Current commercial applications for the *rin* and *nor* mutations involve using these alleles to impart enhanced shelf-life in tomato. The identification of these genes will allow the assessment of additional natural allelic variation in tomato and possibly other fruit crops and inform crop improvement strategies by marker assisted breeding. The genes also offer the possibility of enhancing fruit quality by genetic modification.

In tomato, loci specifically involved in ripening map onto all 12 chromosomes and it is likely that the majority of the ripening-related genes will soon be given map locations in this species. Together with efficient mapping of QTL and single-gene mutant loci, this approach will permit the efficient identification of candidate genes involved in important quality traits (see Giovannoni *et al.*, 1999). In other fruit crops, direct genetic approaches are more difficult. In apple the generation time makes traditional breeding for improved fruit characteristics a long-term investment and it is clearly difficult to test the efficacy of candidate genes by genetic modification. However, a robust molecular map for apple now exists and several important quality traits have been mapped, including crispness and acidity (King *et al.*, 2001). Conservation of regulatory and metabolic

Figure 10.4 Loss of cell-to-cell adhesion in the *Cnr* tomato ripening mutant after pericarp tissue left in water overnight.

functions in fruits and QTL mapping should allow the provision of molecular markers to assist breeding and surveys of natural allelic variation. Also, it may be possible to use the resources from the *Arabidopsis* genome project to identify areas of synteny between genomes and permit isolation of candidate genes (Ku *et al.*, 2000).

The next ten years should be an exciting period for researchers in the fruit ripening area. It is likely to bring key insights into the genetic and molecular basis of the mechanisms underpinning all the major quality traits and a clearer understanding of the place of fruit ripening within the wider context of plant development.

References

Aharoni, A., Keizer, L.C.P., Bouwmeester, H.J., Sun, Z., Alvarez-Huerta, M., Verhoeven, H.A., Blaas, J., van Houwelingen, A.M.M., De Vos, R.C.H., van der Voet, H., Jansen, R.C., Guis, M., Mol, J., Davis, R.W., Schena, M., van Tunen, A.J. and O'Connell, A.P. (2000) Identification of the SAAT gene involved in strawberry flavor biogenesis by use of DNA microarrays. *Plant Cell*, **12**, 647-661.

Ayub, R., Guis, M., Ben Amor, M., Gillot, L., Roustan, J.-P., Latché, A., Bouzayen, M. and Pech, J.-C. (1996) Expression of ACC oxidase antisense gene inhibits ripening of cantaloupe melon fruits. *Nature Biotechnology*, **14**, 862-866.

Baldwin, E.A., Scott, J.W., Shewmaker, C.K. and Schuch, W. (2000) Flavor trivia and tomato aroma: biochemistry and possible mechanisms for control of important aroma components. *HortScience*, **35**, 1013-1022.

Barry, C.S., Blume, B., Hamilton, A., Fray, R., Payton, S., Alpuche-Solis, A. and Grierson, D. (1997) Regulation of ethylene synthesis and perception in tomato and its control using gene technology, in *Biology and Biotechnology of the Plant Hormone Ethylene* (eds A.K. Kanellis, C. Chang, H. Kende and D. Grierson), Dordrecht, Kluwer Academic Publishers, pp. 299-306.

Bate, N.J., Riley, J.C.M., Thompson, J.E. and Rothstein, S.J. (1998) Quantitative and qualitative differences in C-6 volatile production from the lipoxygenase pathway in an alcohol dehydrogenase mutant of *Arabidopsis thaliana*. *Physiologia Plantarum*, **104**, 97-104.

Bennett, M.J., Marchant, A., May, S.T. and Swarup, R. (1998) Going the distance with auxin: unravelling the molecular basis of auxin transport. *Philosophical Transactions of the Royal Society of London* B, **353**, 1511-1515.

Berger, R.G. (1991) Fruits I, in *Volatile Compounds in Foods and Beverages* (ed. H. Maarse), Marcel Dekker, New York, pp. 283-304.

Bleecker, A.B. (1999) Ethylene perception and signaling: an evolutionary perspective. *Trends in Plant Science*, **4**, 269-274.

Boss, P.K., Davies, C. and Robinson, S.P. (1996) Analysis of the expression of anthocyanin pathway genes in developing *Vitis vinifera* L. cv Shiraz grape berries and the implications for pathway regulation. *Plant Physiology*, **111**, 1059-1066.

Bramley, P.M. (1997) The regulation and genetic manipulation of carotenoid biosynthesis in tomato fruit. *Pure and Applied Chemistry*, **69**, 2159-2162.

Brummell, D.A. and Harpster, M.H. (2001) Cell wall metabolism in fruit softening and quality and its manipulation in transgenic plants. *Plant Molecular Biology* (in press).

Brummell, D.A., Harpster, M.H. and Civello, P.M. (1999) Modification of expansin protein abundance in tomato fruit alters softening and cell wall polymer metabolism during ripening. *Plant Cell*, **11**, 2203-2216.

Buttery, R.G. (1993) Quantitative and sensory aspects of flavour of tomato and other vegetables and fruits, in *Flavor Science: Sensory Principles and Techniques* (eds T.E. Acree and R. Teranishi), American Chemical Society, Washington DC, pp. 259-286.

Carey, A.T., Holt, K., Picard, S., Wilde, R., Tucker, G.A., Bird, C.R., Schuch, W. and Seymour, G.B. (1995) Tomato exo-(1-4) β-D-galactanase. Isolation, changes during ripening in normal and mutant tomato fruit and characterisation of a related cDNA clone. *Plant Physiology*, **108**, 1099-1107.

Carpita, N.C. and Gibeaut, D.M. (1993) Structural models of primary-cell wall in flowering plants—consistency of molecular-structure with the physical-properties of the walls during growth. *Plant Journal*, **3**, 1-30.

Chen, A.-R.S. and Chase, T. Jr (1993) Alcohol dehydrogenase 2 and pyruvate decarboxylase induction in ripening and hypoxic tomato fruit. *Plant Physiology and Biochemistry*, **31**, 875-885.

Coombe, B.G. (1976) The development of fleshy fruits. *Annual Review of Plant Physiology*, **27**, 507-528.

Davies, C., Boss, P.K. and Robinson, S.P. (1997) Treatment of grape berries, a non-climacteric fruit, with a synthetic auxin retards ripening and alters the expression of developmentally regulated genes. *Plant Physiology*, **115**, 1155-1161.

Davies, K.M., Bloor, S.J., Spiller, G.B. and Deroles, S.C. (1998) Production of yellow colour in flowers: redirection of flavonoid biosynthesis in *Petunia. Plant Journal*, **13**, 259-266.

Domínguez-Puigjaner, E., Llop, I., Vendrell, M. and Prat, S. (1997) A cDNA clone highly expressed in ripe banana fruit shows homology to pectate lyases. *Plant Physiology*, **114**, 1072-1076.

Eriksson, O., Friis, E.M., Pedersen, K.R. and Crane, P.R. (2000) Seed size and dispersal systems of early cretaceous angiosperms from Famalicao, Portugal. *International Journal of Plant Sciences*, **161**, 319-329.

Fellman, J.K., Mattheis, J.P., Matthinson, D.S. and Bostick, B.C. (1991) Assay of acetyl CoA alcohol transferase in 'Delicious' apples. *HortScience*, **27**, 773-776.

Fernández-Trujillo, J.P., Nock, J.F. and Watkins, C.B. (1999) Fermentative metabolism and organic acid concentrations in fruit of selected strawberry cultivars with different tolerances to carbon dioxide. *Journal of the American Society for Horticultural Science*, **124**, 696-701.

Ferrándiz, C., Pelaz, S. and Yanofsky, M.F. (1999) Control of carpel and fruit development in *Arabidopsis. Annual Review of Biochemistry*, **68**, 321-354.

Ferrándiz, C., Liljegren, S.J. and Yanofsky, M.F. (2000) Negative regulation of the SHATTERPROOF genes by FRUITFULL during *Arabidopsis* fruit development. *Science*, **289**, 436-438.

Fray, R.G. and Grierson, D. (1993) Identification and genetic analysis of normal and mutant phytoene synthase genes of tomato by sequencing, complementation and co-suppression. *Plant Molecular Biology*, **22**, 589-602.

Fray, R.G., Wallace, A. and Fraser, P.D. (1995) Constitutive expression of a fruit phytoene synthase gene in transgenic tomatoes causes dwarfism by redirecting metabolites from the gibberellin pathway. *Plant Journal*, **8**, 693-701.

Fulton, T.M., Grandillo, S., Beck-Bunn, T., Fridman, E., Frampton, A., Lopez, J., Petiard, V., Uhlig, J., Zamir, D. and Tanksley, S.D. (2000) Advanced backcross QTL analysis of a *Lycopersicon esculentum* × *Lycopersicon parviflorum* cross. *Theoretical and Applied Genetics*, **100**, 1025-1042.

Gillaspy, G., Ben-David, H. and Gruissem, W. (1993) Fruits: a developmental perspective. *Plant Cell*, **5**, 1439-1451.

Giovannoni, J.J., Dellapenna, D. and Bennett, A.B. (1989) Expression of a chimeric polygalacturonase gene in transgenic *rin* (ripening inhibitor) tomato fruit results in polyuronide degradation but not fruit softening. *Plant Cell*, **1**, 53-63.

Giovannoni, J.J., Kannan, P., Lee, S. and Yen, H.C. (1998) Genetic approaches to manipulation of fruit development and quality in tomato, in *Genetic and Environmental Manipulation of Horticultural Crops* (eds K.E. Cockshull, D. Gray, G.B. Seymour and B. Thomas), CAB International, Wallingford, pp. 1-15.

Giovannoni, J.J., Yen, H., Shelton, B., Miller, S., Vrebalov, J., Kannan, P., Tieman, D., Hackett, R., Grierson, D. and Klee, H. (1999) Genetic mapping of ripening and ethylene-related loci in tomato. *Theoretical and Applied Genetics*, **98**, 1005-1013.

Giovannucci, E., Ascherio, A., Rimm, E.B., Stampfer, M.J., Colditz, G.A. and Willett, W.C. (1995) Intake of carotenoids and retinol in relation to risk of prostate-cancer. *Journal of the National Cancer Institute*, **87**, 1767-1776.

Given, N.K., Venis, M.A. and Grierson, D. (1988) Hormonal regulation of ripening in the strawberry, a non-climacteric fruit. *Planta*, **174**, 402-406.

Grierson, D., Tucker, G.A. and Keen, J. (1986) Sequencing and identification of a cDNA clone for tomato polygalacturonase. *Nucleic Acids Research*, **14**, 8595-8603.

Grierson, D., Hamilton, A.J., Bouzayen, M., Kock, M., Lycett, G.W. and Barton, S. (1992) Regulation of gene expression, ethylene synthesis and ripening in transgenic tomatoes, in *Inducible Plant Proteins* (ed. J.L. Wray), Cambridge University Press, Cambridge, pp. 155-174.

Griffiths, A., Prestage, S., Linforth, R., Zhang, J.L., Taylor, A. and Grierson, D. (1999) Fruit-specific lipoxygenase suppression in antisense-transgenic tomatoes. *Postharvest Biology and Technology*, **17**, 163-173.

Hamilton-Kemp, T.R., Archbold, D.D., Loughrin, J.H., Collins, R.W. and Byers, M.E. (1996) Metabolism of natural volatile compounds by strawberry fruit. *Journal of Agricultural and Food Chemistry*, **44**, 2802-2805.

Harada, M., Ueda, Y. and Wata, T. (1985) Purification and some properties of alcohol acyltransferase from banana fruit. *Plant Cell Physiology*, **26**, 1067-1074.

Hertog, M.G.L., Hollman, P.C.H. and Katan, M.B. (1992) Content of potentially anticarcinogenic flavonoids of 28 vegetables and 9 fruits commonly consumed in the Netherlands. *Journal of Agricultural and Food Chemistry*, **40**, 2379-2383.

Hobson, G. and Grierson, D. (1993) Tomato, in *Biochemistry of Fruit Ripening* (eds G. Seymour, J. Taylor and G. Tucker), Chapman and Hall, London, pp. 405-442.

Holton, T.A. and Cornish, E.C. (1995) Genetics and biochemistry of anthocyanin biosynthesis. *Plant Cell*, **7**, 1071-1083.

Hong, V. and Wrolstad, R.E. (1990) Characterization of anthocyanin-containing colourants and fruit juices by HPLC photodiode array detection. *Journal of Agricultural and Food Chemistry*, **38**, 698-708.

Huxham, I.M., Jarvis, M.C., Shakespeare, L., Dover, C.J., Johnson, D., Knox, J.P. and Seymour, G.B. (1999) Electron-energy loss spectroscopic imaging of calcium and nitrogen in the cell walls of apple fruits. *Planta*, **208**, 438-443.

Jones, L., Seymour, G.B. and Knox, P.J. (1997) Localisation of pectic galactan in tomato cell walls using a monoclonal antibody specific to (1-4)-β-D-galactan. *Plant Physiology*, **113**, 1405-1412.

Kays, S.J. and Wang, Y. (2000) Thermally induced flavor compounds. *HortScience*, **35**, 1002-1012.

Ke, D., Goldstein, L., O'Mahony, M. and Kader, A.A. (1991) Effects of short-term exposure to low O_2 and high CO_2 atmospheres on quality attributes of strawberries. *Journal of Food Science*, **56**, 50-54.

Ke, D., Xhou, L. and Kader, A.A. (1994) Mode of oxygen and carbon dioxide action on strawberry ester biosynthesis. *Journal of the American Society for Horticultural Science*, **119**, 971-975.

Kennedy, R.A., Rumpho, M.E. and Fox, T.C. (1992) Anaerobic metabolism in plants. *Plant Physiology*, **100**, 1-6.

Kim, I.-S. and Grosch, W. (1981) Partial purification and properties of a hydroperoxide lyase from fruits of pear. *Journal of Agricultural and Food Chemistry*, **29**, 1220-1225.

King, G.J., Lynn, J.R., Dover, C.J., Evans, K.M. and Seymour, G.B. (2001) Resolution of quantitative trait loci for mechanical measures accounting for genetic variation in fruit texture of apple (*Malus pumila* Mill.). *Theoretical and Applied Genetics*, **102**, 1227-1235.

Knee, M. and Hatfield, S.G.S. (1981) The metabolism of alcohols by apple fruit. *Journal of the Science of Food and Agriculture*, **32**, 593-600.

Ku, H-M., Vision, T., Liu, J. and Tanksley, D. (2000) Comparing sequenced segments of the tomato and *Arabidopsis* genomes: large-scale duplication followed by selective gene loss creates a network of synteny. *Proceedings of the National Academy of Sciences of the USA*, **97**, 9121-9126.

Lanahan, M.B., Yen, H.-C., Giovannoni, J.J. and Klee, H.J. (1994) The *Never Ripe* mutation blocks ethylene perception in tomato. *Plant Cell*, **6**, 521-530.

Latrasse, A. (1991) Fruits III, in *Volatile Compounds in Foods and Beverages* (ed. H. Maarse), Marcel-Dekker, New York, pp. 329-387.

Liljegren, S.J., Ditta, G.S., Eshed, Y., Savidge, B., Bowman, J.L. and Yanofsky M.F. (2000) SHATTERPROOF MADS-box genes control seed dispersal in *Arabidopsis*. *Nature*, **404**, 766-770.

Manning, K. (1994) Changes in gene expression during strawberry fruit ripening and their regulation by auxin. *Planta*, **194**, 62-68.

Manning, K. (1998) Isolation of a set of ripening-related genes from strawberry: their identification and possible relationship to fruit quality traits. *Planta*, **205**, 622-631.

Martin, M.J., Motilva, V. and Alarcón de la Castra, C. (1993) Quercetin and naringenin: effects on ulcer formation and gastric secretion in rats. *Phytotherapy Research*, **7**, 150-153.

Mayne, S.T. (1996) Beta-carotene, carotenoids, and disease prevention in humans. *FASEB Journal*, **10**, 690-701.

McCartney, L., Ormerod, A.P. and Gidley, M.J. (2000) Temporal and spatial regulation of pectic (1→4)-beta-D-galactan in cell walls of developing pea cotyledons: implications for mechanical properties. *Plant Journal*, **22**, 105-113.

McGrath, R.B. and Ecker, J.R. (1998) Ethylene signaling in *Arabidopsis*: events from the membrane to the nucleus. *Plant Physiology and Biochemistry*, **36**, 103-113.

Medina Escobar, N., Cardenas, J., Moyano, E., Caballero, J.L. and Munoz Blanco, J. (1997) Cloning, molecular characterisation and expression pattern of a strawberry ripening-specific cDNA with sequence homology to pectate lyase from higher plants. *Plant Molecular Biology*, **34**, 867-877.

Medina-Suarez, R., Manning, K., Fletcher, J., Aked, J., Bird, C.R. and Seymour, G.B. (1997) Gene expression in the pulp of ripening bananas. *Plant Physiology*, **115**, 453-461.

Mitchell, W.C. and Jelenkovic, G. (1995) Characterizing NAD- and NADP-dependent alcohol dehydrogenase enzyxmes of strawberries. *Journal of the American Society for Horticultural Science*, **120**, 798-801.

Moyano, E., Portero-Robles, I., Medina-Escobar, N., Valpuesta, V., Munoz-Blanco, J. and Caballero, J.L. (1998) A fruit-specific putative dihydroflavonol 4-reductase gene is differentially expressed in strawberry during the ripening process. *Plant Physiology*, **117**, 711-716.

Mustilli, A.C., Fenzi, F., Cilliento, R., Alfano, F. and Bowler, C. (1999) Phenotype of the tomato *high pigment-2* mutant is caused by a mutation in the tomato homolog of DEETIOLATED1. *Plant Cell*, **11**, 145-157.

Or, E., Baybik, J., Sadka, A. and Ogrodovitch, A. (2000) Fermentative metabolism in grape berries: isolation and characterization of pyruvate decarboxylase cDNA and analysis of its expression throughout berry development. *Plant Science*, **156**, 151-158.

Ozawa, T., Lilley, T.H. and Haslam, E. (1987) Polyphenol interactions: astringency and the loss of astringency in ripening fruit. *Phytochemistry*, **26**, 2937-2942.

Parker, R.S. (1996) Carotenoids. 4. Absorption, metabolism and transport of carotenoids. *FASEB Journal*, **10**, 542-551.

Pérez, A.G., Sanz, C., Olías, R., Ríos, J.J. and Olías, J.M. (1996) Evolution of strawberry alcohol acyltransferase activity during fruit development and storage. *Journal of Agricultural and Food Chemistry*, **44**, 3286-3290.

Pérez, A.G., Sanz, C., Olías, R. and Olías, J.M. (1999) Lipoxygenase and hydroperoxide lyase activities in ripening strawberry fruits. *Journal of Agricultural and Food Chemistry*, **47**, 249-253.

Pickenhagen, W., Velluz, A., Passerat, J.P. and Ohloff, G. (1981) Estimation of 2,5-dimethyl-4-hydroxy-3(2H)-furanone (furaneol) in cultivated and wild strawberries, pineapples and mangoes. *Journal of the Science of Food and Agriculture*, **32**, 1132-1134.

Pyysalo, T., Honkanen, E. and Hirvi, T. (1979) Volatiles of wild strawberries, *Fragaria vesca* L. compared to those of cultivated berries *Fragaria x ananassa* cv. Senga sengana. *Journal of Agricultural and Food Chemistry*, **27**, 19-22.

Rice-Evans, C.A., Miller, N.J. and Paganga, G. (1997) Antioxidant properties of phenolic compounds. *Trends in Plant Science*, **2**, 152-159.

Robards, K. and Antolovich, M. (1997) Analytical chemistry of fruit bioflavonoids. *Analyst*, **122**, 11R-34R.

Römer, S., Fraser, P.D., Kiano, J.W., Shipton, C.A., Misawa, N., Schuch, W. and Bramley, P.M. (2000) Elevation of the provitamin A content of transgenic tomato plants. *Nature Biotechnology*, **18**, 666-669.

Ronen, G., Carmel-Goren, L., Zamir, D. and Hirschberg, J. (2000) An alternative pathway to β-carotene formation in plant chromoplasts discovered by map-based cloning of *Beta* and *old-gold* color mutants in tomato. *Proceedings of the National Academy of Sciences of the USA*, **97**, 11102-11107.

Salas, J.J. and Sánchez, J. (1999) Hydroperoxide lyase from olive (*Olea europaea*) fruits. *Plant Science*, **143**, 19-26.

Sato-Nara, K., Yuhashi, K.-I., Higashi, K., Hosoya, K., Kubota, M. and Ezura, H. (1999) Stage- and tissue-specific expression of ethylene receptor homolog genes during fruit development in muskmelon. *Plant Physiology*, **119**, 321-329.

Seymour, G.B., Harding, S.E., Taylor, A.J., Hobson, G.E. and Tucker, G.A. (1987a) Polyuronide solubilization during ripening of normal and mutant tomato fruit. *Phytochemistry*, **26**, 1871-1875.

Seymour, G.B., Lasslett, Y. and Tucker, G.A. (1987b) Differential effects of pectolytic enzymes on tomato polyuronides *in vivo* and *in vitro*. *Phytochemistry*, **26**, 3137-3139.

Sheehy, R.E., Kramer, M. and Hiatt, W.R. (1988) Reduction of polygalacturonase activity in tomato fruit by antisense RNA. *Proceedings of the National Academy of Sciences of the USA*, **85**, 8805-8809.

Shibata, Y., Matsui, K., Kajiwara, T. and Hatanaka, A. (1995) Purification and properties of fatty acid hydroperoxide lyase from green bell pepper fruits. *Plant and Cell Physiology*, **36**, 147-156.

Slater, A., Maunders, M.J. and Edwards, E. (1985) Isolation and characterization of cDNA clones for tomato polygalacturonase and other ripening-related proteins. *Plant Molecular Biology*, **5**, 137-147.

Smith, D.L. and Gross, K.C. (2000) A family of at least seven beta-galactosidase genes is expressed during tomato fruit development. *Plant Physiology*, **123**, 1173-1183.

Smith, C.J.S., Watson, C.F. and Ray, J. (1988) Antisense RNA inhibition of polygalacturonase gene expression in transgenic tomatoes. *Nature*, **334**, 724-726.

Smith, C.J.S., Watson, C.F. and Morris, P.C. (1990) Inheritance and effect on ripening of antisense polygalacturonase genes in transgenic tomatoes. *Plant Molecular Biology*, **14**, 369-379.

Smith, D.L., Starrett, D.A. and Gross, K.C. (1998) A gene coding for tomato fruit β-galactosidase II is expressed during fruit ripening. *Plant Physiology*, **117**, 417-423.

Speirs, J., Lee, E., Holt, K., Yong-Duk, K., Steele, N., Loveys, B. and Schuch, W. (1998) Genetic manipulation of alcohol dehydrogenase levels in ripening tomato fruit affects the balance of some flavor aldehydes and alcohols. *Plant Physiology*, **117**, 1047-1058.

Suurmeijer, C.N.S.P., Pérez-Gilabert, M., van Unen, D.J., van der Hijden, H.T.W.M., Veldink, G.A. and Vliegenthart, J.F.G. (2000) Purification, stabilization and characterization of tomato fatty acid hydroperoxide lyase. *Phytochemistry*, **53**, 177-185.

Thomas, B. (1981) *The Evolution of Plants and Flowers*. St Martin's Press, New York, pp. 116.

Thompson, A.J., Tor, M., Barry, C.S., Vrebalov, J., Orfila, C., Jarvis, M.C., Giovannoni, J.J., Grierson, D. and Seymour, G.B. (1999) Molecular and genetic characterisation of a novel pleiotropic tomato-ripening mutant. *Plant Physiology*, **120**, 383-389.

Tieman, D.M. and Klee, H.J. (1999) Differential expression of two novel members of the tomato ethylene-receptor family. *Plant Physiology*, **120**, 165-172.

Vick, B.A. and Zimmermann, D.C. (1980) Oxidative systems for modification of fatty acids: the lipoxygenase pathway, in *The Biochemistry of Plants* (eds P.K. Stumpf and E.E. Conn), vol 9, Academic Press, New York, pp. 53-90.

Wang, C., Chin, C.-K., Ho, C.-T., Hwang, C.-F., Polashock, J.J. and Martin, C.E. (1996) Changes of fatty acids and fatty acid-derived flavor compounds by expressing the yeast Δ-9 desaturase gene in tomato. *Journal of Agricultural and Food Chemistry*, **44**, 3399-3402.

Wilkinson, J., Lanahan, M., Yen, H-C., Giovannoni, J.J. and Klee, H. (1995) An ethylene-inducible component of signal transduction encoded by Never-ripe. *Science*, **270**, 539-544.

Wilkinson, J.Q., Lanahan, M.B., Clark, D.G., Bleecker, A.B., Chang, C., Meyerowitz, E.M. and Klee, H.J. (1997) A dominant mutant receptor from *Arabidopsis* confers ethylene insensitivity in heterologous plants. *Nature Biotechnology*, **15**, 444-447.

Woolley, L.C., James, D.J. and Manning, K. (2001) Purification and properties of an endo-β-1,4-glucanase from strawberry and down-regulation of the corresponding gene, *cel1*. *Planta* (in press)

Yen, H.C., Shelton, B.A., Howard, L.R., Lee, S., Vrebalov, J. and Giovannoni, J.J. (1997) The tomato *high pigment* (*hp*) locus maps to chromosome 2 and influences plastome copy number and fruit quality. *Theoretical and Applied Genetics*, **95**, 1069-1079.

Yoshitama, K., Ishikura, N., Fuleki, T. and Nakamura, S. (1992) Effect of anthocyanin, flavonol co-pigmentation and pH on the color of the berries of *Ampelopsis brevipedunculata*. *Journal of Plant Physiology*, **139**, 513-518.

Index